Assessment of Authentic Performance in School Mathematics

Assessment of Authentic Performance in School Mathematics

Richard Lesh
Susan J. Lamon
Editors

NEW YORK AND LONDON

First Published by 1992
American Association for the Advancement of Science

Published 2009 by Routledge
711 Third Avenue, New York, NY 10017
2 Park Square, Milton Park, Abingdon, Oxfordshire OX14 4RN

First issued in paperback 2016

Routledge is an imprint of the Taylor and Francis Group, an informa business

Copyright © 1992

Library of Congress Cataloging-in-Publication Data

Assessment of authentic performance in school
mathematics / Richard A. Lesh and Susan J. Lamon, editors
    p.   cm.
   Includes bibliographical references and index.
   1. Mathematical ability—Testing. 2. Mathematics—
Study and teaching (Elementary) I. Lesh, Richard A.
II. Lamon, Susan J., 1949–
QA135.5.A797 1992   92-14967
372.7'044'0973—dc20   CIP

The findings, conclusions, and opinions stated or implied
in this publication are those of the authors. They do not
necessarily reflect the views of the Board of Directors,
Council, or membership of the American Association for
the Advancement of Science.

ISBN 13: 978-1-138-96402-0 (pbk)
ISBN 13: 978-0-8058-1877-2 (hbk)

Publisher's Note
The publisher has gone to great lengths to ensure the quality of this reprint
but points out that some imperfections in the original may be apparent.

Preface

This book grew out of a conference sponsored by the Educational Testing Service and the University of Wisconsin's National Center for Research in Mathematical Sciences Education. The conference was held in Princeton, New Jersey, and its purpose was to facilitate the work of a group of scholars who are especially interested in the assessment of higher-order understandings and processes in foundation-level (pre-high school) mathematics. The conference brought together an international team of scholars representing diverse perspectives: mathematicians, mathematics educators, developmental psychologists, technology specialists, psychometricians, and curriculum developers.

Discussions at the conference focused on issues such as the purposes of assessment, guidelines for producing and scoring "real-life" assessment activities, and the meanings of such terms as "deeper and higher-order understanding," "cognitive objectives," and "authentic mathematical activities." International trends were highlighted, as well as current problems, challenges, and opportunities within the United States.

Assessment was viewed as a critical component of complex, dynamic, and continually adapting educational systems. For example, during the time that chapters in this book were being written, sweeping changes in mathematics education were being initiated in response to powerful recent advances in technology, cognitive psychology, and mathematics, as well as to numerous public demands for educational reform. These changes have already resulted in significant reappraisals of what it means to understand mathematics, of the nature of mathematics teaching and learning, and of the real-life situations in which mathematics is useful. The challenge is to pursue assessment-related initiatives that are systemically valid, in the sense that they work to complement and enhance other improvements in the educational system rather than acting as an impediment to badly needed curriculum reforms.

To address these issues, most chapters in this book focus on clarifying and articulating the goals of assessment and instruction, and they stress the content of assessment above its mode of delivery. For example, computer- or portfolio-based assessments are interpreted as means to ends, not as ends in themselves, and assessment is conceived as an ongoing documentation process, seamless with instruction, whose quality hinges upon its ability to provide complete and appropriate information as needed to inform priorities in instructional decision making.

This book is intended for researchers and curriculum developers in mathematics education, for teachers of mathematics, for those involved in the mathematical and pedagogical preparation of mathematics teachers, and for graduate students in mathematics education. It tackles some of the most complicated issues related to assessment, and it offers fresh perspectives from leaders in the field—with the hope that the ultimate consumer in the instruction/assessment enterprise, the individual student, will reclaim his and her potential for self-directed mathematics learning.

We are grateful to the authors of these chapters, who contributed their expertise and energy to this project, and to Gail Guadagnino at Educational Testing Service for typing and preparing the manuscripts.

Richard Lesh
Susan J. Lamon

April 1992

Note: The work described in chapters 2, 3, 4, 10, 11, 12, 13, 14, 15, and 16 was supported in part by the National Science Foundation. Any opinions, findings, and conclusions expressed are those of the authors and do not necessarily reflect the views of the National Science Foundation.

Contributors

Alice Alston
Center for Mathematics, Science, and
Computer Education
Rutgers University
New Brunswick, NJ

Steven Anacker
Cardiovascular Research Institute
School of Medicine
University of California–San Francisco
San Francisco, CA

Merlyn Behr
Dept. of Mathematical Sciences
Northern Illinois University
DeKalb, IL

Alan Bell
Shell Centre University
Nottingham, England

Hugh Burkhardt
Shell Centre University
Nottingham, England

Dan Chazan
Dept. of Teacher Education
Michigan State University
E. Lansing, MI

Robert B. Davis
Center for Mathematics, Science, and
Computer Education
Rutgers University
New Brunswick, NJ

Herbert P. Ginsburg
Dept. of Developmental and
Educational Psychology
Teachers College
Columbia University
New York, NY

Gerald Goldin
Center for Mathematics, Science, and
Computer Education
Busch Campus
Rutgers University
Piscataway, NJ

Brian Gong
Educational Testing Service
Princeton, NJ

Mary S. Kelly
Dept. of Developmental and
Educational Psychology
Teachers College
Columbia University
New York, NY

Susan J. Lamon
Dept. of Mathematics, Statistics, and
Computer Science
Marquette University
Milwaukee, WI

Jan de Lange
Freudenthal Institute
Research Group on Mathematics and
Computing Science Education
University of Utrecht
The Netherlands

Richard A. Lesh
Educational Testing Service
Princeton, NJ

Frank Lester
Journal for Research in Mathematics
Education
Indiana University
Bloomington, IN

Luz S. Lopez
Dept. of Developmental and Educational Psychology
Teachers College
Columbia University
New York, NY

Carolyn A. Maher
Graduate School of Education
Rutgers University
New Brunswick, NJ

Robert J. Mislevy
Educational Testing Service
Princeton, NJ

Swapna Mukhopadhyay
University of Washington
Seattle, WA

Thomas Post
Dept. of Curriculum and Instruction
College of Education
University of Minnesota
Mineapolis, MN

Judah Schwartz
Educational Technology Center
Harvard Graduate School
of Education
Cambridge, MA

Leen Streefland
Freudenthal Institute
Research Group on Mathematics and
Computing Science Education
University of Utrecht
The Netherlands

Malcolm Swan
Shell Centre University
Nottingham, England

Megan Willis
Dept. of Developmental and
Educational Psychology
Teachers College
Columbia University
New York, NY

Takashi Yamamoto
Educational Testing Service
Princeton, NJ

Michal Yerushalmy
Dept. of Education
University of Haifa
Haifa, Israel

Kentaro Yamamoto
Dept. of Developmental and
Educational Psychology
Teachers College
Columbia University
New York, NY

Contents

Preface		v
Contributors		vii
PART I:	**Assessment Objectives**	
Chapter 1	Trends, Goals, and Priorities in Mathematics Assessment Richard Lesh and Susan J. Lamon	3
Chapter 2	Assessing Authentic Mathematical Performance Richard Lesh and Susan J. Lamon	17
Chapter 3	Toward an Assessment Framework for School Mathematics Gerald Goldin	63
Chapter 4	Research and Classroom Assessment of Students' Verifying, Conjecturing, and Generalizing in Geometry Daniel Chazan and Michal Yerushalmy	89
PART II:	**New Items and Assessment Procedures**	
Chapter 5	Balanced Assessment of Mathematical Performance Alan Bell, Hugh Burkhardt, and Malcolm Swan	119
Chapter 6	Assessment of Extended Tasks Alan Bell, Hugh Burkhardt, and Malcolm Swan	145
Chapter 7	Moving the System: The Contributions of Assessment Alan Bell, Hugh Burkhardt, and Malcolm Swan	177
Chapter 8	Assessing Mathematical Skills, Understanding, and Thinking Jan de Lange	195
Chapter 9	Thinking Strategies in Mathematics Instruction: How Is It Possible? Leen Streefland	215

PART III:	**New Perspectives on Classroom-based Assessment**	
Chapter 10	*A Teacher's Struggle to Assess Student Cognitive Growth* Carolyn A. Maher, Robert B. Davis, and Alice Alston	249
Chapter 11	*Assessing Understandings of Arithmetic* Herbert P. Ginsburg, Luz S. Lopez, Swapna Mukhopadhyay, Takashi Yamamoto, Megan Willis, and Mary S. Kelly	265
PART IV:	**New Types of Scoring and Reporting**	
Chapter 12	*Toward a Test Theory for Assessing Student Understanding* Robert J. Mislevy, Kentaro Yamamoto, and Steven Anacker	293
Chapter 13	*Interpreting Responses to Problems with Several Levels and Types of Correct Answers* Susan J. Lamon and Richard Lesh	319
Chapter 14	*Using Learning Progress Maps to Improve Educational Decision Making* Richard Lesh, Susan J. Lamon, Brian Gong, and Thomas Post	343
PART V:	**Difficulties, Opportunities, and Future Directions**	
Chapter 15	*Future Directions for Mathematics Assessment* Richard Lesh, Susan J. Lamon, Frank Lester, and Merlyn Behr	379
Chapter 16	*The Intellectual Prices of Secrecy in Mathematics Assessment* Judah L. Schwartz	427
Index		439

PART I: Assessment Objectives

1  Trends, Goals, and Priorities in Mathematics Assessment

*Richard Lesh and
Susan J. Lamon*

INTRODUCTION

Today, there are strong pressures to move away from traditional multiple-choice or short-answer tests, toward alternative forms of assessment that focus on "real-life" situations, "authentic" mathematics, and "performance" activities. However, in spite of the fact that organizations such as the National Council of Teachers of Mathematics have made significant progress in reaching a national consensus on curriculum and evaluation standards for school mathematics (NCTM, 1989), what we want to move away from is clearer than what we want to move toward in assessment reform. For example, in the first sentence of this paragraph, each of the words in quotation marks tends to be a subject of debate among mathematics educators.

What is meant by real-life situations? or authentic mathematics? or performance activities? The main purpose of this book is to address these kinds of questions in a form that is relevant to priority decision-making issues that arise during the construction of new modes of assessment. Authors in this book were chosen partly because of the leadership roles they have played in reform efforts aimed at high-stakes testing programs in the United States, Great Britain, and the Netherlands. But they were also selected because of their interests and experience in developing materials that contribute to both instruction and assessment in the classroom.

This chapter is divided into three sections. The first section describes the main types of assessment emphasized throughout this book, and several distinctions are described that have influenced mathematics educators' views about the purposes of assessment. For example, some especially relevant distinctions are reflected in similarities and differences among the words *examine, document, assess, evaluate, test, and measure*. The second section emphasizes several ways that current assessment interests developed out of earlier curriculum reform efforts. For example, in past attempts at curriculum reform, it became clear that piecemeal approaches seldom succeeded, especially if the neglected areas involved assessment and teacher education. Such lessons are especially relevant, because it seems unlikely that piecemeal approaches to assessment reform will work any better than they did for curriculum reform in general. Even when attention is focused on assessment, teacher education, program implementation, and the improvement of curriculum materials and instruction must still be taken into account. The third section gives an overview of the remainder of the book—new views about the objectives of instruction and assessment, new types of items and assessment procedures, new perspectives about classroom-based assessment, new types of reports and response interpretation schemes, and future directions for research and development related to assessment.

SOME DISTINCTIONS THAT INFLUENCE VIEWS ABOUT ASSESSMENT

The aim of educational assessment is to produce information to assist in educational decision making, where the decision makers include administrators, policy makers, the public, parents, teachers, and students themselves. None of these consumers of assessment information can be ignored. But even though the high-stakes, accept-or-reject decisions of administrators and other policy makers are important, the authors in this book generally consider the needs of students, parents, and teachers as priorities, because their main goals are to facilitate learning.

In an *age of information*, educational assessment systems must be able to gather information from a variety of sources, not just tests, and they must provide information about individual students, groups of students, teachers of students, and programs for students. Also, the information itself must often include multidimensional profiles of a variety of achievements and abilities, and descriptions of relevant conditions under which individual profiles were developed. Furthermore, the information must be displayed in a form that is simple without being simplistic, and that also meets the needs of a variety of decision makers and decision-making purposes.

No single source of information can be expected to serve all purposes, and no single characterization of students (or groups, or teachers, or programs) is appropriate for all decision makers and decision-making issues. For example, in the assessment of individual students, when the goal is to document developing knowledge and abilities, some of the most useful sources of information involve problem-solving activities in which students simultaneously learn and document what they are learning. But when activities contribute to both learning and assessment, traditional conceptions of reliability must be revised or extended, because performance does not remain invariant across a string of equivalent tasks and the difficulty of a given task depends on whether it occurs early or late in the sequence.

Similarly, when attention shifts from multiple-choice or short-answer tests to project-sized activities (such as those that are emphasized in portfolio forms of assessment), the notion of validity generally needs to be expanded to include at least the following issues:

- *Construct validity:* Are the constructs that are being measured (or described) aligned with national curriculum standards? Do the understandings and abilities that are emphasized reflect a representative sample of those that contribute to success in a technology-based age of information?

- *Decision validity:* Is the information collected, analyzed, organized, aggregated, and displayed in a form appropriate to the entities that are being assessed? Are the results appropriate for the decision-making issues that are priorities to address?

- *Systemic validity:* Does the assessment program as a whole help to induce curricular and instructional changes that foster the development of the constructs that are being monitored?

- *Predictive validity:* Are results of the assessment correlated with performance in other relevant areas (such as success on tests in beginning college courses)?

When attention focuses on instructional decision making by teachers (and others who are familiar with students) rather than on policy decisions by administrators (and others who are not familiar with students), the risks and benefits associated with assessment results tend to change. For example, rapid turn-around times sometimes become more important than high precision or high accuracy or high reliability, and rich and meaningful reports often become more important than those that simply use "g" (general aptitude) as a euphemism for not knowing what is being tested.

In a technology-based society, assessment opportunities are influenced by the fact that reports can achieve simplicity without reducing all information to a single number. Simplicity can often be achieved by using reports that are computer-based, graphics-based, interactive, and inquiry oriented, and that focus on specific questions from specific people in specific situations.

For the alternative approaches to assessment that are emphasized in this book, it is important to underscore the fact that the authors are not simply concerned about developing new *modes* of assessment. They are primarily concerned about changing the *substance* of what is being measured. That is, they are not simply concerned about making minor changes to testing strategies (such as discontinuing the use of multiple-choice items, or focusing on computer-adaptive sequences of questions rather than fixed pencil-and-paper formats). They are concerned about the fact that, when most large-scale, high-stakes standardized tests are evaluated in terms of their alignment with the nationally endorsed *Curriculum and Evaluation Standards for School Mathematics* (NCTM, 1989), the understandings and abilities that are assessed tend to represent only narrow, obsolete, and untypical conceptions about (i) the nature of mathematics, (ii) the nature of real-life situations in which mathematics is useful in our modern world, and (iii) the nature of the knowledge and abilities that contribute to success in the preceding kinds of situations.

In one way or another, nearly every author in this book focuses on fundamental issues that involve clarifying the nature of children's mathematical knowledge; they also focus on developing operational definitions of what it means to "understand" the foundations of elementary mathematics when special attention is given to "deeper and higher-order conceptions" of foundation-level concepts, procedures, and principles. The authors in this book are mainly interested in (i) *examining* students' abilities, (ii) *documenting* their achievements, and (iii) *assessing* their progress. They are not especially interested in *testing*, measuring, and evaluating. Although these words are often used interchangeably, their meanings are not identical.

Examining, Documenting, Assessing, Testing, Measuring, Evaluating

Unless certain distinctions are sorted out with respect to the preceding words, they are likely to cause confusion when readers try to interpret the chapters in this book. Therefore, it is useful to consider the following contrasting definitions.

- *Examining:* To examine something means inspecting it closely.

- *Documenting:* To document something means gathering tangible evidence to demonstrate what occurred.

- *Assessing:* To assess something means describing its current state—probably with reference to some conceptual, or procedural, or developmental landmarks.

- *Testing:* To test something means creating an ordeal (or a barrier, or a filter) to inform decisions about acceptance or rejection, passing or failing.

- *Measuring:* To measure something means specifying both "how much" and "of what" (using some well-specified unit).

- *Evaluating:* To evaluate something means assigning a value to it.

The point of emphasizing the preceding distinctions is that it is possible to examine students, and to monitor their progress, without relying on a test (or other nonproductive ordeals). Also, it is possible to document students' achievements and abilities without measuring them in terms of some hypothesized abstract quantity and without reducing relevant information to a single-number score (or letter grade). Furthermore, it is possible to assess where students are and where they need to go (with respect to well-known landmarks of mathematical understandings and abilities) without assigning values to their current states and without comparing students with one another along a simplistic "good-bad" scale. In fact, *individuals who are "good" in mathematics often have exceedingly different profiles of strengths and weaknesses; learning progress can occur along a variety of paths and dimensions; accurate interpretations of achievements and abilities usually depend on the conditions under which development occurred.*

In current assessment reform efforts, the goal is not simply to produce new kinds of tests. The authors in this book generally have in mind a two-pronged approach to assessment reform: first, to increase the authenticity of tests, where authenticity is measured in terms of alignment with standards such as those published by the National Council of Teachers of Mathematics; and second, and equally important, to increase the credibility and fairness of assessment-relevant information taken from other sources, such as students' extended projects, or teachers' classroom observations.

LESSONS FROM PIECEMEAL APPROACHES TO CURRICULUM REFORM

Leaders in mathematics education have come to realize that piecemeal approaches to curriculum reform seldom succeed. The Mathematical Sciences Education Board writes:

> Few traces remain of the expensive, major curriculum development projects so prominent in the 1960s and 1970s. These free-standing curricula, which were intended to be adopted intact by schools, were naive about the process of change. Teachers were not directly involved in the development, and acceptance of new curricula was viewed as a top-down imposition. [Also] *commonly employed methods of evaluation were themselves obstacles to the teaching of beyond-minimum competencies.* (MSEB, 1990, p. 12; emphasis added)

Similarly, in the National Council of Teachers of Mathematics' recent series, Setting a Research Agenda for Mathematics Education (1989), Romberg describes another reason why many past curriculum reform efforts seldom achieved lasting success:

> In spite of the best intentions of developers and implementors, it was unreasonable to expect that new products or programs would be used as intended in most schools and classrooms. The reason for this is that public schools as they now operate are integrated social systems. Tinkering with parts, such as changing textbooks or the number of required courses, fails to change other components of the system. The traditions of the system force new products to be used in old ways. Current educational practice is based on a coherent set of ideas about goals, knowledge, work, and technology that came from a set of "scientific management" principles growing out of the industrial revolution of the past century. These ideas about schooling need to be challenged and replaced with an equally coherent set of practices in light of the economic and social revolution in which we are now engaged. *Current school mathematics operates within a coherent system; reform will happen only if an equally coherent system replaces it.* (NCTM, 1989, p. 21; emphasis added)

High-stakes tests are widely regarded as powerful leverage points to influence curriculum reform, because such tests tend to be aimed precisely at the infrastructure of schooling. First, tests are used to inform critical policy decisions that mold and shape the education system, and second, tests are used to define, clarify, and monitor goals of the system that is created. Therefore, for better or for worse, it is clear that high-stakes tests strongly influence both what is taught and how it is taught in mathematics education (Romberg, Zarinnia, and Williams, 1989). Such tests are not simply neutral indicators of learning outcomes. When rewards, punishments, and opportunities are at stake, they tend to become powerful components of instruction itself. Consequently, relevant professional and governmental organizations are increasingly making demands, such as "Discontinue use of standardized tests that are misaligned with national standards for curriculum" (Mathematical Sciences Education Board, 1990, p. 21).

Even though mathematics educators have come to realize that piecemeal approaches to curriculum reform are not sufficient, it is not widely recognized that piecemeal approaches to assessment will be equally unlikely to succeed. Recent policy statements from relevant professional and governmental organizations have made significant progress toward clarifying the nature of the most important goals of instruction, and innovative testing programs have produced a number of examples of test items and tests. However, a number of important issues that extend beyond the level of individual problems and isolated objectives need to be addressed. Consider the following:

- Adopting new statements of objectives may do little good if these objectives continue to be expressed as simple unorganized and unweighted lists of rules that convert to test items focusing on either (i) complex strings of low-level facts and skills which continue to be treated as though they should be mastered one at a time and in isolation, or (ii) global heuristics, strategies, or processes which are treated as though they function independently from any substantive mathematical ideas.

- Not using multiple-choice items may do little good if we continue to use problems and scoring procedures that impose artificial constraints by allowing only a single type and level of correct answer, because such constraints tend to trivialize the interpretation and model-refinement phases of problems where deeper and higher-order mathematical understandings tend to be emphasized.

- Gathering assessment information from new types of situations (such as students' project portfolios, or teachers' classroom observations) may do little good if the conceptually rich and instructionally relevant information that is gathered continues to be either collapsed onto a single number line, or left in an unsimplified or uninterpreted form that fails to address the needs of many important types of educational decision makers.

For assessment, just as for instruction, special attention should be focused on components of the system that directly improve the infrastructure of our education system. This goal can only be accomplished by dealing directly with the knowledge and beliefs of teachers, students, parents, administrators, and policy makers, especially their understandings about (i) the nature of mathematics, (ii) the nature of real-life learning and problem-solving situations, and (iii) the nature of abilities that contribute to success when new types of mathematical ideas and tools are used in new kinds of problem-solving situations.

To address such issues, it is useful to shift attention beyond traditional "bottom-up" approaches to assessment that begin by developing new types of objectives and then proceed to introduce new types of test items, tests, and reports. Sometimes it is useful to adopt a "top-down" approach that emphasizes the following questions: (i) What decision-making issues are priorities for educators to address? (ii) What kinds of reports are needed to inform these decisions? (iii) What types of information and data sources- -testing formats, item types, scoring procedures, aggregation techniques- -are appropriate for such reports?

If the bottom-up and top-down approaches are coordinated, then issues that need to be addressed include the following:

- *New levels and types of instructional objectives* need to be emphasized, including deeper and higher-order understandings of cognitive objectives that are not simply complex sequences of behavioral objectives and that are also not simply global process objectives or affective objectives.

- *New levels and types of problem-solving activities* must be emphasized, ranging from clearly defined pure mathematics problems to more complex and open-ended real-life projects in which realistic time, tools, and resources are available.

- *New sources of documentation* for achievement must be considered, including not only innovative new types of tests, but also students' project portfolios, teachers' classroom observations, one-to-one clinical interviews, and computer-based instructional activities.

- *New types of response interpretation procedures* must be developed, including those that go beyond assigning one-dimensional scores or letter grades to identify profiles of strengths and needs for individual students.

- *New data analysis models and procedures* must be developed based on assumptions that are consistent with (or at least not flagrantly inconsistent with) accepted views about the nature of mathematics, learning, and problem solving in real life situations.

- *New types of learning progress reports* must be generated that are simple and yet not simplistic, and that: (i) integrate information from a variety of sources, (ii) focus on patterns and trends in data, and (iii) inform a variety of decision makers and decision-making issues.

- *New decision makers* and *new decision-making issues* must be treated as priorities, where the decision makers include students, teachers, parents, and administrators, and the issues range from program accountability to diagnostic analyses of learning progress for individual students, with emphasis on equity and validity.

AN OVERVIEW OF THE BOOK

This book is divided into five parts. The first part describes some critical conceptual foundations for a new view of assessment in mathematics education. The second part focuses on innovative new items and assessment procedures. The third part describes several emerging new perspectives about classroom-based assessment. The fourth part focuses on examples of some new kinds of reports and response interpretation schemes that will be needed to support these broader views of assessment. Finally, the fifth part shifts attention toward some important future directions for assessment-related research and development and toward some practical matters that can subvert even assessment programs that are based on strong and sound conceptual foundations.

Foundations for a New View of Mathematics Assessment

Opening chapters of this book focus on recent developments in cognitive science, where a great deal of attention has been given to investigations dealing with the nature and development of students' mathematical knowledge and abilities. A prominent theme is that a clear definition of what is being assessed should guide the assessment process.

In chapter 2, Lesh and Lamon describe useful ways to think about the nature of authentic mathematical activities. In particular, they focus on responses to the following kinds of questions: What are examples of important cognitive objectives in mathematics instruction? How are cognitive objectives different from (yet related to) behavioral objectives, process objectives, or affective objectives which have been emphasized in the past? What does it mean to develop deeper or higher-order understandings of elementary cognitive objectives in mathematics? How can instruction and assessment be designed to elicit information regarding students' higher-order understandings?

In chapter 3, Goldin extends his earlier research on task variables and proposes a more comprehensive assessment framework built upon not only task analyses but also idea analyses and response analyses. He also deals with questions of the following type: How can the cognitive processes themselves become the objects of assessment? How are these process objectives linked to content understanding and to the growth of more sophisticated conceptualizations of mathematical ideas?

In chapter 4, Chazan and Yerushalmy focus on the domain of geometry and provide examples in which cognitive processes, such as verifying, conjecturing, and generalizing, are the objects of assessment. They also describe ways to assess higher-order inquiry skills which are some of the main instructional goals of innovative, computer-based instruction with computer-based tools such as The Geometric Supposer.

Innovative New Types of Items and Assessment Processes

The second part of this book highlights recent British and Dutch experiences in innovative, large-scale assessment. The juxtaposition of a specific-to-general approach and a general-to-specific approach to assessment reform also provides a tacit comparison of two research and development paradigms.

In chapters 5 and 6, Bell, Burkhardt, and Swan describe numerous examples of assessment tasks and marking schemes that have been used by examination boards in the United Kingdom, and in chapter 7, they also reflect on assessment-related issues that need to be addressed to effect curriculum change. In particular, they describe problems encountered in their assessment reform efforts and solutions that have proven to be effective in dealing with such problems.

In chapters 8 and 9, de Lange and Streefland give details and examples to illustrate the types of mathematics assessments that have been emphasized in the Netherlands, where the development of well-articulated assessment goals and procedures have been closely linked to instructional-based research and curriculum development. De Lange discusses several creative formats for assessing high-level goals, including free productions, two-stage tasks, and test-tests (tests that consist of designing tests). Streefland explains the nature of realistic instruction designed to facilitate thinking strategies and progressive mathematization.

New Perspectives About Classroom-based Assessment

In contrast to preceding chapters on macroplans for assessment reform, the chapters in Part III focus on assessment-relevant information that is based on classroom observations in American schools and on one-to-one teacher/student interviews. A common theme in this section is that an accurate profile of a student's mathematical understandings depends on the use of multiple techniques that are explicitly designed to overcome the limitations of any one method of capturing students' knowledge and thought processes.

In chapter 10, Maher, Davis, and Alston examine assessment on a microscopic level. That is, one teacher, on the basis of a brief classroom observation, makes judgments about the correctness of children's thinking.

In chapter 11, Ginsburg, Lopez, Mukhopadhyay, Yamamoto, Willis, and Kelly describe how combinations of screening instruments, probes, modified clinical interview techniques, and classroom observations can help teachers assess students' thinking in whole-class situations. They address questions such as, How can standard tests be improved to reflect a broader conception of what it means to think mathematically? How can information from multiple sources be integrated to give a more complete description of diverse aspects of understanding?

New Types of Reports and Response Interpretation Schemes

The fourth part of the book focuses on the interpretation, analysis, and reporting of assessment information based on new types of statistical models built on assumptions derived from modern cognitive psychology.

In chapter 12, Mislevy, Yamamoto, and Anacker discuss several recent advances in measurement and statistics that have been made by researchers who are aiming to connect quantitative models to qualitative differences in student thinking. Questions addressed include the following:

What kind of statistical models seem particularly promising to describe children's mathematical thinking? How can measurement procedures capture distinctions between critical states of understanding?

In chapter 13, Lesh and Lamon use the theoretical perspectives described in earlier chapters to focus on specific problems about ratio and proportion. Questions that are addressed include the following: What are some of common misconceptions about the formulation of "good" problems and "good" responses? What are some specific cognitive objectives of instruction about ratios and proportions? How can reliable scores be produced for problems with multiple solution paths which have different levels of difficulty?

In chapter 14, Lesh, Lamon, Gong, and Post describe one way to represent complex profiles of student abilities and achievement. They use computer-generated "learning progress maps" which succeed in being simple (from the point of view of educational decision makers) because they are graphics-based, interactive, and inquiry oriented, with details that are displayed only when they are requested by individual decision makers. That is, the reports are versatile enough to aggregate and display information in alternative ways to address a variety of decision-making issues.

Future Directions and Practical Concerns

In chapter 15, Lesh, Lamon, Lester, and Behr describe some of the most important assumptions underlying traditional types of standardized testing compared with the types of alternative assessments emphasized in this book. Chapter 15 also describes several priorities for future research (with special attention being given to issues related to equity, technology, and teacher education), and it concludes with specific examples taken from three current closely related projects that were explicitly designed to find practical ways to implement recommendations made in other chapters of this book.

In chapter 16, Schwartz summarizes conclusions reached by a series of recent projects focusing on "The Prices of Secrecy: The Social, Intellectual, and Psychological Costs of Current Assessment Practice" (Schwartz and Viator, 1990).

SUMMARY

It is our hope that this book will provide both general principles and specific examples to help support curriculum reform efforts that

- go beyond testing (for screening) to assessment (for informed decision making);

- go beyond a few discrete assessment events to the seamless integration of instruction and assessment;

- go beyond behavioral objectives to cognitive objectives;

- go beyond multiple choice tests to realistic tasks;

- go beyond right answers to reasoned answers;

- go beyond one-number scores to multi-dimensional profiles; and

- go beyond report cards to learning progress maps.

REFERENCES

Mathematical Sciences Education Board. (1990). *Reshaping school mathematics: A philosophy and framework for curriculum.* National Research Council. Washington, DC: National Academy Press.

National Council of Teachers of Mathematics. (1989). *Curriculum and evaluation standards for school mathematics.* Reston, VA: NCTM.

National Council of Teachers of Mathematics. (1989). *Setting a research agenda for mathematics education.* Reston, VA: NCTM.

Romberg, T.A., Zarinnia, E.A., and Williams, S. (1989). *The influence of mandated testing on mathematics instruction: Grade 8 teachers' perceptions.* Madison, WI: University of Wisconsin-Madison, National Center for Research in Mathematical Science Education.

Schwartz, J.L., and Viator, K.A. (1990). The prices of secrecy: The social, intellectual, and psychological costs of current assessment practice. Cambridge, MA: Educational Technology Center, Harvard Graduate School of Education.

2  Assessing Authentic Mathematical Performance

*Richard Lesh and
Susan J. Lamon*

This chapter addresses the following four questions: What are authentic mathematical activities? What kind of instructional objectives are priorities to address? What kind of problems are particularly useful for examining these priorities? What are some rules of thumb for creating such problems? To explain our answers to these questions, it is necessary to focus on the concept of *models*—in mathematics, in cognitive psychology, and in everyday situations.

CHARACTERISTICS OF AUTHENTIC MATHEMATICAL ACTIVITIES

Authentic mathematical activities are *actual work samples* taken from a representative collection of activities that are meaningful and important in their own right. They are not just surrogates for mathematical activities that are important in "real-life" situations.

To verify the mathematical authenticity of a collection of activities (beyond simply evaluating the authenticity of isolated items), both positively and negatively oriented criteria are relevant. That is, the activities as a whole should require students to use a representative sample of the knowledge and abilities that reflect targeted levels of competence in the field, and at the same time, the activities should avoid narrow, biased, obsolete, or instructionally counter-productive conceptions about the nature of mathematics, the nature of realistic problem-solving situations in which mathematics is useful, and the varieties of mathematical capabilities that are productive in these situations. Stated simply, authentic mathematical activities are those that involve: (i) real mathematics, (ii) realistic situations, (iii) questions or issues that might actually occur in a real-life situation, and (iv) realistic tools and resources. The most important kinds of problem-solving activities that we have in mind have the following characteristics:

- The problem solutions tend to require at least 5 to 50 minutes to construct.

- The contexts might reasonably occur in the students' everyday lives.

- The issues fit the interests and experiences of targeted students.

- The tasks encourage students to engage their personal knowledge, experience, and sense-making abilities.

- The objectives emphasize deeper and higher-order understandings and processes in elementary mathematics.

- The solution procedures allow students to use realistic tools and resources (such as hand-held calculators, pocket computers, notebook computers, consultants, colleagues, or "how-to" manuals).

- The activities generally require more than simply answering a specific question. They involve developing a mathematical model that can be used to describe, explain, manipulate, or predict the behavior of a variety of systems that occur in everyday situations.

- The activities contribute to both learning and assessment.

- The evaluation procedures recognize and reward more than a single type and level of correct response.

CHARACTERISTICS OF PRIORITY INSTRUCTIONAL OBJECTIVES

As noted in chapter 1, the revolution in mathematics education of the past decade resulted from powerful advances in technology, cognitive psychology, mathematics, and mathematics education, together with dramatic changes in demands for a competitive work force. Behavioral psychology (based on factual and procedural rules) has given way to cognitive psychology (based on models for making sense of real-life experiences), and technology-based tools have radically expanded the kinds of situations in which mathematics is useful, while simultaneously increasing the kinds of mathematics that are useful and the kinds of people who use mathematics on a daily basis.

In response to these trends, professional and governmental organizations have reached an unprecedented, theoretically sound, and future-oriented new consensus about the foundations of mathematics in an age of information (see, for example, National Council of Teachers of Mathematics [NCTM], 1989). To address the new goals of mathematics instruction, alternative assessment programs are being demanded, created, and refined from California to Connecticut, from Chicago to Houston, and from Australia to the Netherlands.

A hallmark of most new programs is a focus on "authentic performance" rather than on simply measuring some undefined factor. In general, new programs emphasize that it is not sufficient merely to replace multiple-choice items with fill-in-the-blank counterparts. Realistic applied problems are not created by just starting with an abstract algebraic (or arithmetic, or geometric) sentence and replacing the abstract symbols with the names of real objects. Unfortunately, clarity about goals for instruction does not necessarily result in equally clear operational definitions (that is, procedures and criteria) for measuring the extent to which the goals are being met. In spite of the enormous progress that has been made in specifying curriculum and evaluation standards for school mathematics (NCTM, 1989), what is not desirable continues to be far clearer than what is. In debates among leaders in current mathematics education reform movements, a great many issues remain to be resolved about the nature of "real" mathematics, realistic problems, realistic solutions, and realistic solution processes, as well as what it means to have a deeper or higher-order understanding of an elementary mathematical idea. For example:

- According to recent reports from professional and governmental organizations, harsh criticism has been aimed at the kinds of content-by-process matrices that have been used in traditional forms of assessment and instruction. But what is missing from such matrices? And how can alternative objectives frameworks avoid these deficiencies?

- In mathematics education today, it is fashionable to be a "constructivist." But what is it that students are expected to "construct"? Similarly, twenty years ago, mathematics laboratories focused on concrete "embodiments" to help students understand foundation-level mathematical concepts and principles. But what was it that the concrete materials were supposed to embody? When "discovery learning" was emphasized, what was it that students were supposed to discover? Surely the answer in each case must include something more than simply facts and procedural rules.

- As a reaction to instruction and assessment that focuses on decontextualized abstractions, cognitive scientists often emphasize the importance of *situated knowledge* (Greeno, 1988, 1988b, 1987). But what is an example of a situated understanding that is not simply a specific fact or rule that fails to generalize to other situations?

- In the past, college mathematics courses for preservice elementary teachers have often been characterized by superficial treatments of advanced (college level) topics rather than deeper or higher-order treatments of elementary (K-8) topics. But what does it mean to have a deeper or higher-order understanding of an elementary mathematical idea?

- When students are encouraged to use technology-based tools in instruction and testing, these tools tend to be "capability amplifiers" which are both conceptual and procedural in nature. For example, in realistic problem-solving situations, when students use tools such as pocket calculators (with graphing and symbol-manipulation capabilities) or notebook computers (with word processors and spreadsheets and other modeling or simulation tools), the tools often introduce new ways to think about givens, goals, and possible solution paths—in addition to providing new ways to get from givens to goals. They are much more than new ways to achieve old goals using old mathematical ideas. Furthermore, the psychological characteristics of a student-without-tools may be quite different from those of a student-with-tools. Mathematics educators have only recently begun to do extensive research on real-life problem solving involving these new breeds of students with amplified abilities. What ideas and processes should be emphasized, in teaching and testing, when students have access to powerful technology-based tools?

- The pendulum of curriculum reform tends to swing back and forth between basic skills and general problem-solving processes. While each extreme has some merit, both also have some serious shortcomings. For example, the atomization and fragmentation that tend to accompany an emphasis on discrete basic skills tend to undermine long-term learning. On the other hand, when attention focuses mainly on content-independent problem-solving processes, questions like, "Where is the mathematics?" often arise.

- To avoid such pendulum swings, it is important for teaching and testing to emphasize *cognitive objectives* (Greeno, 1976, 1980, 1988) that are not simply *behavioral objectives* (factual or procedural rules) and that are not simply reducible to general *process objectives* (content-independent rules). But how can cognitive objectives be stated in a form that is both clear and precise from the point of view of assessment, and instructionally sound and meaningful from the point of view of teachers and students?

According to the theoretical approach that will be emphasized in this chapter, all of the preceding questions are variations on a single theme, and a single idea is the key to answering all of them. For example, consider the question, "What characterizes a cognitive objective that distinguishes it from a behavioral objective or a global process objective?" Our answer is a *model* (that is, a complete functioning system for describing, explaining, constructing, modifying, manipulating, and predicting our increasingly complex world of experiences). In other words, to answer the questions that were stated at the beginning of this chapter, a primary goal will be to clarify what it means to base the most important cognitive objectives of mathematics instruction on the construction of mathematical models.

THE NATURE OF COGNITIVE MODELS

A principle that is a cornerstone of modern cognitive science is illustrated in Figure 1—that is, humans interpret experiences by mapping them to internal models. For example, what a person "sees" or "hears" in a given situation is filtered, organized, and interpreted by the cognitive models that he or she has constructed, based on past experience. Therefore, two people often interpret a single situation in quite different ways. If a person has only developed primitive models that fit a given situation, then the way this person thinks about the situation will tend to be relatively barren and distorted. In such cases, as suggested in Figure 1, the information that is available may not be noticed, and patterns that are not appropriate are likely to be perceived even though they are not objectively given.

Figure 1. Humans Interpret Experiences By Mapping to Internal Models

In language-oriented areas of artificial intelligence research, frames of reference are often referred to as scripts, or frames, or as other types of representational systems, such as Schank's stories (1991). In mathematics, there has been a long history of using a variety of representation systems to describe, explain, or predict experiences in real or possible worlds. These representational systems are usually referred to simply as models, because they are used to model structurally rich phenomena. No matter which of the preceding terms is used, the point is that humans generate interpretations that are influenced by both external data and internal models, and in many problem-solving/decision-making situations, the information that is relevant is based on hypothesized patterns and regularities beneath the surface, not just on calculations or deductions based on isolated pieces of data.

THE IMPORTANCE OF MODELS IN AN AGE OF INFORMATION

The principle that is illustrated in the figure above has also been described in a number of recent popular publications. For example, the best-selling book, *Megatrends 2000*, describes the principle in the following way:

> We are drowning in information and starving for knowledge....
> Without a structure, a frame of reference, the vast amount of data
> that comes your way each day will probably whiz right by you.
> (Naisbitt and Aburdene, 1990, p. 13).

In many real-life problem-solving and decision-making situations, an overwhelming amount of information is relevant, but this information often needs to be filtered, weighted, simplified, organized, or interpreted before it is useful. Sometimes needed information may not be provided, yet a decision may need to be made anyway, and made within specified time limits, budget constraints, risks, and margins for error. Models are needed to provide meaningful patterns that can be used (i) to make rapid decisions based on strategically selected cues, (ii) to fill holes or go beyond a minimum set of information, (iii) to provide explanations of how facts are related to one another, or (iv) to provide hypotheses about missing (or hidden or disguised) objects or events that may need to be actively sought out, generated, or (re)interpreted.

The essence of an age of information is that the models humans develop to think about the world also mold and shape that world. This is why, in professions ranging from business to engineering to law to music, many of the patterns and regularities that exist in the world are not simply preordained laws of nature, they are model-based products of human constructions. In fact, as the year 2000 approaches, many of the most important systems that humans must learn to understand (that is, construct, analyze, explain, manipulate, predict, and control) are businesses, communication networks, social systems, and other systems that are based on models that are themselves constructed by humans. Furthermore, many of the most important characteristics of these systems are based on patterns and regularities beneath the surface, not just on surface-level perceptions.

Increasingly, problem solving and decision making require model construction, model refinement, and model adaptation. In fields ranging from business to engineering, and from the arts to the sciences, many of the most important goals of education involve the construction of models that provide conceptual and procedural amplifiers for interacting with our increasingly complex worlds of experience.

FACILITATING THE CONSTRUCTION OF POWERFUL MODELS

Many professional schools, such as our nation's leading business and engineering schools, furnish a wealth of examples of teaching models for thinking about the world. The value of such models is easy to recognize if we imagine a well-educated modern business manager or engineer sent back in time. Modern professionals would often appear to be unusually intelligent compared with their counterparts in earlier centuries. Their enhanced capabilities would not be the result of higher general intelligence. Instead, their enhanced capabilities would result from their use of the powerful, elementary-but-deep models and tools that our culture enables students to construct.

In fields where the most important goals of instruction are associated with the construction of models for making (and making sense of) complex systems, *case studies* are often used to help students construct models that have proven to possess the greatest power and utility. In business schools, for example, students often use spreadsheets or similar tools to construct models to explain trends or patterns in systems which can then be used as prototypes for making sense of other structurally related situations.

Models develop from concrete to abstract, from intuitions to formal systems, and from situated knowledge to decontextualized understandings. To encourage their development, the situations that are used resemble the concrete embodiments which are familiar to mathematics educators from their experiences in mathematics laboratory forms of instruction (Dienes, 1957; Lesh, Post, and Behr, 1987).

THE NATURE OF "REAL" MATHEMATICS

In a series of recent reports from the National Council of Teachers of Mathematics (1989), the Mathematics Association of America (Steen, 1988), the American Association for the Advancement of Science (1989), and the Mathematical Sciences Education Board (1990, 1990a), the mathematics education community has reached a new consensus about the nature of real mathematics (Ernest, 1991). The key characteristics that distinguish mathematics from other domains of knowledge can be summarized as follows:

> Mathematics is the science and language of pattern.... As biology is a science of living organisms and physics is a science of matter and energy, so mathematics is a science of patterns.... To know mathematics is to investigate and express relationships among patterns: to discern patterns in complex and obscure contexts; to understand and transform relations among patterns; to classify, encode, and describe patterns; to read and write in the language of patterns; and to employ knowledge of patterns for various practical purposes.... Facts, formulas, and information have value only to the extent that they support effective mathematical activity. (Mathematical Sciences Education Board, 1990, p. 5)

From this perspective, a simplified view of mathematics learning and problem solving would involve the construction, refinement, or elaboration of models (Figure 2), plus (i) mappings from "real-world" situations into a "model world," (ii) transformations within the model world, and (iii) explanations or predictions from the model world back into the real world situation.

- Doing "pure" mathematics means investigating patterns for their own sake, by constructing and transforming them in structurally interesting ways, and by studying their structural properties.

- Doing "applied" mathematics means using patterns as models (or structural metaphors, or quantitative structures) to describe, explain, predict, or control other systems—with refinements, extensions, and adaptations being made to these models when necessary.

Figure 2. A Simplified View of Mathematical Modeling

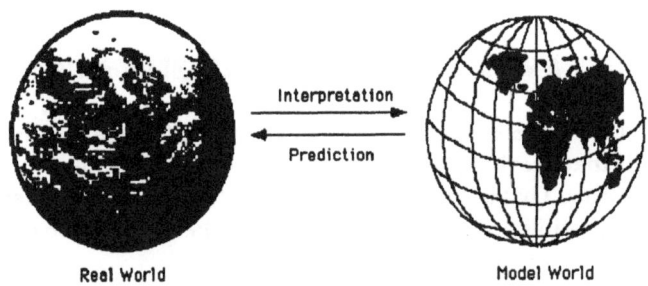

Real World　　　　　　　　　　Model World

Of course, this description of mathematical problem solving is too simplistic for many purposes. For example, when humans use a given mathematical model to describe/explain/predict/control a given learning or problem-solving situation, their models tend to be partly internal and partly external. Also, several unstable and possibly conflicting models are often used in sequence and/or in parallel, with each model emphasizing or deemphasizing somewhat different aspects of the situation or of the underlying abstract system. Furthermore, a given (abstract) model tends to be embedded simultaneously within a variety of interacting notation systems (for example, involving written symbols, spoken language, manipulatable concrete models, static pictures or diagrams, or real-life systems, prototypes, or structural metaphors) each of which again emphasizes or deemphasizes different aspects to the modeled situation. Therefore, solutions to realistic problems often involve (i) parallel and interactive uses of several distinct notation systems and/or problem interpretations, (ii) partial mappings between components of the modeled situation and corresponding components of the model and/or notation system, and (iii) constructing, adapting, extending, integrating, differentiating, and/or refining of a series of models that gradually become more complete, accurate, and sophisticated. However, for the purposes of this chapter, the simplified portrayal of mathematical modeling is sufficient. The essential points are the following:

- Many of the most important mathematical "objects" that students are expected to study are not simply isolated factual and procedural rules or general problem-solving processes. They are models (or patterns, or structural metaphors) that are useful for building and making sense of real and possible worlds.

- Mathematical models are complete functioning systems, which consist of: (i) *elements* (for example, quantities, ratios of quantities, shapes, coordinates), (ii) *relationships* among elements within the system, (iii) *operations* or *transformations* on elements in the system, and (iv) *patterns* that govern the behavior of the relations, operations, and transformations.

AN EXAMPLE OF A MATHEMATICAL MODEL IN ELEMENTARY MATHEMATICS

Cartesian coordinate systems are examples of elementary mathematical models that provide powerful ways to describe (or think about) real or possible worlds. That is, by imposing rectangular coordinate systems on the world, it is possible to use equations and numbers to describe whole new classes of situations, locations, or relationships.

An important point to notice about elementary-but-deep models is that it often takes the genius of someone like René Descartes to introduce the models on which the conceptual system is based. Yet today, it is relatively easy to help average-ability middle schoolers construct these powerful models. Furthermore, if these middle schoolers could be sent back in time to a period prior to the birth of René Descartes, they would often appear to be geniuses, because, in a wide variety of structurally complex situations, they would be able to generate descriptions, explanations, and predictions that would seem miraculous to people in ancient civilizations.

Some other important points to notice about Cartesian coordinate systems include the following:

- Cartesian coordinate systems are capability amplifiers that function in much the same way as pocket calculators. But while pocket calculators tend to emphasize procedural capabilities, Cartesian coordinates tend to emphasize conceptual amplifiers. Still, both are capability amplifiers that involve conceptual and procedural components.

- Even though Cartesian coordinate systems are often represented using diagrams that look something like game boards for checkers, mathematicians often work with Cartesian systems without using such diagrams; the diagrams are only embodiments in which the abstract system is embedded. The model is the underlying abstract system itself.

- Even though Cartesian coordinate systems are linked to a variety of rules for comparing, ordering, organizing, combining, and transforming data, their primary function is not really data processing. Their primary function is to provide a description or interpretation for making sense of underlying patterns and regularities, and for expressing them in a form that is generative and easily manipulatable.

- Even though Cartesian coordinate systems are obviously human constructions that are not inherent parts of nature, educated citizens in the 1990s often have difficulty remembering what the world was like before these powerful conceptual systems were such familiar parts of our cultural heritage. In fact, today, many people even think it's obvious to view the world within a four-dimensional framework in which time is the fourth dimension. In science and mathematics, things that are "obvious" have evolved dramatically from one era to another as a function of the conceptual models that we use to describe and explain our experiences.

AN EXAMPLE OF A MATHEMATICAL MODEL IN ELEMENTARY ARITHMETIC

In elementary arithmetic, there are many elementary-but-deep models of a type similar to Cartesian coordinate systems. While the elements in Cartesian coordinate systems are *coordinates* of the form (n,m), the elements in other mathematical models may be mathematical entities ranging from *signed numbers*, to *ratios*, to *proportions*, to *functions*, to *vectors*, to *shapes*, to *sequences*. Some of the models that underlie arithmetic become so familiar to citizens of modern societies that they seem to be part of nature. However, the names of our number systems are strong reminders of the historic obstacles that had to be overcome before the underlying models associated with them gained acceptability. For example:

- In the beginning, there were *natural numbers* {1,2,3,4,...}. The invention of *zero* occurred much later. *Negative numbers* were looked upon negatively, and *fractions* were considered to be unacceptable in polite society.

- Later in history, as fractions and negatives began to seem sensible (or rational), they were were described as being *rational numbers*, as opposed to *irrational numbers* (like √2 and π) that just didn't make any sense to most people.

- As more time passed, even irrational numbers were included in the set of *real numbers*, even though *imaginary numbers* (such as *i* or √-1) continued to be treated as unreal.

- Eventually, *complex numbers* proved to be very useful to describe important events in nature, even though they included numbers that had previously been called imaginary.

In the history of mathematics, it took centuries to construct the notation systems and underlying models that citizens of twentieth century societies take for granted. Furthermore, before these models were constructed, earlier civilizations were often severely limited in the kinds of economic, social, technological, or scientific systems that they were able to create, and they were often similarly hampered in their attempts to make sense of many kinds of patterns and regularities that occurred in their worlds or experience.

Figure 3 suggests that many students in our schools today have limitations similar to those that were experienced by citizens of ancient cultures because what they see or hear is filtered, organized, and interpreted by the cognitive models that they have constructed, and because their primitive models are only able to produce barren and distorted interpretations of their experiences. For example, extensive research on the development of students' knowledge reveals that a large share of the American population is often extremely restricted in their reasoning abilities if they must make judgments about systems that involve more than directly observable counts and measures (Lesh, Behr, and Post, 1987).

Figure 3. Humans Interpret Underlying Patterns & Regularities By Mapping To Mathematical Models

Even for middle school teachers whose relevant computational abilities tend to be flawless, reasoning abilities are often extremely restricted if the emphasis shifts from systems whose elements are directly perceivable quantities to systems whose elements involve higher-order entities such as ratios, rates, or functions (where the elements involve relationships among several quantities). For example, consider the tasks shown in Figure 4, which are typical of those for which success rates have been consistently below 50 percent for large samples of 8th graders or adults, including middle school teachers (see, for example, Lesh, Post, and Behr 1988; Post, Behr, Lesh, and Harel, 1991).

Figure 4

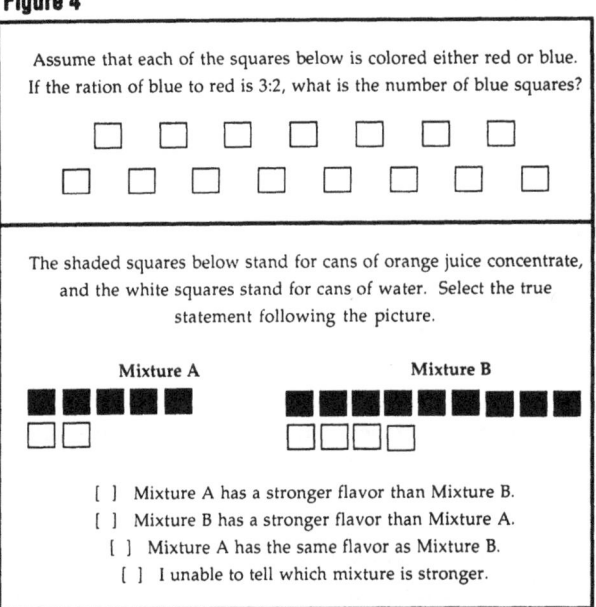

Many more examples could be given that emphasize conceptual rather than procedural proficiency, and similar results could be cited based on more realistic ethnographic observations in real-life settings (see, for example, Saxe, 1991; Lave, 1988) or on clinical interviews involving concrete materials (Lamon, 1990; Lesh, Landau, and Hamilton, 1983). The results of such studies reveal how weak the conceptual foundations are underlying many students' procedural facility. It is important to note that some students who perform poorly on the conceptually oriented tasks may be highly capable individuals. But, like intelligent citizens of early cultures, they have simply never had the opportunity or the need to construct mathematical models whose elements involve anything more sophisticated than simple counts and measures.

Unfairness on tests is easy to recognize if some students have access to powerful electronic tools that are unavailable to other students. But unfairness of another sort occurs when some students go into tests equipped with powerful conceptual models that other students have never encountered. The situation is further exacerbated when deficiencies with respect to such capability amplifiers are portrayed as differences in native abilities.

DEEPER AND HIGHER-ORDER UNDERSTANDINGS OF A CONCEPTUAL MODEL

To clarify what it means to have a deeper or higher-order understanding of a mathematical model, it is useful to return to fields such as business and engineering where there is a long history of treating acquisition of models as a key goal of instruction. In such fields, it is common for the most powerful models to be accompanied by (i) technical language to facilitate communication about basic models and the systems they describe, (ii) specialized notation systems to expedite the construction and manipulation of particularly important models and systems, (iii) diagrams or descriptions that focus attention on holistic characteristics of the system-as-a-whole, and (iv) formulas, computation tools, spreadsheet programs, or other simulation and modeling tools that can be used to generate hypotheses, descriptions, or predictions in typical decision-making situations. However, it is clear that understanding the underlying models involves far more than simply being able to remember and execute rules within the notation systems, diagrams, rules, and tools. In fact, in realistic settings, when the purpose is to get from specifically stated givens to clearly identified goals, the relatively simple job of executing the relevant procedures is regarded as a low-level clerical task; if "data crunching" is necessary, powerful technology-based tools tend to be used. Therefore, in our nation's leading schools of business and engineering, expert job interviewers whose goal is to hire individuals capable of higher-level skills, including on-the-job problem solving and decision making, tend to focus on questions that emphasize the following types of deeper and higher-order understandings:

- Students are asked to interpret standard and nonstandard situations using traditional models that have the greatest power and usefulness.

- Students are asked to construct new models, or sort out and integrate existing models, to determine (i) which kind of information should be gathered, (ii) how the data should be interpreted, quantified, and analyzed, and (iii) whether trial results are sensible and useful.

- Students are asked to analyze or critique competing interpretations of a given situation, for example, by detecting significant model-reality mismatches using standard models, or by evaluating risks, benefits, and underlying assumptions associated with alternative models and suggesting appropriate modifications or extensions to the model.

- Students are required to perform tasks in which fact/skill-level understandings and abilities must be embedded within flexible, adaptable, and well-integrated systems of knowledge rather than treating lower-level definitions, facts, and rules as rigid and isolated pieces of dogma.

In such situations, it is usually easy to recognize that (i) every model deemphasizes and simplifies some aspects of reality in order to emphasize and clarify other aspects, and (ii) every model is based on some assumptions that do not completely fit the realities in the problem situation. Therefore, when models are important cognitive objectives of instruction, one of the main goals of assessment is to probe the nature of the interpreting models that individual students have constructed to determine their accuracy, complexity, completeness, flexibility, and stability when they are used to generate descriptions, explanations, and predictions in a variety of problem-solving settings and for a variety of purposes under differing conditions. For example, when explanations are generated, the quality of responses depends on the following kinds of criteria: (i) How much information was noticed? (ii) How well (and how flexibly) was the perceived information organized? (iii) How sophisticated, or complex, or rich were the relationships that were noticed? (iv) Were observations and subjective relationships perceived that were not objectively given?

To assess the types of models that individual students have constructed, and to assess the stability of these models in a variety of situations and conditions, assessment must go beyond testing the *amount* of information that a student notices in a given situation; it must also assess the nature of the *patterns* of information that are noticed and identify valid and invalid assumptions that are made about underlying regularities. Similarly, the goal of instruction is not simply to get students to master more factual and procedural rules, but rather to help them construct powerful models that provide conceptual/ procedural amplifiers in priority types of problem-solving and decision-making situations.

DEEPER AND HIGHER-ORDER UNDERSTANDINGS IN ELEMENTARY MATHEMATICS

Exploring similarities between the kind of models that occur in elementary mathematics and the kind that are emphasized in professional schools (or in other disciplines) has helped to identify the following six types of deeper and higher-order understandings:

- *Students should use models to interpret real-life situations.* That is, they should go beyond school math problems (in which the givens, goals, and available solution steps are clearly specified) to also deal with real-life problems in which models must be constructed to generate descriptions, explanations, and predictions.

- *Students should think about underlying models.* Students should go beyond thinking *with* a given mathematical model to also think *about* the model as a complete functioning system, for example, by investigating the accuracy, precision, and goodness of fit of the descriptions that are generated for given problem-solving situations, and by investigating structural properties of the model (or system as a whole).

- *Students should explore similarities and differences among alternative representation systems associated with a given model.* Because most mathematical models can be embodied within a variety of alternative notation systems (involving spoken language, written symbols, static graphics, manipulatable concrete materials, or real-life prototype experiences), students should go beyond executing factual and procedural rules within a given notation system and should also investigate (i) translations from one notation system to another, and (ii) strengths and weaknesses associated with alternative embodiments.

- *Students should think about thinking.* That is, they should think about the processes that are needed to construct and refine an adequate model, and plan, monitor, and assess the construction process.

- *Students should think about systems of knowledge.* They should go beyond learning lists of isolated facts and rules to also develop well organized and clearly differentiated systems of knowledge. In particular, they should go beyond constructing isolated models to also develop coherent systems of models, for example, by investigating similarities and differences among alternative models in a variety of problem-solving situations.

- *Students should think about the nature of mathematics and assess their own personal capabilities.* Beyond constructing and investigating mathematical models, students should also form accurate and productive beliefs about (i) the nature of mathematical models (for example, concerning their power and limitations in real-life problem-solving situations), and (ii) the nature of their own personal problem-solving and model-constructing capabilities.

When cognitive objectives of instruction include the construction and modification of models for describing, explaining, constructing, modifying, manipulating, and predicting patterns and regularities that govern the behavior of complex systems, the six categories described above represent distinct types of deeper and higher-order understandings associated with models and modeling. Furthermore, as Table 1 shows, they also correspond nearly one-to-one with the most interesting new categories of objectives emphasized in the *Curriculum and Evaluation Standards for School Mathematics* published by the National Council of Teachers of Mathematics

Table 1. Relationships Between NCTM Standards and Model-Based Knowledge and Abilities

New Categories of NCTM Objectives		Higher-Order Understandings of Cognitive Objectives		Model-Based Process Objectives
mathematics as problem solving	<=>	think about "real life" situations	<=>	analyze & interpret problem situations
number sense & estimation	<=>	^		
mathematical structure	<=>	think about models	<=>	generalize & extend overall solutions
mathematics as communication	<=>	think about representation systems	<=>	translate within and between modes of representation
mathematics as reasoning	<=>	think about thinking	<=>	plan & execute solution steps
			<=>	monitor & assess intermediate & final results
mathematics as connections	<=>	think about systems of models	<=>	identify similarities and differences
	<=>	think about real world applications	<=>	translate and interpret between models and the real world
mathematical disposition	<=>	think about mathematics and personal capabilities		

(NCTM, 1989). The table also shows how the seven categories of NCTM objectives correspond with model-based process objectives that we have identified in past publications (Lesh, 1990).

ACTIVITIES THAT ENCOURAGE THE DEVELOPMENT OF MATHEMATICAL MODELS

This section contains several examples of model-eliciting activities in real-life situations (Problems 1-4). Each example has been used with middle school students through adults, with problem solvers working either individually or in three-person teams. Each activity takes at least 30 to 60 minutes to complete. The solutions involve constructing interpretations that are based on ratios or proportional reasoning (A/B=C/D). No artificial restrictions are placed on time, tools, or resources; in particular, calculators and computers are available. The goal is not simply to produce answers to specific questions. Instead, the responses involve producing a description, or an explanation, or a prediction which requires students to document explicitly how they are thinking about the situations. Then the descriptions and explanations that students construct are used as models to interpret other structurally similar situations. In other words, the problems are more like case studies in professional schools than they are like traditional kinds of textbook word problems.

Problem 1

THE BANK ROBBERY PROBLEM

Students were shown the following newspaper article.

Two gunmen held up the Second National Bank, around ten thirty this morning. One was a middle-aged man in a grey suit, who pointed a large handgun at a frightened teller and demanded cash. Another man pulled a shotgun out of a laundry sack and held it on the customers, while tellers filled the sack with money, most of it in small bills. The two perpetrators escaped before police arrived.

"It was awful!" exclaimed Louise DiChello, head teller for Second National. "I pushed the alarm button. Where are the police when you need them?"

Responding to charges of slow response to emergency calls, Police Commissioner Tyrone Campbell claimed that the bank's silent alarm did not go off during the robbery. "If the hardware fails, the police get the blame."

Bank President George Bromley stated that, according to his accounting, the amount stolen was close to a million dollars. "The bank's insurance will cover the loss. But it's sad to see thieves and thugs get away with crimes like this!"

Anyone having information about today's robbery should call the CrimeStoppers HotLine 1-800-STOP*IT.

Problem A: Could the events really have happened as told in the newspaper story? Could two robbers escape with a million dollars in small bills in a laundry sack as reported? - - - Analyze the situation. What suggestions would you offer for solving this crime? Write a note to the detectives on the case explaining your reasoning. Give details so they will understand.

Problem B: Imagine you are a detective investigating the crime. An employee of the bank tells you that Bank President Bromley has been suffering financial setbacks recently. She wonders if the robbery was fake. You wonder about the large sum reported stolen, and do some calculations to deduce whether it would have been physically possible to carry out a million dollars in small bills as described. You figure an American dollar bill weights about 1 gram. If 50% of the stolen money was $1 bills, 25% was $5 bills and 25% was $10 bills, estimate how much the loot would have weighed. Would it have been possible for 2 robbers to carry this amount out of the bank?

Problem 2

THE CD TOSSING GAME

For a popular carnival game, a player tosses a coin onto a gameboard that looks like a checker board. If the coin touches a lines on the gameboard the player loses. If not, the player wins! Players get three throws for a dollar!

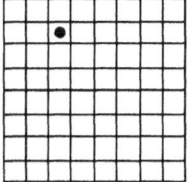

You've been asked to design a similar game for a fund raising carnival at your school. For prizes, a local record store will sell up to 100 CD's to your class for $5 per disc. You can choose any discs you want just as long as the regular price of the CD is less than $20. To make the game more fun, you've decided to let players throw old scratched CDs rather than coins. Two sizes of CDs can be used (3 inch discs and 5 inch discs). So, the cost of three throws can depend on which size a player chooses to throw.

All plans for games must be approved by the carnival planning committee. You want to make as big of a profit as possible. But, if too few people win, people won't want to play the game. Write a plan to submit to the carnival planning committee that includes details about the size of the game board, the cost of throws, the chances of winning, and an estimate of the expected profits.

Problem 3

THE SEARS CATALOG PROBLEM

Problem: Fred Findey began teaching here at the high school 10 years ago. He and his new bride rented an apartment at 318 Main Street for $315 per month, and he also bought a new VW Rabbit for $6,200. His starting salary was $16,300 per year. This year, Fred's sister, Pam, also began teaching at the high school. Pam, too, just got married. In fact, she rented the very same apartment as her brother did 10 years ago, only now the rent is $610 per month. She also bought a new VW Rabbit that sold for $13,700. Using this information, and these (*see items below*) newspapers and catalogs, write a letter to the School Board recommending (and justifying) how much you think Pam should get paid.

note: Students were given: (i) a calculator; (ii) two Sears catalogs -- one current, and the other from 10 years ago, and (iii) two newspapers – one current, and the other from 10 years ago.

Problem 4

DON'T DRINK & DRIVE HOTLINE

To prepare for the problem, show students the following newspaper article.

Drunk Driving Laws for the New Year

The new drunk driving law will go into effect at 12 midnight January 1, meaning New Year's Eve celebrators should have designated non-drinking drivers in their group. A driver who is found to have a blood alcohol level of 0.05% or more will be subject to a $500 fine and will risk prosecution and a possible prison term.

New Year's Eve party-goers should be aware of a few facts. A 12 ounce can of beer, a 5 ounce glass of wine and a 1 ounce glass of hard liquor should each be counted as one drink. The effect of one drink on a heavy person is much more than the effect of one drink on a light person.

Experts point out that once a person has stopped drinking, the blood alcohol level drops about 0.015% every hour. For example, a 140 pound person with a blood alcohol level of 0.10% would have a blood alcohol level of 0.085% after one hour and a blood alcohol level of 0.070% after two hours.

While a person can legally drive at blood alcohol levels of 0.04% or lower, experts emphasize that impairment is still possible, and the best policy is to not drink and drive.

Most restaurants and banquet halls are offering free soft drinks to designated drivers and free taxi rides home upon request. For party-goers who prefer to avoid driving or riding with others, overnight packages are available at most local hotels.

Blood Alcohol Concentration (%)
Within One Hour

Body Weight	Number of Drinks				
	1	2	3	4	5
100	0.04	0.09	0.15	0.20	0.25
120	0.03	0.08	0.12	0.16	0.21
140	0.02	0.06	0.10	0.14	0.18
160	0.02	0.05	0.09	0.12	0.15
180	0.02	0.05	0.08	0.10	0.13
200	0.01	0.04	0.07	0.09	0.12

Problem: Your community has decided to start the **Don't Drink & Drive Hotline**. As part of this group effort your class has been asked to develop a method so that hotline workers can quickly estimate a caller's blood alcohol level. Develop useful tools (for example, tables, graphs, computer software, etc.) that will be helpful to hotline workers and describe how to use your tools/method for scenarios like those below.

Test Your Materials: After developing your tools/method for estimating a caller's blood alcohol level, test your materials by role playing a telephone call. Have someone (a friend, a classmate, a parent, etc.) read one of the scenarios below and pretend to call you. You should pretend to be the hotline employee who answers the call. Ask questions to gather the information that you need and use your quick-and-easy-to-use method to estimate the blood alcohol level of the caller.

> SCENARIO I. Jake ate lunch between 12:00 and 1:00 and had two glasses of beer. He weighs 240 pounds. It is now 3:00. Pretend that you are Jake and are calling the DDD Hotline.
>
> SCENARIO II. Shortly after arriving at a friend's house at 6:00, Regina had a cocktail. Dinner was served from 7:00 to 9:00. With dinner she drank two glasses of red wine. Regina weighs 112 pounds. It is 11:00. Pretend that you are Regina and are calling the DDD Hotline.

To create effective model-eliciting activities, one of the main goals is to create problems that encourage (and do not *dis*courage) sense-making based on students' personal knowledge and experience. A second rule is to focus on problems that encourage the construction and investigation of elementary mathematical models that are likely to have the greatest power and utility (short-term and long-term) for the students who are involved. A third rule is to create tasks in which students go beyond (unconscious) thinking with the models to also (consciously) think about them, for example, by constructing them, by modifying and adapting them for a variety of purposes, and by investigating their structural properties in a variety of meaningful situations. A fourth rule is to ensure that an appropriate range of problem types, response types, and interpretation possibilities are represented, and, in particular, to ensure that realistic types of givens, goals, tools, settings, or procedures are not neglected. A fifth rule is to avoid problems that have only a single level and/or type of correct response.

The last rule is especially important because, in general, the only way to create problems with only a single correct answer is to eliminate the phases of problem solving that focus on processes such as problem interpretation, or response justification, or the testing and refinement of hypotheses about underlying patterns and regularities. In other words, to eliminate the possibility of more than a single correct answer, it is usually necessary to eliminate exactly those phases of problem solving in which attention is focused on the underlying mathematical structure of the problem and on deeper and higher-order mathematical understandings of the mathematical structure.

To create activities that encourage students to construct significant mathematical models, authors should ask themselves, "What kinds of situations create the need for people to create models, whether they are working in mathematics, in science, in business, or in everyday life?" Answers to this question include the following:

- *Models are needed when it is necessary to make predictions based on underlying patterns or regularities,* for example, to anticipate real events, to reconstruct past events, or to simulate inaccessible events.

- *Models are needed when constructions or explanations are requested explicitly,* for example, for describing hypothesized patterns and regularities, or for describing decision-making situations in which too much or not enough information is available.

- *Models are needed when it is necessary to justify or explain decisions,* for example, by describing underlying assumptions, conditions, and alternatives.

- *Models are needed to resolve interpretation mismatches,* for example, between hypotheses and reality or between two competing interpretations, predictions, or explanations of a given situation.

- *Models are needed when it is necessary to recreate and critically analyze conclusions or descriptions generated by others.*

In general, to develop effective model-eliciting activities, one of the major goals is to create meaningful contexts in which students will recognize the need for a model. Then model construction tends to follow naturally.

RULES OF THUMB FOR WRITING EFFECTIVE MODEL-ELICITING ACTIVITIES

Twenty years ago, when authors of this chapter first began to conduct research on problems that people today refer to as *authentic performance activities,* we usually cited reality as our primary criterion for distinguishing "good" problems from "bad" (see, for example, Bell, Fuson, and Lesh, 1976). We still consider realism to be a praiseworthy goal, though realism is a principle that is not so simple and straightforward to implement. To see why, consider the following:

- Problems that are real for authors or teachers often have little to do with a middle schooler's reality. For example, a teenager's main interests often center around fanciful situations, or around "what if" distortions of the real world, rather than around reality in the more traditional sense.

- A topic that is timely (or "hip") one year often is treated as old fashioned the next, and one student's reality is often quite different from another's. For example, a rural middle-class white male's reality tends to be considerably different from that of an inner-city, lower-SES black female.

- Videodisc portrayals of real-life situations sometimes encourage students to suspend their reality judgments in the same way that they do when they watch movies and television. Explorations with concrete materials (such as Cuisenaire rods) frequently turn out to be as abstract as if they had been done with written symbols.

- Computer-based explorations sometimes become very real to students, even though they often make no reference at all to objects or events in students' everyday lives.

- Problems that focus on skills for low-level employment often turn off the very students who were expected to consider them relevant. On the other hand, pure mathematics activities (for example, involving pattern exploration in number theory) often give students a realistic view of mathematics as it really is for a research mathematician, even though the experiences might seem far removed from the students' everyday life.

Over the years, another problem attribute proved to be even more fundamental than "reality" for describing of the kind of problems that we want to emphasize. That is, the real purpose of emphasizing realistic problems was to encourage students to construct and investigate powerful and useful mathematical ideas (that is, models and principles associated with models) based on extensions, adaptations, or refinements of their own personal knowledge and experience. Therefore, we refer to our most effective problems as *model-eliciting activities* (Lesh and Kaput, 1988) because their solutions involve constructing, transforming, investigating, modifying, integrating, differentiating, or using mathematical models or patterns. For example, typical solutions to the Sears Catalog Problem, described in the preceding section, illustrate the kinds of mathematical ideas students invent in model-eliciting situations.

When we first began to gather information about students' solutions to the Sears Catalog Problem, the main model (or reasoning pattern) that we expected students to construct had to do with proportional reasoning of the form $A/B=C/D$. In fact, most of the students we observed did indeed end up thinking about the problem using some type of proportional reasoning model. But a high percentage went far beyond an interpretation of the problem based on simple ratios, proportions, or linear equations. For example, to find a useful way to think about the situation, students often invented ideas such as weighted averages, trends, interpolation, extrapolation, data sampling, margins of error, or others that their teachers thought were too sophisticated for youngsters to learn.

In their solutions to the Sears Catalog Problem, students also invented surprisingly sophisticated ways to deal with the following kinds of issues:

- *Data sampling.* For example, how many, and which, items should be considered? Which should be ignored? What should be done about unusual cases (such as the fact that the cost of pocket calculators decreased, while the cost of most other items increased)? How should the data be classified or organized? What kinds of patterns and relationships (for example, additive, multiplicative, exponential) should be hypothesized?

- *The quantification of qualitative information.* For example, what weights should be assigned to various kinds of information? How can information be merged that is based on different kinds of quantities or units of measure?

- *Conditional results.* For example, because of equity issues, or risks and benefits associated with alternative answers, final decisions about salaries should depend on additional information about conditions in the past and perhaps on assumptions pertaining to the present and future.

In fact, we found that, when model-eliciting activities are used to encourage students to make sense of problem-solving and decision-making situations based on their own personal knowledge and experiences, students who had been labeled average or below average often emerged as extraordinarily talented, because they routinely invented (or significantly extended, modified, or refined) mathematical models that went far beyond those that their teachers believed they could be taught. (Lesh and Akerstrom, 1982; Lesh and Zawojewski, 1987)

AN EXAMPLE: TEACHERS SOLUTION OF A MODEL-ELICITING PROBLEM

Many K-8 teachers have not had much experience working on project-sized, model-eliciting problems. Therefore, before they try to create such problems for their own students, and before they try to assess students' responses to such problems, it is useful for them to participate (as students) in a few problems that were explicitly designed to fit their own everyday decision-making activities.

The Math Placement Problem (Problems 5A-5D) was designed especially to help teachers gain firsthand experience with realistic model-eliciting problems. It was also designed to focus on two questions that are special concerns for many teachers: (i) What understandings and abilities should be emphasized when students are allowed to use technology-based tools? (ii) What understandings and abilities should be emphasized when students are allowed to work in teams?

Problem 5A

THE MATH PLACEMENT PROBLEM

Problem A: Imagine that you are the teacher at a middle school that has developed its own performance assessment program. The tests are not multiple-choice, and students' achievement scores reflect not only test performance but also teachers' classroom observations and evaluations of students' work portfolios. You've been given the following assignment.

Your school offers three sections of 9th grade mathematics classes. Your assignment is to work with the school counselor and another teacher to develop a policy for assigning students to one of these three sections. Write a letter to the principal describing the policy that you recommend for assigning students to the available courses. Then, demonstrate how your policy should be applied to the students whose test scores are shown in the following data sheet.

Math and Reading Achievement (Grade Level Equivalent Scores)

	3rd Grade		4th Grade		5th Grade		6th Grade		7th Grade		8th Grade	
	Math	Read	Math	Read	Math	Read	Math	Read	Math	Read	Math	Read
Al	2.1	3.0	2.9	3.5	3.3	5.0	3.6	6.2	5.9	7.8	8.6	8.2
Barb	3.8	3.1	3.8	3.1	4.5	4.8	4.8	5.5	5.9	6.1	5.8	6.1
Carl	4.8	5.0	5.7	6.2	6.8	7.2	7.6	8.0	8.8	9.1	10.8	9.2
David	4.8	4.9	5.0	5.8	5.5	7.8	6.1	9.6	7.5	10.8	8.8	12.6
Edith	5.0	5.9	6.8	7.1	8.0	8.8	10.2	11.0	10.8	12.1	11.0	12.2
Fran	5.0	5.3	5.8	5.9	6.6	6.6	7.2	7.3	7.5	7.8	8.0	8.1
Greg	1.5	2.3	2.6	3.5	4.2	4.5	4.8	5.0	7.0	7.5	8.5	8.8
Hank							2.3	1.5	5.8	6.0	9.8	8.8
Ida	3.3	3.1	4.5	4.8	5.6	5.5	6.8	6.9	7.8	8.1	8.9	9.2
Jan	5.6	5.1	7.9	7.0	9.0	7.5	9.6	8.0	10.8	8.3	8.0	7.8

COMMENTS FROM PREVIOUS TEACHERS

AL	Al works hard. He always turns in his homework, and he even comes in after class for help. But, math has been difficult for him. - - - The projects in his portfolio are not inspired; but they show his dedication.
BARB	Poor attendance. Often late for class. Since her mother died two years ago, Barbara has had to take care of her younger brothers and sisters at home. Her homework is rarely finished.
CARL	Charles has consistently worked hard and is a very productive contributor to class discussions. He seems to know a great deal more than he has been able to demonstrate on tests.
DAVID	David is gaining confidence in himself. His success in sports seems to be rubbing off on other activities. - - - David is a leader.

EDITH	Edith is a gifted student. Recently, however, she seems to have lost interest in her work. Although she finishes assignments on time, she doesn't seem to devote much time to them.
FRAN	Fran is working to the best of her ability, but she needs work on basic skills. She spends too much time on low level skills, and still low level mistakes tend to hurt her performance on tests and projects.
GREG	Greg's projects are the best in the class. He is also good at creative types of problem solving – especially in geometry. Greg's immature behavior has interfered with his progress this year and has occasionally disrupted the whole group. He is the class clown.
HANK	Hank only spoke Spanish when he moved here three years ago. He has shown tremendous improvement this year, but his previous training has been rather weak so he still has a lot to learn. If his remarkable improvement continues he could become the best in the class.
IDA	Ida has a lot of natural ability but she hasn't worked up to her potential because of failure to complete assignments and to pay attention in class.
JAN	Jan has discovered boys. She can be a good student when she wants to; lately, she seems to have lost interest. She often falls asleep in class.

Problem 5B

Problem B: After writing a trial policy for sorting students into your school's three levels of mathematics courses, the other teachers you've been working with got a new idea. They transformed the students' scores as shown below. Look at what they did. Then rewrite your policy in any way that you believe is appropriate to take into account the new information.

Math and Reading Scores Compared to Actual Grade Level

	3rd Grade		4th Grade		5th Grade		6th Grade		7th Grade		8th Grade	
	Math	Read	Math	Read	Math	Read	Math	Read	Math	Read	Math	Read
Al	-0.9	0.0	-1.1	-0.5	-1.7	0.0	-2.4	0.2	-1.1	0.8	0.6	0.2
Barb	0.8	0.1	-0.2	-0.9	-0.5	-0.2	-1.2	-0.5	-1.1	-0.9	-2.2	-1.9
Carl	1.8	2.0	1.7	2.2	1.8	2.2	1.6	2.0	1.8	2.1	2.8	1.2
David	1.8	1.9	1.0	1.8	0.5	2.8	0.1	3.6	0.5	3.8	0.8	4.6
Edith	2.0	2.9	2.8	3.1	3.0	3.8	4.2	5.0	3.8	5.1	3.0	4.2
Fran	2.0	2.3	1.8	1.9	1.6	1.6	1.2	1.3	0.5	0.8	0.0	0.1
Greg	-1.5	-0.7	-1.4	-0.5	-0.8	-0.5	-1.2	-1.0	0.0	0.5	0.5	0.8
Hank							-3.7		-4.5	-1.2	-1.0	1.8	0.8
Ida	0.3	0.1	0.5	0.8	0.6	0.5	0.8	0.9	0.8	1.1	0.9	1.2
Jan	2.6	2.1	3.9	3.0	4.0	2.5	3.6	2.0	3.8	1.3	0.0	-0.2

Problem 5C

Problem C: The school counselor you've been working with has found some other ways to simplify (and graph) the data you were given. For example, an average score was calculated across all six grades; and the following graph was drawn. Again, modify your policy in any way that you believe is appropriate to take into account all of the information that is now available.

Math Scores (Only) For All Grade Levels

	3rd	4th	5th	6th	7th	8th	Average
Al	-0.9	-1.1	-1.7	-2.4	-1.1	0.6	-1.1
Barb	0.8	-0.2	-0.5	-1.2	-1.1	-2.2	-0.7
Carl	1.8	1.7	1.8	1.6	1.8	2.8	1.9
David	1.8	1.0	0.5	0.1	0.5	0.8	0.8
Edith	2.0	2.8	3.0	4.2	3.8	3.0	3.1
Fran	2.0	1.8	1.6	1.2	0.5	0.0	1.2
Greg	-1.5	-1.4	-0.8	-1.2	0.0	0.5	-0.7
Hank				-3.7	-1.2	1.8	-1.0
Ida	0.3	0.5	0.6	0.8	0.8	0.9	0.7
Jan	2.6	3.9	4.0	3.6	3.8	0.0	3.0

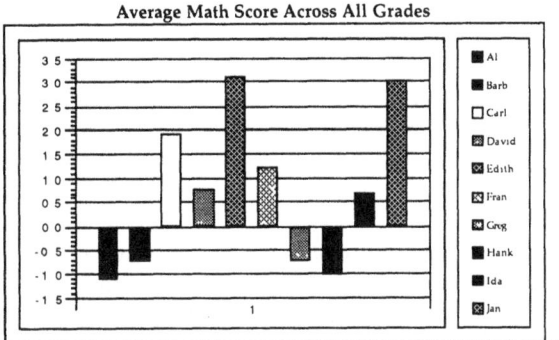

Problem 5D

Problem D: While playing around with the graphing capabilities of the spreadsheet you've been using to graph the data shown in problem 3, the computer constructed the following graph. Does it show anything new that you should consider? Again, modify your policy in any way that you believe is appropriate to take into account all of the information that is now available.

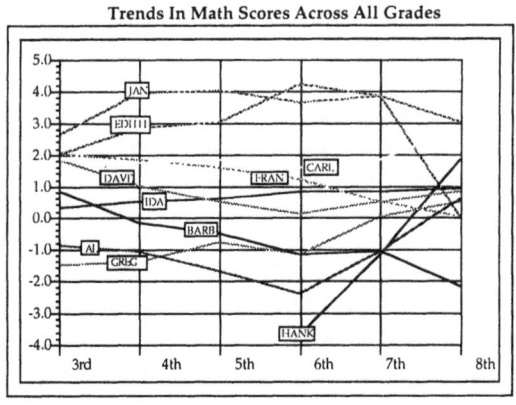

One Solution to the Math Placement Problem

One group of teachers used the available analyses and graphs to support the recommendation that the school should abandon its previous the policy of sorting students into low, middle, and high ability groups. Instead, the recommendation was made that three equivalent math groups should be created, with "difficult" students being distributed equally in the three sections.

To help teachers recognize the strengths and needs of each student, the teachers also recommended that executive summaries of the preceding graphs and analyses should be made available to teachers.

Because solutions to the Math Placement Problem involve using powerful computer-based tools, problem solvers no longer needed to be preoccupied with routine calculation and graphing skills. Instead, they are able to focus on conceptually oriented activities such as constructing appropriate and useful ways to think about the information that was available, or such as analyzing and interpreting results produced by computers or colleagues.

An important attribute of the Math Placement Problem is that the givens involve more than just *pieces* of data. They also involve *patterns* of data; and, they involve more than a single type of qualitative or quantitative information. Furthermore, the goal involves more than simply producing a specific answer to a particular question. It involves creating a policy that can be applied to a whole class of specific cases. In fact, appropriate responses involve more that just making decisions (or giving answers). They also involve justifying and explaining decisions (or answers); they involve constructions in which students explicitly describe the data, patterns, and relationships that their interpretations of the problem take into account. In other words, students use the computer-based tools to construct responses in much the same way that geometry tools are used to construct geometric objects. In this way, the quality of responses depends on constructions themselves as well as on decisions based on the constructed information.

Because the Math Placement Problem involved a familiar situation for the teachers who worked on it, they noticed that many of their own most important decision-making activities involved interactions with members of their groups and with technological tools (for example, spreadsheets). They also noticed that people who were good at dealing with the problem situation were not necessarily those who were best at textbook word problems because different kinds of knowledge, abilities, skills, and personalities were often needed. For example, in the Math Placement Problem, when they worked together in three-person teams, and when they used computer-based tools, different people generally assumed different roles (information gatherer, manager/coordinator, data cruncher, reality checker), which often shifted throughout the problem-solving process. Therefore, many of the most important abilities that were emphasized involved coordinating and communicating plans, processes, and results among people playing different roles.

When problem solving focuses on the development of models for thinking about realistic decision-making issues (rather than on linking together factual and procedural rules for getting from givens to goals), technology-based tools generally make it easier for students to focus on higher-order mathematical activities by minimizing distractions associated with computational drudgery. Technology-based tools not only change the computational and procedural demands associated with problem-solving activities, they also change the conceptual demands, for example, by creating new possibilities for selecting, organizing, and interpreting information and by forcing students to externalize (and be explicit about) interpretation frameworks that otherwise may have remained internal.

AN EXAMPLE: TEACHERS' IMPROVED VERSIONS OF EXISTING PROBLEMS

To help teachers create realistic problems that focus on authentic mathematical modeling activities, it is useful to begin by analyzing (and trying to improve) problems from exemplary resources such as the projects that are described throughout this book. For example, Problem 6A was taken from the NCTM's 1989 *Curriculum and Evaluation Standards for School Mathematics*. A transcript of one teacher's report and revised problem follows. (Note: The woman submitting this report was actually a softball player.)

Problem 6A

The Original Softball Problem

The table gives the record for Joan Dyer's last 100 times at bat during the softball season. She is now coming up to bat. Use the data to answer the following questions:

What is the probability that Joan will get a home run? What is the probability that she will get a hit? How many times can she be expected to get a walk in her next 14 times at bat?

Home Runs	9
Triples	2
Doubles	16
Singles	24
Walks	11
Outs	38
Total	100

One Teacher's Analysis Of The Softball Problem

Critique: On the surface, this problem appears to be embedded in a real-world situation: Joan is coming up to bat, and the problem description gives some data about her prior performances. But, in a real situation, it wouldn't be sensible for someone (other than a math teacher) to want to know the answer to the questions as they are stated (concerning probability of a home run, or the probability of a hit). In fact, simply computing this "probability" using the intended rule depends on ignoring common sense and/or practical experience. In reality, the probability depends on who is pitching (Are they left handed or right handed?), on field conditions or the weather, and on a lot of other factors that people who play softball are aware of (Who is in a slump or on a streak? Who is good under pressure). Furthermore, since we don't know who is asking the question, or why, we can't know what to take into account, or how accurate the answer needs to be, or what the risks or benefits might be. Therefore, the criteria for judging the quality of answers are not implicit to the situation; and, solutions must be judged according to whether they conform to the calculation that was expected, rather than according to whether they succeed in any practical or meaningful sense of the word.

Overall, I do not think such a problem is realistic, or that it promotes authentic performance. Real softball players would actually have to turn off their "real life" knowledge and experience.

Analysis: In this problem, it is not particularly necessary to have Joan coming up to bat right now. In fact, for the third question we might wonder what coming to bat, at this particular moment in time, has to do with expectations about walks for the next 14 times at bat, which may not even happen in the same game. Is there some reason, now that Joan is coming up to bat, to want to know her expected number of walks for the next 14 times at bat? Why not the next 15 times at bat? Is it possible that having Joan at bat in the first place was only the awkward result of an well-intentioned effort to create a life-like context (for a problem with essentially one appropriate solution path that leads to a single "right" answer)? Or was it a superficial gesture to suggest that even girls play sports? If not for either of these reasons, why it is there?

Possible Improvements: Here is a suggestion for improving this problem. Notice that the "math answer" is not an end in itself. It is a means to an end (or a tool for informing actions, decisions, and judgments). If Joan really is coming up to bat now, the mathematics should address a plausible question that might occur in that situation. I think that the revised item asks a more authentic question in the sense that it asks for a decision that might be required in the context.

Problem 6B

The Improved Softball Problem

You are the manager of a softball team. It is the bottom of the ninth inning, two outs gone, and no one is on base. Your team is one run behind. You plan to send in a pinch hitter in hopes of scoring the tying run. Your possibilities are Joan, Mary, and Bob. Their batting records are given in the table below. Who would you choose to bat? Explain your reasoning.

	Joan	Mary	Bob
Home Runs	9	15	6
Triples	2	5	3
Doubles	16	11	8
Singles	24	34	18
Walks	11	20	12
Outs	38	85	36

The main difference between the original softball problem and the modified softball problem is that the original asked a "school question" but didn't provide any clues about the real-life issues or decisions that the response was intended to inform. By contrast, the modified problem (6B) calls for a realistic response to a situation that might really occur in the lives of the people who were asked to work on the problem. It is the problem solver's responsibility to identify and address relevant mathematics questions.

One of the main reasons for emphasizing realistic decision-making situations is to encourage students to make sense of the problems based on extensions of their own personal knowledge and experiences; embedding problems in such contexts tends to be the only way for sensible judgements to be made about such things as: (i) how accurate the answer needs to be, (ii) what the consequences of making an error might be, and (iii) how quickly a response must be generated.

In real-life situations, people's three-second answers tend to be quite different from their five-minute answers, or their sixty-minute answers; high-stakes answers tend to be quite different from low-stakes answers. This is why one of the most important abilities associated with real-life problem solving involves sizing up problems in appropriate ways; it is also why some of the most important kinds of estimation skills are used to support these sizing-up processes.

THE NATURE OF REALISTIC SITUATIONS IN WHICH MATHEMATICS IS USEFUL

The problems in the preceding sections illustrate a number of important characteristics of realistic situations in which mathematical models are useful. This section will give a brief summary of some of these characteristics as they relate to givens, goals, solution paths, and response assessments.

Concerning the nature of realistic data sources: In realistic problem-solving and decision-making situations (such as the Sears Catalogue Problem or the Math Placement Problem), judgments often must be based on patterns or trends in data, not just isolated pieces of information, and on hypothesized regularities beneath the surface of things, not just on information that is given by direct perception. Also, the relevant information often involves several different types of qualitative and/or partly-quantified information that must be quantified or coded in appropriate ways before relevant calculations can be made. Furthermore, an overwhelming amount of information may often be relevant; this information may need to be filtered, weighted, simplified, organized, or interpreted before it is useful. Some relevant information may not be available, yet a decision may be needed anyway, within specified time limits, budget constraints, and margins for error.

To create model-eliciting activities of the type described in this chapter, problems must somehow create the need for the relevant models, so that meaningful patterns can be used to (i) base rapid decisions on a restricted set of data, (ii) fill holes or go beyond a minimum set of information, (iii) provide explanations of how facts are related to one another, or (iv) provide hypotheses about missing or hidden objects or events that may need to be actively sought out, generated, or (re)interpreted.

Concerning the nature of realistic demands on knowledge: Realistic problems tend to emphasize the use of organized systems of knowledge, not just isolated facts, skills, or bits of information. In fact, model-eliciting problems generally cannot be solved using a single computational formula. Instead, relevant knowledge usually must be integrated from a variety of topics. For example, in mathematics education research on children's natural solutions to real problems, it has been shown that solutions often draw on mixes of ideas and procedures from a variety of topic areas ranging from arithmetic, to measurement, to statistics, to geometry, to ideas and procedures that are not taught in schools (Carpenter, Moser, and Romberg, 1982; Confrey, 1990; Fuson, 1988; Lesh and Akerstrom, 1982; Steffe, 1988). Similar observations have also been made by educational anthropologists who have studied the everyday problem solving behaviors of tailors, carpenters, cooks, shop keepers, shoppers, or other ordinary folks (Carraher, Carraher, and Schliemann, 1985; Rogoff and Lave, 1984; Saxe, 1991). In both higher mathematics and everyday problem solving, solutions to real-life problems often reside at the borders, intersections, and fringes of traditional curriculum categories.

In general, the most effective model-eliciting problems can be interpreted/quantified/modeled in several alternative ways, and the interpretations are based on models that are created by extending or refining students' real-life knowledge and experiences.

Concerning the nature of realistic tools and resources: In real-life situations, few problems occur in an isolation booth where the only available tools are pencils and paper. In fact, in most real-life problem-solving situations, the tools and resources that are available include not only pocket calculators but also computers, resource books, and colleagues or consultants.

When such tools and resources are available, new types of conceptual capabilities, generally quite different from the pencil-and-paper computational abilities traditionally emphasized in schools, become important. For example, when a graphics-oriented spreadsheet is used in problems such as the Math Placement Problem, trial results are often fast and easy to generate and what-if explorations easy to conduct. Therefore, attention can shift away from the production of results toward higher-order processes such as (i) analyzing the appropriateness of alternative assumptions or interpretations of data, (ii) planning, monitoring, and assessing procedures that are executed, (iii) conducting explorations about alternative levels and ways of collecting, organizing, or coding available information, or (iv) testing and revising interpretations and trial solutions to adjust the precision, accuracy, risks, and benefits associated with hypotheses and predictions.

In general, when students have access to realistic tools and resources, the natural tendency is for deeper and higher-order model construction processes to emerge. The kinds of skills that are emphasized go far beyond number crunching to involve processes such as quantifying qualitative information, estimating quantities and measures, drawing informative diagrams or graphs, generating symbolic descriptions (using written or spoken language), or generating sequences of commands to be executed by a computer, a colleague, or an assistant. In fact, the kinds of skills that emerge as important often involve far more than simply getting from the givens to the goals that are specified by others. For example, relevant skills often involve giving commands and finding useful ways to think about givens and goals.

To create problems that are model-eliciting and that also involve explicitly documenting the model that is elicited, computer-based tools are often quite useful. They make it possible for students to produce complex constructed responses in which they explicitly reveal how they are thinking about the problem situations. For example, in situations similar to the Math Placement Problem, model-eliciting problems should be stated so that the spreadsheets and graphs that students produce will provide direct evidence about (i) what information is being considered, (ii) how the information is being interpreted, (iii) what relationships or patterns are being taken into account, and (iv) what operations or transformations are considered to be appropriate. In other words, the spreadsheet and graphs can sometimes explicitly reveal what models (or systems) students use to interpret/describe/explain problem situations.

Concerning the nature of realistic settings: In real-life situations that are relatively complex, people often work in groups in which different group members have different interests, experiences, and expertise. When diverse groups work together, certain higher-order capabilities tend to be emphasized such as (i) partitioning problems into smaller pieces that can be attacked by different people, (ii) partitioning solution processes into different roles and functions so that people with different expertise and tools can work together cooperatively, (iii) defining questions, givens, and goals so that outside help can be sought, and (iv) communicating, planning, monitoring, and explaining results to others inside or outside the group. Also, because information must be shared among members of the group, there tend to be both the need and the opportunity to externalize internal reasoning processes and to think about thinking. In particular, activities tend to be emphasized that the NCTM's *Curriculum and Evaluation Standards for School Mathematics* refer to under headings such as "mathematics as communication" or "mathematical connections."

To create the kind of model-eliciting problems described in this chapter, it is often useful to design the situations in such a way that students assume a variety of different roles. Sometimes this can be done by assigning one student to be the manager, while preparing ahead of time for other students to be proficient in the use of a potentially relevant tool (such as a spreadsheet) or collection of relevant background information.

Concerning the nature of realistic solution processes: In realistic situations that require the construction of a model, several modeling cycles may be needed to create an adequate way to think about (or describe) the givens, goals, and solution paths. For example, in cases similar to the Math Placement Problem, each cycle may involve goal clarification, question refinement, trial solution evaluation, data (re)interpretation, and a variety of other sense-making activities. That is, students go beyond simply using models that already exist: they construct new models by modifying, extending, integrating, or refining existing models. In fact, even when it is possible to use an available model without modification, some noncomputational activities generally emerge as important, such as (i) mapping data from the real world into the model world (for example, by filtering, interpreting, parsing, coding, and organizing available information), (ii) carrying out explorations within the model world (for example, by formulating questions or hypotheses that can be verified or rejected), or (iii) mapping from the model world into the original problem situation (for example, to evaluate explanations, hypotheses, trial results, or predictions).

To create the kind of model-eliciting problems described in this chapter, it is important for students to go beyond thinking *with* a given model (graph, table, or symbolic description) to thinking *about* the model and its underlying assumptions (Campione, Brown, and Connell, 1989).

Concerning the nature of realistic goals: When people shop for groceries, purchase automobiles, or engage in other everyday decision-making situations, the goals often involve far more than simply producing an explicitly requested mathematical answer to someone else's well-formed question. Justifying or explaining decisions is often as important as simply making them. To decide whether to believe conclusions cited in newspaper editorials and advertisements, for example, it is important to ask insightful questions and to critically analyze, critique, or explain answers produced by others.

At a time when newspapers such as *USA Today* are beginning to look more and more like spreadsheets with graphics, studies of adult literacy confirm that it is naive to assume that the only kind of mathematical objects that ordinary people find useful are whole numbers. For example, even to do nothing more than intelligently read the sports pages of a newspaper, the kinds of mathematical products that may need to be generated include (i) *numbers* that are based on patterns or trends in data rather than on isolated and explicitly stated facts; (ii) *estimates* that may involve *ratios*, or *rates*, or *functions* involving relationships among several quantities, rather than only quantities or measures that can be observed directly; (iii) *questions* or *hypotheses* or *predictions* that can be verified or rejected; (iv) *rules* to explain patterns or regularities that are embedded in diagrams or written/spoken statements; (v) *organizational schemes* based on tables, graphs, rankings, or coordinate systems that clarify, highlight, or make some factors easy to see, at the cost of distorting, disguising, or eliminating other factors; (vi) *graphs, arithmetic sentences, equations, diagrams,* or other *concrete or symbolic description or representations* that explain (or make sense of) a given situation; (vii) *statements* that are logically deduced from known facts, rather than being calculated; (viii) *probabilities* that describe the occurrence of given nondeterminant events; (ix) *statistics* that summarize or describe information about collections or patterns of data rather than simply reporting the measures of pieces of data; or (x) *shapes* (or *measures*, or *coordinates*, or *constructions*) that fit specified conditions.

Furthermore, when such objects are recognized as legitimate and useful products of mathematical activities, the kind of problem-solving and decision-making goals that are addressed include (i) *optimizing* the results of given processes, (ii) *simplifying* or *modularizing* the procedures needed to produce given results, (iii) *finding detours* (or alternative ways) to use under alternative conditions, (iv) finding fair ways to *partition* given quantities, (v) *diagnosing* or *correcting errors* in other peoples' results or conclusions, (vi) *specifying parameters* that result in desirable outcomes, (vii) *comparing, choosing,* or *ranking* different kinds of objects or events, or (viii) *manipulating* a given system in ways that presumably improve its functioning even though the results of alternative policies might not be possible to observe.

Concerning the nature of realistic assessments of results: In realistic problem-solving and decision-making situations, students seldom need to rely on the judgment of an external authority (such as a teacher, or textbook author, or test maker) to tell them whether their answers are acceptable. Consider the case where the goal is to generate a plan, or a description, or an explanation of a real-world situation. A variety of different types and levels of responses are usually possible (for example, using verbal rules, symbolic equations, diagrams or graphs, or concrete models). Yet, in most cases, there still tend to be clear and objective criteria to determine whether responses are good enough for a given purpose. For example, the quality of a response usually depends on factors such as (i) the type and amount of information taken into account, (ii) judgments about risks and benefits associated with alternatives, (iii) the problem-solver's awareness of possible sources of errors (for example, due to over-simplifications or assumptions), (iv) hypothesized trends and patterns that explain regularities beneath the surface of things, (v) awareness of conditions that might result in different opportunities or constraints, (vi) resources that are or are not available (for example, time, money, tools), and (vii) purposes and preferences of students themselves. Also, the quality of responses often depend on answers to questions such as, Is it more important to generate answers quickly or with high degrees of accuracy? Can trial answers be tested and revised? Are overestimates preferable to underestimates? Should different answers be given under alternative conditions?

To create effective model-eliciting problems, one of the key tricks is to ask a real question that is neither too vague nor too specific. Productive goals for model-eliciting problems should be similar to productive goals for businesses or government or adult-level, on-the-job projects. They should clarify how you will know when you are done, and when you are expected to be done (date and deliverable). The criteria for evaluating the quality of work should also be clearly understood. Contrast President Kennedy's goal to "put an American on the moon by 1970" to President Bush's goal of "making American students first in the world in mathematics and science by the year 2000." In the first case, the criteria for success are clear, while in the latter, they are not.

HOW MATHEMATICAL MODELS DIFFER FROM OTHER MATHEMATICAL SYSTEMS

In the research literature on mathematics learning, problem solving, and instruction, one issue that strongly influences both the observations that are made and the conclusions that are reached has to do with the "grain size" and the type of objects under investigation. For example:

- If mathematics is thought of as (nothing more than) a collection of condition-action rules, then thinking tends to be equated with information processing, and learners tend to be treated as information processing units. Also, distinctions between experts and novices tend to be characterized in terms of the presence or absence of specific factual and procedural rules, and students' misconceptions are often interpreted as being similar to "buggy" computer programs (see, for example, Brown and Van Lehn, 1980).

- If mathematics is thought of as being based on a relatively small number of general cognitive structures and reasoning patterns, then the most important aspects of thinking tend to focus on the construction and adaptation of these structures (see, for example, Piaget and Beth, 1966; Steffe, Cobb, and von Glasersfeld, 1988) and misconceptions tend to be explained by determining which of several alternative conceptual systems students are using.

- If attention focuses on entire conceptual fields, such as those associated with multiplicative relationships, or additive relationships, or exponential relationships (see, for example, Vergnaud, 1988), then attention tends to focus on meanings that are derived from organized systems of ideas rather than being derived from additive combinations of isolated ideas or procedures.

This chapter is based on the notion that to study the nature of students' mathematical knowledge (plus the understandings and abilities that are needed to use this knowledge in everyday situations), the most appropriate unit of analysis is at the level of mathematical models rather than at the level of (i) information processing rules, (ii) general Piagetian-style cognitive structures, or (iii) entire conceptual fields.

Mathematical models are closely linked to each of these three perspectives, yet they are also distinct from each. For example, a collection of factual and procedural rules is associated with any given mathematical model, yet understanding the model means far more than simply learning a list of isolated rules. Also, because real-life problems seldom fall into neat and tidy disciplinary categories, the mathematical models that students construct generally emphasize the kind of integrated knowledge that is emphasized by those who focus on conceptual fields (see, for example, Vergnaud, 1988). Finally, all mathematical models are complete, functioning mathematical systems (of the type referred to in Table 2) and all mathematical systems are potential models for describing and explaining real or possible worlds. Yet in real-life situations, the models that students construct tend to be somewhat different from both general cognitive structures (of the type described by Piaget) and general mathematical structures (as typically described in mathematics textbooks).

How are the cognitive/mathematical models that have been described in this chapter different from other (more "pure") types of mathematical systems? *Our answer is that mathematical models are situated mathematical systems.* On the one hand, they provide powerful structural metaphors for describing, constructing, explaining, predicting, and controlling real-life situations. On the other hand, they can also be explored without reference to external events.

Whereas pure mathematical systems are abstractions that are distinct from the notation systems in which they are embedded—as well as being independent of the mind of any particular human—mathematical models do not function in the abstract. That is, in nontrivial situations, mathematical models are always embedded in representational systems (written symbols, spoken language, diagrams or pictures, concrete or manipulatable materials, or real-life prototype experiences); and, for a given student, potential models (that is, mathematical systems) do not really become actualized until they are used to model some external phenomenon. Furthermore, in the minds of the humans who construct them, the meanings of mathematical models are always influenced by the purposes for which they are constructed and by situations in which they are constructed. Therefore, the kind of mathematical models emphasized in this chapter are usually more concrete and more closely linked to students' specific problem-solving experiences than the general systems typically described in mathematics textbooks.

Table 2

Basic Mathematical Systems Underlying Ideas in Precollege Textbooks			
Elements	Relations	Operations	Representations, Notations, Definitions, Computations, and Transformations
simple counts	$=, \neq, <, >$	$+, -, \times, \div, \rfloor$	base ten numeration
composite counts	$=, \neq, <, >$	$+, -, \times, \div$	24 eggs => 2 dozen
simple measures	$=, \neq, <, >$	$+, -, \times, \div$	feet, inches, centimeters
derived measures	$=, \neq, <, >$	$+, -, \times, \div$	7cm => .7dm = .07 m
very large/small numbers	$\approx, =, \neq, <, >$	$\wedge, +, -, \times, \div$	scientific notation
signed numbers	$=, \neq, <, >$	$+, -, \times, \div$	$-n, |n|, \pm$
coordinates (locations)	$|p-q|, \Theta$	$+, -, \times, \div$	$|p-q|$
fractions	$=, \neq, <, >$	$+, -, \times, \div, /$	2/3

decimals & percents	=,≠,<,>	+,−,x,÷	%, $
ratios	≈,=,≠,<,>	•	a:b,
rates	=,≠,<,>	+,−,x,÷	miles-per-hour
vectors	≈,=,≠,<,>	•,*,+,−,x,÷	(a,b)
matrices & operators	≈,=,≠,<,>	•,*, ⊗	|a,b|
real numbers	≈,=,≠,<,>	^,+,−,x,÷	π,√,e
complex numbers	≈,=,≠,<,>	^,+,−,x,÷	i
probabilities (measures of the frequency of events)	≈,=,≠,<,>	•,+,−,x,÷	P(A), P(A/B), P(A')
statistics (measures of sets of data)	≈,=,≠,<,>	+,−,x,÷	graphs,
sets (& elements of sets)	≡	∪, ∩, ∈,	Ω,∅
logical propositions	=>,−>	^,v,¬	∀, φ, ∃, ∴, →:, ∃
programming commands	=>,−>	^,v,¬	∀, φ, ∃, ∴, →:, ∃
shapes in a plane	symmetric ≅, ≑, ≡	∪, ∩, ∠,⊥	constructions, reflections, translations, expansions
shapes in 3-D space	symmetric, ≅, ≈, ≡	∪, ∩,	constructions, reflections, translations, expansions
algebraic expressions	=,≠,≈	^,+,−,x,÷	substitute, simplify, commute, distribute, etc
algebraic equations	=>, =	+,−,x,÷	transform
algebraic functions	≈,=>	o, *, +,−,x,÷	simplify
trigonometric expressions	=,≠,≈	^,+,−,x,÷	SIN, COS, TAN
trigonometric functions	≈,=>	o, *, +,−,x,÷	simplify
exponential & logarithmic expressions	=,≠,≈	^,+,−,x,÷	LOG, e
exponential & logarithmic functions	≈,=>	compose, +,−,x,÷	simplify
sequences	≈,=,<,>	+,−,x,÷	∞, —>,
series	≈,=,<,>	+,−,x,÷	Σ,Π,∞, —>, ...
continuously changing quantities (derivatives)	=,≈	+,−,x,÷	∂y/∂x,Δ
accumulating quantities (integrals)	=,≈	+,−,x,÷	∫, Σ

In Table 2, which gives basic mathematical systems, all of the mathematical systems (including the kind of mathematical models emphasized in this chapter) consist of three kinds of entities: (i) elements, (ii) relations among elements, and (iii) operations on elements. Consequently, one convenient way to distinguish one mathematical system from another is simply to name the elements, for example, simple counts, simple measures, ratios, rates, signed numbers, and so on. Another way to distinguish one mathematical system from another is to list the factual and procedural rules that govern the behavior of elements within the system. Nonetheless, the real mathematical objects are neither the elements nor the representations, notations, definitions, computations, and transformations that describe their behaviors. The real mathematical objects are the underlying patterns that the symbol systems describe. Doing mathematics means investigating the structural properties associated with these patterns (or models) and constructing and adapting these patterns (or models) for a variety of situations and purposes.

The mathematical systems described in Table 2 tend to be both too large and too small to serve as the proper unit of analysis for most situations in learning and instruction. They are too small, for example, because nontrivial real-life problems seldom fall within neat disciplinary topic areas. Therefore, to describe or explain most real-life problems, it is usually necessary to construct a model that is based on combinations and adaptations of several pure mathematical systems. The systems are too large, because it is only at an extremely advanced stage of conceptual development that students become able to think in flexible ways using conceptual models that are as large and complex as (for example) the system of rational numbers. In general, the past decade of cognitive science research has revealed that mathematics learning and problem solving are far more situated and piecemeal than earlier researchers had recognized (Greeno, 1988).

Finally, these systems are sometimes simultaneously too large and too small, because, when students and educators are forced to work prematurely with mathematical systems that are as large and complex as those in the table, they often lose sight of the forest because of all of the trees. In particular, they tend to lose sight of the underlying patterns (or models) that are intended to be described, and instead focus on simply knowing and executing isolated factual/precedural rules within the system.

From the point of view of mathematics learning and instruction, substantial effort must be made to focus on underlying systems that are sufficiently small and concrete so that students do not lose sight of the underlying systems-as-a-whole, while, at the same time keeping these systems sufficiently large to have impressive power, utility, and generalizability. But, how large is too large? And how general is too general? The answers depend on the level of development of individual students. In principle, a student might someday reach a level of development when virtually all of the systems referred to in the table have been integrated into a single supersystem. For example, elements in one system (for example, the natural numbers) are often embedded within another system (for example, fractions, or signed numbers) which are in turn included within still other systems (for example, real numbers, complex numbers, vectors, or matrices). On the other hand, as this process of subsumption takes place, the meaning of symbols such as + and = often differ considerably from one level to another (or from one situation to another), so that, if a single integrated system could ever be said to exist in the mind of a given student, this system would have to involve notions such as "levels and types of equivalence" and "levels and types of combination" which apply under different conditions or in different situations.

SUMMARY

To close this chapter, we will review answers to some of the main questions that were posed at the beginning of the chapter.

Question: What are the most important objectives that students should learn (and that assessment instruments should measure) in mathematics instruction?

Answer: According to the point of view that was adopted in this chapter, the most important goals of modern mathematics education are to help students construct models that provide powerful conceptual/procedural amplifiers for making sense of their increasingly complex worlds of experience. For example, within most mathematics courses, or at any given grade level, there tend to be no more than ten to twenty basic models that underlie nearly all of the specific concepts and procedures that students are expected to learn. It is the development of these models that should be emphasized in both assessment and instruction.

Unlike earlier periods when students were expected to demonstrate their knowledge and ability by showing how many facts and skills they knew, increasingly, the main mark of intelligence is considered to involve the ability to analyze, manipulate, synthesize, and critically interpret information in the interest of real-life problem solving. Specific facts and skills are associated with each of the preceding models, but these models themselves are not merely condition-action rules. They are complete, functioning, systems-as-a-whole whose properties are not simply derived from their parts and whose purposes are not simply to provide rules for getting from givens to goals during the process of answering questions posed by others. Instead, their purposes usually involve describing, explaining, predicting, manipulating, and controlling real or possible worlds of experience.

Question: What types of activities are particularly promising for assessing the preceding dimensions of understanding?

Answer: For a given student, to assess the extent to which a given model has been constructed (or to assess the level of development of the model), one kind of activity that has proven to be especially useful is a model-eliciting activity in which students go beyond simply giving an answer to construct a response using available information, objects, tools, and procedures. Often, these constructed response activities tend to be quite similar to activities involving the construction of geometric figures. For example, in the Math Placement Problem described earlier, teachers began to construct a response by explicitly selecting, filtering, coding, and organizing relevant data within a spreadsheet; next they recoded and reorganized the information in a variety of ways; then they constructed graphs and reasoning paradigms to support their constructed decision-making strategy. Finally, they looked for other structurally similar problem-solving situations in which the model they had constructed might provide insightful explanations, predictions, or interventions. Consequently, to evaluate the quality of a given person's response to the Math Placement Problem, it is necessary to evaluate the model that was constructed, and the construction process itself, in addition to evaluating isolated answers or specific decisions that were based on the model.

Question: How can the construction of significant models be facilitated?

Answer: For fields in which the most important goals of instruction focus on the construction of models for describing, predicting, and controlling the behavior of complex systems, case studies are often used as model-eliciting activities to help students develop powerful structural metaphors to make sense of actual or anticipated worlds of experience. For example, in business schools, this development process often involves the following three phases:

- *The model-development phase:* Using a variety of conceptual frameworks and technology-based tools, students construct models to describe, explain, manipulate, and predict the behavior of structurally rich systems. That is, they are case studies that provide prototypes (or structural metaphors) for interpreting other important problem-solving situations.

- *The model exploration phase:* Models that have been constructed are investigated for their own sake (as in a pure math activities) to extend their power and utility by focusing on underlying patterns and regularities.

- *The model application phase:* Models that have been refined and elaborated are applied to new problem-solving situations which could not have been dealt with adequately in the absence of the newly constructed model. That is, students actively look for new situations that can be described or explained using the model they have constructed.

After these three phases, similarities and differences are often explored comparing the original problem, the pure math model, and the final application. This process is depicted in Figure 5.

Figure 5. Three Stages In A Modeling Approach To Mathematics Instruction

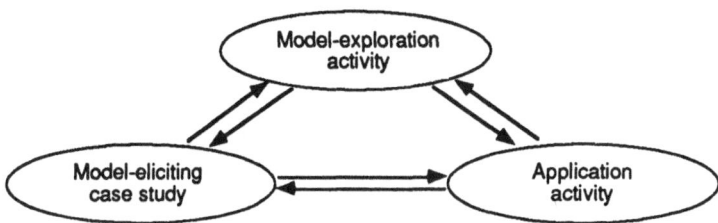

The preceding modeling approach to instruction is similar to instructional approaches that Dienes (1957) advocated for use in mathematics laboratories. As Figure 6 suggests, Dienes' instructional techniques focused on use of concrete manipulatable materials which served as embodiments of targeted mathematical systems. That is, students explored two or three sets of structurally rich materials, then they investigated similarities and differences among structurally isomorphic activities with these materials.

Figure 6. Dienes' Multiple Embodiment Approach To Mathematics Instruction

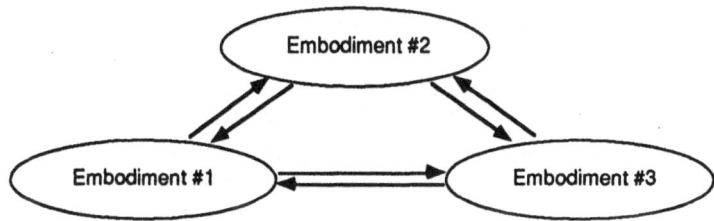

The preceding two approaches to instruction are clearly quite similar. For example, both focus on mathematical systems-as-a-whole; both emphasize the use of concrete materials; both concentrate on mathematical systems that provide conceptual foundations for the most important understandings and processes that are priorities for children to learn. The main difference between the two approaches is that Dienes' principles focus on pure mathematics activities (in which the symbols are concrete objects rather than abstract notation), whereas two of the three stages of our modeling approach focus on applied mathematics activities (which are explicitly based on children's' real-life experiences), even though all three stages are aimed at demonstrating the power of pure mathematical activities.

Our modeling approach to instruction is deliberately consistent with the way modern mathematicians think about their own activities in mathematics and mathematical modeling. Also, it was explicitly created to be consistent with three of the most basic principles of modern cognitive science, namely, (i) humans interpret their experiences by mapping them to internal models, (ii) these internal models must be constructed, and (iii) constructed models result in situated knowledge that is gradually extended and decontextualized to interpret other structurally similar situations. Moreover, our modeling approach to instruction is also consistent with a constructivist philosophy about how human knowledge develops.

REFERENCES

American Association for the Advancement of Science. (1989). *Science for all Americans*. Washington, D.C.: AAAS.

Bell, M., Fuson, K., and Lesh, R. (1976). *Algebraic and arithmetic structures: A concrete approach for elementary school teachers*. New York: Free Press.

Brown, J.S., and Van Lehn, K. (1980). Repair theory: A generative theory of bugs in procedural skills. *Cognitive Science, 4*, 379-426.

Campione, J, Brown, A., and Connell, M. (1989). Metacognition: On the importance of understanding what you are doing. In R. Charles and E. Silver (Eds.), *The teaching and assessing of mathematical problem solving*. Hillsdale, NJ: Lawrence Erlbaum Associates.

Carpenter, T.P., Moser, J., and Romberg, T.A. (Eds.). (1982). *Addition and subtraction: A cognitive perspective*. Hillsdale, NJ: Lawrence Erlbaum Associates.

Carraher, T., Carraher, D., and Schliemann, A.D. (1985). Mathematics in the streets and the schools. *British Journal of Developmental Psychology, 3*, 21-29.

Confrey, J. (1990, April). Origins, units, and rates: The construction of a splitting structure. A paper presented at the annual meeting of the American Educational Research Association. Boston.

Dienes, Z. (1957). *Building up mathematics*. London: Hutchinson Educational Ltd.

Ernest, P. (1991). *The Philosophy of mathematics education*. Bristol, PA: The Falmer Press, Taylor and Francis Inc.

Fuson, K.C. (1988). *Children's counting and concepts of number*. New York: Springer-Verlag.

Greeno, J.G. (1976). Cognitive objectives of instruction: Theory of knowledge for solving problems and answering questions. In D. Klahr (Ed.), *Cognition and instruction*. Hillsdale, NJ: Lawrence Erlbaum Associates.

Greeno, J.G. (1980). Some examples of cognitive task analysis with instructional implications. In R.E. Snow, P-A. Federico, and W.E. Montague (Eds.), *Aptitude, learning and instruction: Vol. 2. Cognitive process analyses of learning and problem solving*. Hillsdale, NJ: Lawrence Erlbaum Associates.

Greeno, J.G. (1983). Conceptual entities. In A. Stevens and D. Gentner (Eds.), *Mental models*. Hillsdale, NJ: Lawrence Erlbaum Associates.

Greeno, J.G. (1987). Instructional representations based on research about understanding. In A.H. Schoenfled (Ed.), *Cognitive science and mathematics education*, Hillsdale, NJ: Lawrence Earlbaum Associates; pp. 61-88.

Greeno, J.G. (1988). The situated activities of learning and knowing mathematics. In M. Behr, C. Lacampagne, and M. Wheeler (Eds.), *Proceedings of the tenth annual meeting of the International Group for the Psychology of Mathematics Education*. DeKalb, Illinois.

Greeno, J.G. (1988a). For the study of mathematics epistemology. In R.I. Charles and E.A. Silver (Eds.), *The teaching and assessing of mathematical problem solving*. Reston, VA: National Council of Teachers of Mathematics; pp. 23-31.

Greeno, J.G. (1988b). Situations, mental models, and generative knowledge. In D. Klahr and K. Kotovsky (Eds.), *Complex information processing: The impact of Herbert A. Simon*. Hillsdale, NJ: Lawrence Erlbaum Associates.

Lamon, S.J. (1990, April). Ratio and proportion: Cognitive foundations in unitizing and norming. Paper presented at the 1990 annual meeting of the American Educational Research Association, Boston.

Lave, J. (1988). *Cognition in practice: Mind, mathematics and culture in everyday life*. New York: Cambridge University Press.

Lesh, R. (1990) Computer-based assessment of higher-order understandings and processes in elementary mathematics. In J. Kulm (Ed.), *Assessing higher-order thinking in mathematics*. Washington, DC: American Association for the Advancement of Science.

Lesh, R., and Akerstrom, M. (1982). Applied problem solving: Priorities for mathematics education research. In F. Lester and J. Garofalo (Eds.), *Mathematical problem solving: Issues in research*. Philadelphia: Franklin Institute Press; pp. 117-129.

Lesh, R., Behr, M., and Post, T. (1987). Rational number relations and proportions. In C. Janvier (Ed.), *Problems of representation in teaching and learning mathematics*. Hillsdale, NJ: Lawrence Erlbaum Associates.

Lesh, R., and Kaput, J. (1988). Interpreting modeling as local conceptual development. In J. DeLange and M. Doorman (Eds.), *Senior secondary mathematics education*. OW&OC: Utrecht, Netherlands.

Lesh, R., Landau, M., and Hamilton, E. (1983). *Conceptual models in applied mathematical problem-solving research*. In R. Lesh and M. Landau (Eds.), Acquisition of mathematics concepts and processes. New York: Academic Press; pp. 263-343.

Lesh, R., Post, T., and Behr, M. (1987) Dienes revisited: Multiple embodiments in computer environments. In I. Wirszup and R. Streit (Eds.), *Developments in school mathematics education around the world.* Reston, VA: National Council of Teachers of Mathematics; pp. 647-680.

Lesh, R., Post, T. and Behr, M. (1988). Proportional reasoning. In J. Hiebert and M. Behr (Eds.), *Number concepts and operations in the middle grades.* Reston VA: The National Council of Teachers of Mathematics; Vol. 2, pp. 93-118.

Lesh, R., and Zawojewski, J. (1987). Problem solving. In T. Post (Ed.), *Teaching mathematics in grades K-8: Research-based methods.* Boston: Allyn and Bacon.

Mathematical Sciences Education Board. (1990). In Lynn Arthur Steen (Ed.), *On the shoulders of giants: New approaches to numeracy.* National Research Council. Washington, DC: National Academy Press.

Mathematical Sciences Education Board. (1990a). *Reshaping school mathematics: A philosophy and framework for curriculum.* National Research Council. Washington, DC: National Academy Press.

Naisbett, J., and Aburdene, P. (1990). *Megatrends 2000: Ten new directions for the 1990s.* New York: William Morrow.

National Council of Teachers of Mathematics. (1989). *Curriculum and evaluation standards for school mathematics.* Reston, VA: NCTM.

Piaget, J., and Beth, E. (1966). *Mathematical epistemology and psychology.* Dordrecht, Netherlands: D. Reidel.

Post, T., Behr, M., Lesh, R., and Harel, G. (1991). Intermediate teachers' knowledge of rational number concepts. In E. Fennema, T. Carpenter, and S. Lamon. (Eds.), *Integrating research on teaching and learning mathematics.* Albany, NY: State University of New York Press.

Rogoff, B., and Lave, J. (Eds.). (1984). *Everyday cognition: Its development in social context.* Cambridge, MA: Harvard University Press; pp. 67-94.

Saxe, G.B. (1991). *Culture and cognitive development: Studies in mathematical understanding.* Hillsdale, NJ: Lawrence Erlbaum Associates.

Schank, R.C. (1991). *Tell me a story: A new look at real and artificial memory.* New York: Charles Scribner's Sons, Macmillan.

Steen, L.A. (1988). *Calculus for a new century: A pump, not a filter.* (MAA notes No. 8). Washington, D.C.: Mathematical Association of America.

Steen, L.A. (1988). The science of patterns. *Science, 240,* 611-616.

Steffe, L.P. (1988). Children's construction of number sequences and multiplying schemes. In J. Hiebert and M. Behr (Eds.), *Number concepts and operations in the middle grades.* Reston, VA: National Council of Teachers of Mathematics; Vol. 2, pp. 119-140.

Steffe, L.P., Cobb, P., and von Glasersfeld, E. (1988). *Construction of arithmetical meanings and strategies.* New York: Springer-Verlag.

Vergnaud, G. (1988). Multiplicative structures. In J. Hiebert and M. Behr (Eds.), *Number concepts and operations in the middle grades.* Reston VA: The National Council of Teachers of Mathematics; Vol. 2, pp. 141-161.

3 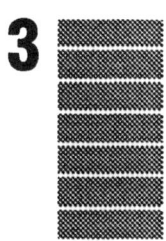 Toward an Assessment Framework for School Mathematics

Gerald A. Goldin

INTRODUCTION

There is increasing recognition that the methods currently used most widely by schools for assessing student mathematics achievement are having a substantial negative impact on meaningful learning. Often it is assumed that the situation can be improved by replacing tests that measure low-level skills, computational algorithms, and routine problem-solving with new instruments containing more sophisticated, nonroutine problems. Ideally, with an appropriate pool of test items, it is suggested that "teaching toward the test" would no longer compromise the goals of the assessment and that a student's successful performance would unquestionably reflect a deep mathematical understanding.

This chapter argues against this approach and stresses the need for a sound *cognitive model* as the basis of a framework for assessing meaningful mathematics learning and understanding in schools. Exploring in detail a few mathematical assessment items illustrates how the outcomes of *any* assessment—traditional or nontraditional—depend on the teacher's prior understanding of what is being assessed. Particular cognitive processes cannot be identified with a mathematics problem that elicits them, nor can they be assumed to be necessary to solve the problem. It follows that reform of assessment involves much more than the creation of new instruments. What is needed is not only an appropriate cognitive model, but also an understanding among teachers, school administrators, students, and the general public of mathematical *processes* as well as content, and of how new methods of assessment are intended to address these processes.

Based on the examples explored here, we shall discuss an assessment model entailing several different kinds of broadly applicable cognitive representational systems drawn from an earlier model for problem-solving competence (Goldin 1987, 1988) together with domain-specific capabilities organized into conceptual schemes.

Background

The past fifteen years have seen some substantial changes in the intellectual leadership of mathematics education in the United States. Among other positive developments has been a transformation in the prevailing educational perspective on what it means to learn and to understand mathematics.

By the mid-1970s, a behaviorist view had come to predominate in school mathematics. Influenced by the claims of behavioral psychologists to scientific rigor, some educators became advocates for the position that vocabulary purporting to describe students' mental states or cognitive processes should be discarded from the lexicon (Mager, 1962; Sund and Picard, 1972). Many schools rewrote their curricular goals accordingly. *Understanding* in mathematics (and in other school subjects) became virtually *identified with performance*—or, more precisely, with the student's achievement of sets of performance objectives, most often expressed as the reliable and rapid attainment of correct answers to mathematical problems of various types. This view might not have become so prevalent had it not had, as a source of wide political support, its extraordinary compatibility with the "back to basics" reaction against the "new mathematics" movement of the 1960s (a movement which itself had had mixed results in fostering mathematical understanding in the majority of students). The behavioral objectives approach lent itself well to reliance on standardized skills tests in mathematics, both to define the goals of instruction (basic skills), and to provide the assessment of success in achieving those goals.

The results on balance have been quite negative, even by performance objective measures. Not only did insightful mathematics learning virtually disappear from many classrooms, but a substantial number of children did not acquire or retain even the skills they were taught as rote procedures. Today, as a partial consequence, our society confronts a real crisis in mathematics teaching and learning. But the prevailing view of mathematical understanding has changed. Meaningful mathematical understanding is now widely seen by many educators (as well as cognitive theorists) as entailing a complex system of elaborately constructed cognitions, developed over time, involving not only overt concepts and procedures but considerable tacit knowledge that can be brought to bear without the conscious awareness of the teacher or the student (for example, Lesh and Landau, 1983; Davis, 1984).

Obviously the current thinking requires substantial, fundamental revision in several aspects of mathematics education, including not only the objectives of instruction and the preparation of teachers, but also our assessment techniques—at least if we desire to observe in some systematic way the nature of the student understandings that are the outcomes of learning. The objectives of mathematics education have been addressed in some detail in a number of recent public documents (for example, NCTM, 1989; NRC, 1989). Attention has also been given to issues in mathematics teacher development through national publications (NCTM, 1990; Davis, Maher, Noddings, 1990), and through many innovative regional institutes for teachers. To date, in 1992, there have been only isolated and preliminary efforts toward the fundamental reform of assessment techniques.

DEVELOPING NEW ASSESSMENT TECHNIQUES: OPPORTUNITIES AND DANGERS

It can of course be maintained that the assessment of a student's understanding is fundamentally a matter of qualitative judgment based on long-term, personal interaction and is therefore best accomplished by the individual classroom teacher. Perhaps it is a mistake to try to assess understanding systematically or to develop a framework for doing so. This possibility should not be dismissed out of hand—while leaving assessment to the classroom teacher seems a radical idea in the United States, it is generally the method of choice in, for example, the Federal Republic of Germany. There are, however, some strong arguments for taking the present opportunity to develop new instruments and modes of assessment, provided the appropriate cognitive and teacher development foundations are laid.

First, with the setting of goals for mathematics education transcending mere computational speed and accuracy, a well-designed system for measuring understanding descriptively could be a useful resource. It could in principle not only provide feedback to society as to how well educational institutions are achieving their more ambitious objectives, but also inform and consequently enhance the classroom teacher's judgments —enabling the teacher to build on individual students' strengths while addressing cognitive obstacles more effectively.

Second, new assessment frameworks may permit us to replace our current system of tests with a much better alternative. *This is an urgent necessity for real mathematics education reform.* To the degree that present assessment techniques rely on routine computational, algorithmic, or word problem tasks, or can easily be misconstrued as doing so, they continue to function as powerful inhibitors of teaching for mathematical understanding in schools. School boards, administrators, and teachers are reluctant to devote class time to conceptual understanding, exploratory activity, construction of mathematical meanings, or mathematical invention, because they (often correctly) perceive these activities as untested and irrelevant to the test scores that are the short-term bottom line in mathematical achievement. In the long run, of course, the damage shows up in many forms, including low scores even on the standard test items, because students have not developed adequate conceptual foundations.

An immediate example is provided by some of the K-8 national mathematics testing series released in the United States in 1991. As a consequence of the current movement toward mathematics education reform, these now include more complex, nonroutine topics and problems, and they develop some important mathematical ideas both computationally and conceptually. While textbook publishers deserve their share of past responsibility for overemphasizing rote computation in mathematics, the 1991 series offer a potentially significant opportunity for change at the classroom level. Unfortunately, experience suggests that, in many schools, despite national recommendations to the contrary, conceptually based activities will be culled from the curriculum and treated as "enrichment"— for occasional use only, with selected students—as if the understanding and doing of mathematics (as opposed to the rote learning of skills) were not central to our educational goals, but merely an optional add-on. This anticipated outcome is made more likely by the fact that, due to traditional format restrictions, complex problem strategies and conceptual understanding are by and large deemphasized in textbook chapter and unit tests, that are the only immediately provided assessment methods in the K-8 series. Furthermore, items included on textbook tests that might be considered as addressing higher-level or deeper understanding are easily misconstrued as intending to measure lower-level skills or memorized terminology. An assessment framework accessible to teachers could be just what is needed to overcome such difficulties.

But a note of caution must also be sounded. There are dangers as well as opportunities in developing new instruments for assessment that, in practical use, can be harmful as well as helpful. *Assessment instruments must never substitute for the teacher's own understanding of mathematics, nor for the teacher's own model of student understanding.* A teacher who does not adequately comprehend a mathematical idea, or have a good cognitive model of what it means for a student to understand a mathematical concept, will not be able to compensate successfully by teaching toward the test, especially when the test is a nontraditional instrument designed to assess higher-order or deeper understandings. And teachers or administrators who do have such understandings must be able to use the information gained from new instruments to enhance and supplement their perspectives, not to replace them.

Having agreed at last to attend to the inner cognitions of students, mathematics educators must recognize that the assessment is not the set of goals—it is only a means of gathering information related to, but different from, the set of goals. It would be a poor doctor who thought that the objectives of medical treatment were mainly to obtain satisfactory readings on descriptive instruments—thermometers, stethoscopes, and so on—without a physiological model that distinguished health from disease. Such a practitioner, seeking only to alleviate the symptoms of illness, might succeed in the short run in relieving symptoms, but would fail spectacularly in achieving healthy patients. In developing an assessment framework for mathematical understanding, this medical analogy can help us take account of the following general principle: *No matter how sophisticated the mathematical problems we may pose* (so that they seem to *require* higher-level thinking, strategic problem solving, and/or conceptual understanding for their solution), *it is conceivable to devise—and to teach—practical, rule-based, noninsightful procedures for solving them.* Under these conditions successful performance does not reflect understanding. Indeed, in such circumstances the purpose of the assessment is defeated just as surely as lowering a patient's fever through aspirin, or an ice bath, prior to a medical examination would defeat the purpose of taking her temperature. In schools, such misuses of assessment are most likely to occur when educators do not themselves understand what is actually being assessed, or why. These are major dangers in developing a new assessment framework, and they must be carefully avoided.

INGREDIENTS OF AN ASSESSMENT FRAMEWORK

The following are some of the perspectives for which I wish to argue in approaching the issue of a useful assessment framework:

- Assessment that is no longer limited to discrete, low-level mathematical skills requires not only an idealized, structured model for a mathematical content domain, but a sound cognitive model for describing the capabilities to be assessed. While such models can become very complex, it is important that the resulting assessment framework be simple enough to be useful at the same time that it is sufficiently detailed to reflect what is involved in doing the kind of mathematics we want to encourage.

- A useful cognitive model for this purpose can be developed based on two dimensions, which cut across each other: (i) mathematical conceptual schemes, reflecting the organization of domain-specific cognitive capabilities, and (ii) several different kinds of cognitive representational systems, which characterize capabilities for broader problem solving and transfer to novel situations. A framework for assessment based on such a model can be *reflective* in taking account of the student's analysis of his or her own cognitive strategies. It also provides the basis for a more *descriptive* assessment, so that we can obtain useful profiles of a student's mathematical development as we evaluate insightful problem solving and depth of understanding.

I shall also try to illustrate these ideas with a few examples.

The Need for an Explicit Cognitive Model

There are several reasons why it is important to set forth a cognitive model as a prerequisite to creating an assessment framework. The first, as mentioned above, is the need to prevent the measure of achievement—the test or the assessment framework—from being identified or taken as synonymous with the central goals of the curriculum. In part this happens because we have no other *independent* characterization of those goals. To be specific, consider a nonstandard, exploratory problem about egg timers (Problem 1).

4-minute timer

7-minute timer

Problem 1. You have two egg-timers, in which fine sand runs from one compartment to another in a fixed interval of time (see illustration). It takes exactly 4 minutes for the sand to run through one timer, and 7 minutes for it to run through the other.

What other intervals can be timed? Can you use these timers to measure an interval of exactly one minute? Explore different pairs of egg-timers.

Let us ask what the purpose would be of such a problem activity in a teaching lesson. What might we be assessing if this problem were used in an individual interview, as a group problem-solving activity, within a student project, or on a test?

As mathematics educators interested in higher-order and deeper understandings, we might expect that such a problem is intended to develop or assess a student's ability and willingness to do at least some of the following (listed in no particular order):

- Explore a new concrete situation.

- Try a number of specific procedures.

- Visualize the outcomes of particular sequences.

- Organize and record the outcomes of trials in a useful way.

- Make appropriate use of addition and subtraction of numbers, in the specific context of comparing and concatenating intervals of time.

- Modify the order of steps in particular sequences.

- Make conjectures from special cases.

- Investigate a conjecture systematically.

- Put together a strategy for systematic study of the problem.

- Arrive at a mathematical generalization based, in this case (tacitly or explicitly), on the fact that the numbers 4 and 7 are relatively prime (involving structures of multiplication and division).

- Generalize spontaneously from one problem to a family of related problems.

- Try to understand why a conjecture might be true (intuitive precursor of the idea of mathematical proof).

- Proceed with an investigation without fear of being wrong.

- Be influenced by the problem's internal logic, rather than by the perceived expectations of the teacher.

- Participate constructively in group problem-solving interactions (depending on the mode in which the problem is addressed).

- Have fun with such an exploration, and so on.

Depending on which cognitive objectives are highlighted, the context or mode of presentation of the problem may change. A list such as the above is based on at least a tacit model for the kinds of things that constitute mathematical understanding or sophistication. Without such a model, a teacher even with the best of intentions in preparing students to solve such a problem during an assessment, can do many things that defeat the preceding objectives. For example, the teacher may

- show the student a number of ways to use the timers in sequence (so that there is no need to explore spontaneously),

- demonstrate a number of specific procedures (so that the student tries them only in imitation of the teacher),

- decide the outcomes of particular sequences for the students (so that the students need not anticipate, or visualize, or reason to consequences),

- set up a chart for the students to record the outcomes of trials (so that the students themselves only enter numbers), and so on, until finally, the students have no sense of fun or accomplishment in the problem-solving, and see only a routinized method for solving a class of "egg-timer problems" along with other problem types such as "money problems," "rate problems," and so on. Perhaps a student, having been thus prepared, will solve the problem correctly during the assessment—but what is measured will be recall of a demonstrated, rule-based procedure rather than any of the desired cognitive processes.

A second reason to have a cognitive model in advance of developing an assessment framework is to be able to discuss the goals of instruction (and consequently the goals of the assessment) explicitly, and to the extent feasible to decide these goals consciously rather than tacitly. For example, we have heard much about the need for real-life problem solving in the curriculum, as though the immediacy and verisimilitude of application were the most important criteria for what should be taught. But are they? No one doubts that one goal of mathematics education should be to develop the student's ability to treat real situations mathematically, but is it justified to declare this the most important or cognitively valuable goal? Might not an exclusive focus on real-life mathematics problems preclude the student's achievement of mathematical understanding, of abstraction from specific situations, of transfer to new situations, or of insight into the beauty and simplicity of mathematical reasoning? Or are these envisioned as automatic, but incidental consequences of an emphasis on real-life mathematics? The new textbooks and their accompanying tests, even as they have increased the complexity and conceptual depth of their treatments, have also sought to place more emphasis on realistic problems in realistic contexts—but these may or may not encompass what we mean by higher-level or deeper understanding. Having a sound cognitive model as the foundation of an assessment framework can help us avoid oversimplified interpretations of what it means to do mathematics.

A third reason for a cognitive model is that new approaches to assessment can now make use of new technology—for example, computer environments can be designed for conducting individualized assessments, with built-in elaborate contingencies based on student responses. The fact that such schemes require major commitments to structured programs suggests that it would be a good idea to invest some time in the design elements, to determine just what it is we want to assess before we set about building computer-based assessment systems. An adequate, accessible cognitive model would seem to be a prerequisite. This is especially important because the American public now tends to have little conceptual understanding of mathematics and to have a highly procedural/algorithmic orientation toward what it means to do mathematics. Parents and policymakers may tend to place their faith in new assessment schemes merely because they are high-tech, and again—without efforts to the contrary—we may see the assessment procedure defining by default the cognitive goals of instruction.

In short, we need a good way to characterize desirable cognitions in mathematics, a characterization that captures the essential cognitions that we as mathematicians and educators hope to develop in our students. Then, and only then, should we seek to develop an assessment framework that describes the extent to which these cognitions have actually been developed. At the classroom level, we need teachers who understand what the goals of instruction are, why they are what they are, and how certain kinds of assessment items are expected to measure particular instructional goals.

TRADITIONAL AND NONTRADITIONAL ASSESSMENT ITEMS

To make more concrete what we mean by a cognitive model, let us next consider some further assessment items, comparing a traditional with a nontraditional question. Again, most of the analysis that follows is independent of whether the questions are posed on standardized tests, occur within student projects, are presented in individual interviews, or form the basis of group problem-solving activities.

Understanding, Traditionally Assessed

Problem 2 is a routine story problem of a type that, in one form or another, might have been given to middle school students at any time during the past century to assess their understanding of school mathematics. It is expected that (with pencil and paper) the student will calculate the two products [4 x $3.79 and 3 x $4.85], subtract the latter result [$14.55] from the former [$15.16], and obtain the answer [$.61].

Problem 2. Mixed nuts cost $3.79 per pound, while cashews cost $4.85 per pound. How much more does a 4-pound bag of mixed nuts cost than a 3-pound bag of cashews?

On the surface, such an item does appear to assess the student's understanding of multiplication and subtraction in a real-life context, as well as his or her ability to perform and make use of routine arithmetic computations. Undoubtedly this belief accounts for such problems having survived generations of mathematics education reforms. When we look more deeply, though, we can see that this characterization of what is assessed is too simple. On the one hand, solving the problem insightfully may involve, for example, (i) the student's having some kind of *broad heuristic strategy* (which some might call "metacognitive") for structuring such story problems, including the ability to extract the wanted and given information, identify the goal information, and so on; (ii) the student's having the ability to interpret the problem statement semantically, to visualize the problem elements in the cultural context of purchases being made; (iii) the student's being able to regard specific aspects of the problem situation (not merely the words) as calling for certain arithmetic operations, so that the "cost of the mixed nuts" can be obtained by multiplication, the "cost of the cashews" by a second multiplication, and "how much more" by subtraction of the latter product from the former; (iv) the student's not only having the ability to carry out the desired computations correctly, using standard algorithms, but being able to monitor the meaning of the computations and the reasonableness of the results through estimation and contextual reasoning; so that (for instance) an answer of $1,455 would immediately be deemed inappropriate for the purchase of a bag of cashew nuts; and (v) the student's being able to organize the information, keeping track of goals and subgoals along the way, so that (for example) after each arithmetic operation is performed, the student knows or can determine how the numerical answer relates to the problem's goal structure and to which semantic element of the problem situation it corresponds.

These procedures are, in and of themselves, sufficiently complicated so that, if we accept the preceding as a partial description or "model" of the needed problem-solving processes, it must be acknowledged that the problem does require higher-order cognitive activity or deeper understanding. In fact, many students experience considerable difficulty with such problems, suggesting the complexity of the processes they do in fact involve. However, despite the complexity of the requisite cognitive processes in this description, there are some deeply unsatisfactory characteristics of such problems used as assessment items, which we now mention.

First, many would point out that this task is not exactly the real-life, practical problem it pretends to be. In an actual pricing situation, it is more probable that an estimated answer rather than an exact answer would be called for and calculated. Were a precise answer needed, it is likely that a calculator would be at hand to obtain it more easily than through a conventional calculation. In this case it would be necessary (in some states of the United States) to consider additional complications such as the sales tax. Furthermore, we are not told why the answer is needed, and it is difficult to conjecture a real-life situation in which the precise difference in these two prices would be the problem goal.

Second, the foregoing partial description of procedures suggests numerous important, *tacit* understandings which may or may not be brought to bear by the student. These are addressed only incidentally by the problem. *Learning or assessment based solely on the problem-as-posed does not tell us what failed and what succeeded.* It might be well to make some of the possibilities overt. For example, the problem anticipates that a 4-pound bag of mixed nuts costs more than a 3-pound bag of cashews. Does the student (i) take this for granted (and simply subtract the smaller product from the larger, or the second product from the first), or (ii) monitor the calculation and verify when the time comes that the cashews really cost less, or (iii) check this assumption by estimation at the outset? What are the consequences of each possibility? If (iii) occurs, for instance, the student might obtain 4 x $4 = $16 while 3 x $5 = $15, and conclude that indeed the mixed nuts cost more, but not a lot more; while a sophisticated student might even note that since the cost-per-pound of the mixed nuts was rounded upward in this estimation by more than the cost-per-pound of the cashews, one should not have total confidence in the outcome of the estimate. This is just one of the large number of tacit capabilities which, taken together, may characterize understanding of the mathematics of the task, but which success on the task itself does not assess. Within certain modes of learning or assessment, such as tutoring or individual interviews, it may actually be possible to measure such outcomes; but it is necessary, then, to introduce a questioning or interview procedure that seeks to elicit them.

Third, and perhaps most important, there is the entrenched expectation that the problem is "routine." This term has two distinguishable meanings. In the sense that problems like this one are frequently used for assessment and are standard content in most textbooks, to be routine is not a negative characteristic—after all, if ideal, nonroutine problems were to be substituted in textbooks and assessment instruments, they would quickly become routine in this sense. But there is a more intrinsic sense of routine that pertains to this problem—namely, *the expectation that its solution involves only the straightforward application of previously learned computational rules, in semantic situations where there is a standard one-to-one correspondence between the called-for computation and the situational entity* (see below). There are no new mathematical constructions anticipated here and no difficult decisions are anticipated as to the operations that are called for or what they mean. On the contrary, the implication of routine is that any such constructions or difficult decisions occurred long ago and have become automatic; if not, this would bespeak a deficiency in the student's preparation to solve the problem.

A Tacit Model

Because of the routine nature of Problem 2, it is thus possible to bypass much of its cognitive complexity—and some teachers and books tend to do so. For example, the problem can be addressed without developing a mature concept of the intensive quantity "cost per pound," if the student is merely taught to interpret the phrase "4 pounds at $3.79 per pound" as a purely syntactic instruction to multiply. [The general syntactic form here is "x A's at y B 's per A," where x and y are numbers and A and B are nouns describing objects, units of measurement (including money), and so on.] We see this bypass attempt at its worst in the so-called key words approach to story problems—an approach that has often been deplored but that survives in many classrooms. It survives because, unfortunately, it works to obtain answers to the routine problems most often used in classrooms. A tacit (and highly inadequate) cognitive model underlying such approaches to the problem is illustrated in Figure 1.

Figure 1. A tacit, naive cognitive model.

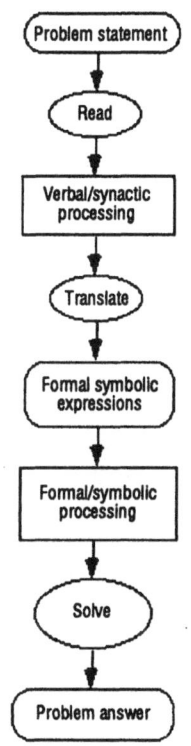

The compartments with rounded corners in Figure 1 refer to external configurations (the verbal problem statement; formal expressions or equations generated by the solver; the written answer to the problem). The rectangular boxes stand for internal systems of representation (verbal processing, involving reading the problem and identifying relevant syntactic expressions; formal/symbolic processing, involving structured arithmetical computations). The ovals refer to internal processes that interface between the internal systems and the external configurations, or between two internal systems. The strategy depicted is sequential—read the problem, translate each part of the problem statement into an arithmetical procedure ("words" to "symbols"), and then perform the necessary arithmetic.

By making this frequently held tacit model explicit and detailed, it becomes apparent how, without a more adequate cognitive model to the contrary, many teachers of mathematics would plausibly see what is being assessed here as the conversion of word problems about cost to formulas (translation of modes of external representation), and the performance of computations, rather than the understanding of mathematical concepts, heuristics, problem-solving strategies, or applications. This is another instance of the above-mentioned general principle that no matter how sophisticated the problem, one can devise a noninsightful procedure for solving it.

For these reasons the exclusive focus on such problems as the epitome of what it means to do mathematics at the middle school level is a very bad idea, despite the potential complexity of the cognitions that might be brought to bear.

Comparison with Nontraditional Assessment Items

When a nontraditional problem (Problem 3, for example) is introduced in the context of assessment, mathematics educators frequently respond with approval or disapproval, making some implicit judgments about the item's validity and at the same time projecting their assumptions about what it would take to solve the problem. The purpose of the examples that follow is not to present good or bad items, but to make some of those assumptions explicit—that is, to discuss what might be measured by certain kinds of nonroutine items, compared to routine assessment items such as Problem 2. We shall see that the cognitions to be assessed are very sensitive to the conditions of the assessment. Without a clear characterization of what we want to measure, there is a great deal of ambiguity. We shall see that the question to pose first is not, "What does the problem assess?" but "What are we *trying* to assess through the problem?" The answer is not obvious from the problem itself.

Problem 3. Let the symbol @ stand for the *average* or *mean* of two numbers. For example, we shall write 6 @ 8 = 7, because 7 is the mean of the pair 6 and 8. Is the operation @ commutative? Is it associative? Explain why or why not.

One possibility, in line with the recommendations of many educators, is that through Problem 3 (and others like it), one might assess conceptual understanding of commutativity and associativity by asking that these concepts be transferred to a new and unfamiliar context. If this is indeed the assessment goal, however, the student should not have been prepared for the problem (for example, as an examination question) through prior exploration in class of the properties of @ as an operation. Were the student to know before seeing the problem that the mean can be treated as a binary operation, perhaps even having seen introduced a symbol such as @ to stand for the operation, having learned that it is commutative but not associative, and having seen these properties illustrated for the mean with examples and counter examples, then we are not assessing transfer to a new context at all, but only how well the student learned what was taught about the operation in this context. Thus, to accomplish the stated goal of the assessment, we are to a limited extent testing the student on material that has not been taught—something widely regarded as unfair, and whose purpose is not well understood by the public.

But the issue is not limited to what the student has not seen. For the problem to accomplish its assessment goal, it must also be understood that the student has some prior understanding of "the average or mean of two numbers." If not, the problem as posed may never address the transfer of the concepts of commutativity and associativity to a new context, because the intended context is itself not understood. It must also, of course, be assumed that the student has some prior knowledge of commutativity and associativity in other contexts, such as addition or multiplication. Thus what the problem actually assesses depends sensitively on the student's prior preparation; for the intended purpose, there is a fine line between prior preparation that is necessary and that which is impermissible. These conditions are, at least to some extent, controlled directly by the classroom teacher; thus *what is accomplished by an assessment depends on the teacher's understanding of the intent of the assessment.*

Suppose that the prerequisites are in fact met: the student has some understanding of the mean, but has not seen its properties as a binary operation discussed. Then a variety of complex cognitive processes can occur. The student might employ any of a number of heuristic strategies. One approach is to make trials to determine whether @ is commutative: try 6 @ 8 [which is (6 + 8)/2 = 14/2 = 7], and compare the result with 8 @ 6 [which is (8 + 6)/2 = 14/2 = 7]. The decision to compare 6 @ 8 with 8 @ 6 through calculation would seem to require syntactic transfer of the commutative property, as normally stated for the binary operations + or x, to the new operation @, using a procedural notion of binary operation, that is encouraged by the calculation of 6 @ 8 in the problem statement. Several further trials might convince the student that commutativity holds, but this conclusion may be regarded (correctly) as a conjecture, or (incorrectly) as a proven fact, depending on the student's understanding (at a level of mathematical logic more sophisticated than is generally appreciated) of what constitutes an unfounded conjecture, what constitutes a conceptual reason for a mathematical pattern, and what constitutes proof. At some point, the student might reason by modeling the general on the particular (another heuristic strategy) and ask, Why is the operation commutative? We are dividing by two in all cases, in whichever order; thus @ is commutative "because a sum like 6 + 8 is always the same as 8 + 6"; that is, because addition is commutative for the pair 6 and 8, and "the same argument" applies to any other pair of numbers.

Similarly, associativity could be investigated by making trials, for example, by comparing (6 @ 8) @ 2 with 6 @ (8 @ 2), by choosing three different numbers more or less at random to test and test again, or by making syntactic transfer of the meaning of associativity from a familiar domain. The first calculation gives 7 @ 2 = 4.5; the second gives 6 @ 5 = 5.5; thus, the answers are different. If the student understands the role of a counterexample in disproving a conjecture made for all numbers, then it is demonstrated that the operation @ is not associative.

We see that the problem, if used as intended, can assess not only the transfer of the concepts of commutativity and associativity to the new situation, but also the use of some problem-solving heuristics and the understanding of mathematical reasoning at a fairly sophisticated level. And indeed, such capabilities are an important, but often tacitly disregarded, part of the conceptual understanding of structural properties in mathematics such as commutativity and associativity.

As was mentioned, some prior acquaintance with the mean of two numbers is assumed in this problem, but thus far only its procedural interpretation using formal notational symbols has been discussed. Of course, understanding the mean of 6 and 8 encompasses far more than the procedure of adding and dividing by two. There is descriptive knowledge about the mean, for example, the anticipation that the result is a number between the original two. There is visualization of the mean, such as its semantic interpretation as an intermediate height, or as the midpoint of a segment joining the two original numbers on a number line; there is its interpretation as a center of mass, and so on. Such representations can help with the problem: If @ is seen as specifying the midpoint of a line segment, for instance, the commutative property can be understood as the assertion that the midpoint is independent of the order in which we specify the end points. Similarly, understanding of commutativity and associativity can involve descriptive knowledge, for example, the idea that if one of these properties holds for a few generic examples it is probably true, but that examples involving identity elements are not generic. It can involve imagery, for example, visualizing the commutativity of addition as describing the reversal of two rods glued together, automatically preserving the total length. The problem posed can thus assess transfer of understanding to a new domain, but the understanding whose transfer is assessed may be, but is not necessarily, more than procedural/notational.

Minor variations of a problem like this one can lead to very different conclusions about what is being assessed. After the knowledge needed to solve the problem (or that is potentially helpful in doing so) is analyzed into several components, it is instructive to consider what happens when we provide students with elements of one or another component within the question itself. For example, the following knowledge components related to Problem 3 might be identified:

- The calculation of @ (that is, a procedure for finding the mean);

- The definition of commutativity and associativity for a familiar operation;

- The definition of these properties for @ (entailing transfer of the statement of each property);

- The understanding that to suggest a possible property of an operation is to make a conjecture that can be investigated, not to state a fact or convention that must have been previously learned (so that the answer "I don't remember," or "We haven't had that yet," is seen as inappropriate to this question, while it might be appropriate to say "I don't remember" if asked to state the commutative property);

- The trying of special cases (use of a heuristic process that may not be necessary but can be helpful here);

- The seeking of mathematical reasons for a pattern;

- The modeling of the general on the particular (a potentially helpful process);

- The construction of related representations of the operations or of the properties (for example, the mean as the midpoint of an interval); and so on.

The variation shown in Problem 3a seeks to remove its dependence on prior acquaintance with the mean by including an explanation. Let us consider what else happens: The change in the problem suggests a change in our assumptions about prior knowledge, that is, the assumption here may be that while the student has prior acquaintance with the commutative and associative properties in connection with operations such as addition and multiplication, there is no prior acquaintance at all with the mean. If this assumption is true, solving the problem then requires more than the transfer of the concepts of commutativity and associativity to a new domain. It requires, and assesses, the student's ability to construct the new domain from the given verbal description of a new operation; and if this ability is undeveloped, the assessment of transfer will never take place. This represents a significant change in the originally stated objective.

Problem 3a. Let the symbol @ stand for the *average* or *mean* of two numbers. This is found by adding them and dividing their sum by 2. For example, 6 @ 8 = 7, because 6 + 8 is 14, and 14 divided by 2 is 7. Is the operation @ commutative? Is it associative? Explain why or why not.

In the next variation of the problem (3b), we also remove the need to recall the statements of the commutative and associative properties for a familiar operation. The result is to change the tacit assumptions still further. In revising the problem this time, we have done more than assist the student with recalling statements of the commutative and associative properties.

Without some explicit statement to the contrary, we have also modified drastically the assumptions about prior knowledge. From Problem 3b, it might be inferred that because commutativity and associativity are defined in the problem statement, we are not trying to assess anything about their prior conceptual understanding at all, but to assess the student's ability to construct mathematical interpretations of these concepts from new definitions, to transfer the interpretations to a newly constructed domain, and to use appropriate heuristic reasoning techniques.

Problem 3b. The operation of addition (+) is *commutative* because when two numbers are added, their sum is the same in either order. For example, 6 + 8 = 14 and 8 + 6 = 14. Addition is also *associative*, because when three numbers are added it does not matter which pair is added first. For example, (6 + 8) + 2 = 14 + 2 = 16, while 6 + (8 + 2) = 6 + 10 = 16.

Now let the symbol @ stand for the *average* or *mean* of two numbers. This is found by adding them and dividing their sum by 2. For example, 6 @ 8 = 7, because 6 + 8 is 14, and 14 divided by 2 is 7.

Is the operation @ commutative? Is it associative? Explain why or why not.

Problem 3c removes some of the dependence on knowledge about conjectures, proofs, and counter examples. In this version, the student is again asked to interpret a prior understanding of commutativity and associativity in an unfamiliar operational domain (@), but instead of having to decide whether @ obeys these properties, using knowledge about the role of conjectures, proofs, and counter examples, the result is provided and the reasoning process is structured for the student within the problem statement. The student must only try special cases to fulfill the stated conditions.

Problem 3c. Let the symbol @ stand for the *average* or *mean* of two numbers. For example, we shall write 6 @ 8 = 7, because 7 is the mean of the pair 6 and 8. Give an example (using two numbers) which illustrates the commutative property of the operation @. Give an example (using three numbers) to show that @ is not associative.

Another variation of the problem (3d) stresses the use of a representation other than formal mathematical notation. Under the right conditions, a version such as Problem 3d assesses first a particular capability of the student—the ability to construct a number-line representation for the new operation (@)—and then assesses the transfer of prior conceptual understanding of commutativity to the new context in relation to that representation. This is distinct from the (rather less definitive) assessment of the student's spontaneous decision to construct such a representation, which might or might not take place in the course of solving other versions of the problem.

Problem 3d. Let the symbol @ stand for the *average* or *mean* of two numbers. This is found by adding them and dividing their sum by 2. For example, 6 @ 8 = 7, because 6 + 8 is 14, and 14 divided by 2 is 7.

Draw a number line and, using two numbers as an example, show the meaning of @ with a diagram on the number line. Then explain what your picture suggests about whether @ is or is not commutative.

The above problem variations illustrate two things: First, beyond the problem statement itself, *it is the assumed conditions of prior learning which in a major way determine the kinds of cognitive processes that solving the problem elicits, and which the item thus assesses.* Second, *whether or not a problem assesses a particular capability depends on whether the student actually reaches the point in the problem where it would be appropriate to make use of the capability.* Even if simple recall of a concept (such as the mean) is part of what the item is intended to assess, the assessment of other aspects of the student's understanding may be contingent on a positive outcome for this one. In short, there are always contingency structures implicit in assessment by means of complex tasks.

These observations apply to many modes of problem presentation. In some modes—individual interviews, computer-based assessment, and possibly group problem-solving—there is the possibility that specific heuristic suggestions or hints can be provided to the problem solver along the way. Then, whether or not particular cognitive capabilities are assessed need not be entirely dependent on the student's prior successful exercise of other capabilities on the same problem. Some of the many different things that the above problem variations assess could, for example, be observed within a single well-designed clinical interview through a series of carefully structured questions. In such an interview, the most important design principle is to permit the student, at each stage, maximum latitude for "free" problem solving—providing minimal suggestions when an impasse is reached, so that spontaneous cognitive processes can occur and be observed.

Let us summarize the major points of similarity and difference between the variations of the nontraditional question, Problem 3, and the previously discussed traditional one, Problem 2. What is assessed by the routine problem depends, like the nonroutine problem, on prior knowledge; and it too has a contingency structure such that the opportunity for successful application of some capabilities depends on other capabilities. But in the routine problem, the prior preparation of the student is assumed to be as direct as possible; little that is new is expected to be constructed; and the structure of capabilities (if regarded at all) is usually simplified as much as possible, as in Figure 1. In the nonroutine case, we expect new constructions, but until we introduce a cognitive model, what they might be is only implicit. Depending on the processes that we want to assess, it is possible create numerous problem variations, even to provide suggestions along the way, ensuring that whether certain capabilities are assessed is independent of the student's successful use of others. But some teachers, familiar with more routine problems, may not expect new constructions, or may be unable to specify the requisite capabilities. And if a nonroutine assessment item is to be used for evaluation purposes, there may be considerable pressure on the teacher to provide direct experiences, which can defeat its purpose. With these comparisons in mind, we now turn to the issue of creating an assessment framework.

TOWARD A FRAMEWORK FOR ASSESSMENT BASED ON A COGNITIVE MODEL

An assessment framework must be much more than a collection of routine or nonroutine problems of various degrees of complexity in various content domains. It should enable us to describe the capabilities that are to be assessed, to make explicit the conditions of the assessment, and to explain how the problem items are intended to be used to elicit and measure the student's cognitive processes. Keeping in mind our earlier discussion, the framework should have the following characteristics:

- It should be based on an independent characterization of the understanding that we want to assess, so that we can infer cognitive capabilities from behaviors without identifying abilities with behaviors.

- It should be descriptive, that is, capable of informing us what the individual student can and cannot do, and capable of describing heuristics, representations, or concepts that are partially developed. We need to move toward a picture that lets us see each student's emerging capabilities and how they are structured.

- It should be reflective, allowing the student not only to grapple with mathematical discovery and conceptual constructions but to reflect on these processes. Thus, part of what is assessed should be the student's own self-descriptions, and how the student places mathematical activity contextually in his or her own life.

Systems of Cognitive Representation

To specify what is meant by mathematical thinking, that is, to describe independently the understanding that we want to assess, requires that we consider ways in which information can be represented internally by problem solvers. We have previously discussed five kinds of cognitive representational systems brought to bear during mathematical problem solving; they are sufficiently different from each other to deserve separate mention (Goldin, 1987; 1988): (i) verbal/syntactic, (ii) imagistic, (iii) formal notational, (iv) executive/heuristic, and (v) affective. These systems, together with some important processes that interface among them, are depicted in Figure 2. Such a model provides a way to organize and to characterize some of the capabilities comprising mathematical understanding.

Before discussing briefly each type of representational system, let me stress that the goal I propose is not to assess these separately and discretely. Typically, any skilled problem-solving activity in mathematics entails an interplay of several systems of cognitive representation. Thus, the assessment goal is to provide opportunities for many such systems of representation to be brought to bear, not only so that specific competencies within particular systems can be observed, but to gain wholistic information about the student's abilities in coordinating various representations, and in appropriately recasting mathematical ideas from one internal system of representation into another.

To treat cognitive representational systems as components of an assessment model relates directly to how we see the purpose of mathematics education. It presupposes that we are not trying only to teach sets of problem-specific skills, but to develop broad, powerful cognitive systems that can enable the student to grapple with new situations as they arise, to represent them internally in a variety of ways, and to think mathematically by making use of the representations.

Let me next comment briefly on Figure 2, and in doing so compare it with the more naive model in Figure 1. Figure 2 omits external configurations entirely (to keep the diagram to manageable size). The five rectangles refer to internal systems of representation, and the ovals to processes that interface between them. It is helpful here to think of a representational system as consisting of configurations of a certain kind, together with higher-level structures that determine the configurations that are possible and the ways in which they can be processed. The processes in the ovals describe how configurations in one system of representation can evoke, influence, or result in new constructions of those in another.

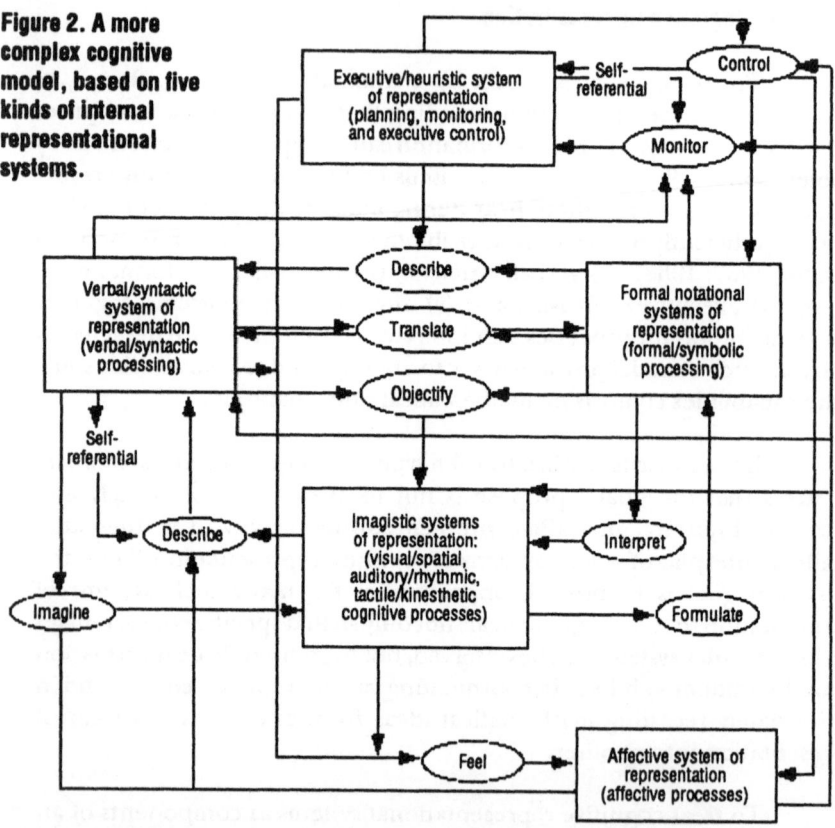

Figure 2. A more complex cognitive model, based on five kinds of internal representational systems.

Verbal/syntactic system of representation. First we have verbal configurations, equipped with syntactic structure. Certainly an important aspect of mathematical understanding involves competencies associated with the syntactic processing of ordinary language, ranging from identification of declarative information and questions or goal statements in a problem, to recognition of various kinds of mathematical phraseology. In discussing Problem 2, for example, we criticized the attempt to characterize the mathematical reasoning as purely syntactic translation (for example, from a phrase taking the form "x A's at yB's per A," into an instruction to multiply); but of course, the recognition and processing of syntax remains important. In a capable solver, however, many other cognitive processes should be in play, prior to and in addition to translation into formal symbolic procedures.

Imagistic system of representation. Imagistic representation includes the construction and processing of visual/spatial configurations, tactile/kinesthetic configurations, and so on. For example, when we discussed various understandings of the mean (the operation @ in Problem 3), one possibility included its visualization as an intermediate height. There could also be a kinesthetic component to such a representation (for example, the student might imagine indicating the intermediate height with an outstretched hand). Construing the mean as the location of the center of mass of two equally heavy objects at given locations suggests a kinesthetic configuration, in which the student imagines balancing the two masses with a finger on the midpoint of the segment connecting them. A conceptual understanding of "x A's at yB's per A" undoubtedly involves concrete, imagistic models of extensive and intensive quantities, of units or groupings of objects, and so on. Imagistic configurations can be evoked by words, by symbols, by heuristic processes (for example, the "draw a diagram" strategy), or by affect. It is certainly a mistake to omit them from an assessment framework.

Formal notational system of representation. Formal notational processing refers to systems of mathematical symbols and their use (for numeration, for arithmetical operations, for algebraic representation, and so on). Figure 2 is intended to suggest not only calculation and computation (for which we can envision cognitive processes within the rectangle), but also the interpretation of notational configurations imagistically, the monitoring of formal procedures (for meaning, for reasonableness, and so on), the description of formal processes verbally, and so forth, as represented by the oval compartments. Traditional assessment schemes in mathematics have placed a lot of emphasis on formal notational processing, but relatively little on the interface between formal notational and other systems of representation.

Executive/heuristic system of representation. Executive decisionmaking and control, monitoring of the student's own problem-solving processes, and heuristic planning are considered as another cognitive representational system. Here we include complex, well-formulated strategies such as trying special cases, means-ends analysis, or drawing diagrams, as well as more vaguely defined strategic decisions. When we speak of a reflective framework for assessment, the intent is to include as a goal the description not only of students' heuristic planning and problem organization capabilities, but also of certain self-referential or metacognitive capabilities—their concomitant monitoring of their own mathematical reasoning, their introspections in relation to their own problem solving, and their discussions and retrospective analyses of their planning processes.

Affective system of representation. Finally, it is important that an assessment framework include the assessment of affect—the feelings or emotional states that occur during mathematical activity, and the consequences of those feelings. This should not be limited to long-term beliefs and attitudes about mathematics, but should address the changing states of affect during mathematical activity: Does the student begin with a sense of curiosity and anticipation of mathematical discovery, or with worry and self-doubt? Does anxiety impede the student, and if so, what is its source? Does the student make use of feelings of frustration constructively, to suggest new heuristic processes (such as trying a simpler problem), or are they a signal to give up? How much fun does the student have with mathematics? Clearly, the assessment of affect must involve a great deal of sensitivity to the individual, making use of modes of interaction very different from those traditional in mathematical assessment.

Conceptual Understanding in Specific Content Domains

Cognitive representation does not take place in the abstract, but always in a contextual domain of knowledge, sometimes referred to as the mathematical content in the curriculum. For purposes of assessment, it is useful to see conceptual understanding in any particular content domain as involving configurations and processes that cut across many representational systems. For example, Problem 2 involved the multiplication and subtraction of multidigit numbers in the context of money and making purchases; such an activity would conventionally be taught as a story problem exercise, addressing that content. However, we saw that the representational capabilities that the problem assesses are not usually made overt. The problem content can be formulated in complex imagistic and heuristic ways, or its representation can be limited to formal notational computation. For Problem 1, on the other hand, we listed a number of broad capabilities, which could now be spelled out in considerably more detail using the model in Figure 2. Only two of these, (addition and subtraction of numbers, and the properties of numbers that are relatively primal) refer specifically to what is commonly called mathematical content. It is clear that the emphasis is not on these but on assessing a rather wide variety of heuristic capabilities. In Problem 3, we saw how setting out to assess mathematical content (commutative and associative properties) at a deeper level of understanding leads into issues of heuristic and imagistic representation.

Structured collections of domain-specific capabilities, organized into sets of related configurations in several different cognitive representational systems and accompanying information-processing action-sequences, give descriptive meaning to what are sometimes called schemes. Thus, the desired assessment framework can be visualized as a kind of Cartesian product of domain-specific mathematical content with the cognitive processes in several representational systems depicted in Figure 2. Traditional mathematics tests have tended to emphasize the former while ignoring the latter; an assessment framework that achieves both is clearly necessary.

Directions for Research and Development

A high priority should be to develop a working model of such an assessment framework in a particular content domain (perhaps for a standard curricular topic at the elementary or middle school level, such as whole number multiplication) that can be tried on a small scale and then extended. First, the cognitive model underlying the assessment framework should be developed carefully, including precursor representations and several different kinds of cognitive representation of the central concepts. Then the framework needs to be implemented with a variety of methods for assessing the key cognitive components (not in isolation, but in combination), such as structured interviews, concrete models, group problem solving, creative projects and portfolio evaluations, as well as pencil-and-paper tests and (perhaps) contingency-based interactive computer environments. A useful system must be able to provide accurate descriptions of what students can do, as well as to identify weaknesses or inadequately developed capabilities; the issue of the reliability or repeatability of the techniques also needs to be addressed.

Ultimately, it will be not only our willingness to join broad, adequately complex cognitive models with the more domain-specific mathematical content traditionally tested, but also our commitment to develop understanding of new assessment goals and methods in teachers, administrators, parents, and the general public, that will enable us to assess meaningful mathematics learning effectively.

NOTE AND REFERENCES

The author thanks the Exxon Education Foundation for its support to the Rutgers University Center for Mathematics, Science, and Computer Education for research on assessment of mathematical understanding.

Davis, R.B. (1984). *Learning mathematics: The cognitive science approach to mathematics education.* Norwood, NJ: Ablex.

Davis, R.B., Maher, C.A., and Noddings, N., (Eds.). (1990). Constructivist views on the teaching and learning of mathematics (JRME Monograph No. 4). Reston, VA: National Council of Teachers of Mathematics.

Goldin, G.A. (1987). Cognitive representational systems for mathematical problem solving. In C. Janvier (Ed.), *Problems of representation in the teaching and learning of mathematics.* Hillsdale, NJ: Lawrence Erlbaum Associates; pp. 125-145.

Goldin, G.A. (1988). The development of a model for competence in mathematical problem solving based on systems of cognitive representation. In A. Borbas (Ed.), *Proceedings of the Twelfth International Conference of PME*. Veszprem, Hungary: OOK Printing House; Vol. 2, pp. 358-365.

Lesh, R., and Landau, M. (Eds.) (1983). *Acquisition of mathematics concepts and processes*. New York: Academic Press.

Mager, R. (1962). *Preparing instructional objectives*. Palo Alto, CA: Fearon.

National Council of Teachers of Mathematics (1989). *Curriculum and evaluation standards for school mathematics*. Reston, VA: NCTM.

National Council of Teachers of Mathematics (1990). *Professional standards for teaching mathematics*. Reston, VA: NCTM.

National Research Council. (1989). *Everybody counts: A report to the nation on the future of mathematics education*. Washington, DC: National Academy Press.

Sund, R. and Picard, A. (1972). *Behavioral objectives and evaluational measures: science and mathematics*. Columbus, OH: Merrill.

4  Research and Classroom Assessment of Students' Verifying, Conjecturing, and Generalizing in Geometry

*Daniel Chazan and
Michal Yerushalmy*

When evaluating the effect of an educational innovation, it is important to test for the goals that the innovation has set out to accomplish. In the Supposer approach to teaching geometry (described below), in addition to teaching the content of high school geometry, teachers try to inculcate in their students mathematical and scientific inquiry skills, beliefs, and attitudes that are helpful in solving inquiry problems. These skills develop in students throughout a year-long course. We would like to be able to assess the success of this approach in teaching students to be good inquirers. The task is an extremely difficult one.

We begin this chapter with a short description of the approach we favor for teaching high school geometry. We then provide a rough outline, which we developed with a group of teachers, of the types of higher-order skills (as well as beliefs and attitudes) involved in exploring an inquiry problem. Having provided this background, we concentrate on students' verifying, conjecturing, and generalizing skills. We first present a research instrument designed to compare generalizations created by students. After presenting this paper-and-pencil test, we present a more thorough analysis of the verifying, conjecturing, and generalizing skills used by competent explorers of inquiry problems. This analysis derives from sessions with classroom teachers as well as considerations suggested by the research literature on induction and thinking skills (Gentner and Gentner, 1983; Holland, Holyoak, Nisbet, and Thagard, 1986; Kuhn, Amsel, and O'Loughlin, 1988; and Nickerson, Perkins, and Smith, 1985). Finally, we suggest some classroom methods for assessing students' progress in developing verifying, conjecturing, and generalizing skills and a grading scheme for students' reports designed to encourage the skills we wish to foster.

ONE WAY TO USE THE SUPPOSERS

The Geometric Supposers (Schwartz, Yerushalmy, and Education Development Center, 1985) are computer programs that allow users to start with an initial shape (for example, a triangle), create geometric constructions (for example, draw all the altitudes), and make measurements of the diagrams that result from the constructions. The programs also store a record of users' activities as a procedure which then can be repeated on a new initial shape (for example, any other triangle, see Figure 1). This repeat feature allows users to test the generality of the conclusions they reach about the results of a particular construction. (See Yerushalmy and Chazan, 1990, for a more detailed description of the software and the ways it supports students in using diagrams.)

Figure 1.
Altitudes in three different triangles.

With paper and pencil, one can do all that these programs do, but not as quickly nor as accurately. This difference in speed and accuracy makes feasible an approach to the teaching of Euclidean geometry only theoretically possible with pencil and paper. (For a comprehensive description of this approach, see Chazan and Houde, 1989.) In this approach, student exploration becomes an important part of the course. Classes no longer meet only for teacher presentations to the whole group or for review of homework problems. Teachers pose open-ended inquiry problems to students that lead to fruitful exploration. (For an examination of such problems, see Yerushalmy, Chazan, and Gordon, 1988.) These, in contrast to Schoenfeld's (1988) description of traditional five-minute exercises, are usually explored for one or more classroom periods, written about for homework, and then discussed.

Students explore the problems with the aid of the Supposers, usually in pairs in a computer lab, and generate conjectures. Some of the conjectures are true and some are false. Some of the false conjectures are easily modified to be true. In class discussions, students share their conjectures and present arguments in favor of their ideas. As students are introduced to mathematical proofs and as their facility with deductive proof develops, they are expected to present deductive arguments (formal or informal) for their statements.

In this approach, students do not sit down to prove statements that they know are true and that they know have been proven year after year in geometry classes. Some of the statements that they try to prove may not turn out to be true; others may not be present in their textbooks and may be unfamiliar to their teachers (Kidder, 1985). With this approach, there is a new goal for students and a new standard for student performance: students should become competent explorers of open-ended problems. This new goal requires that students know how to work together and break down a large task, generate hypotheses, use the computer to get feedback about their hypotheses, formalize their hypotheses, generalize their hypotheses, change and extend a problem, and argue for their conclusions.

Supposer Inquiry Skills: An Outline

We explored this new standard with a group of teachers as part of a three-year study conducted under the auspices of the Harvard Educational Technology Center (ETC). Throughout the first two years of monthly meetings, the teachers in the group referred to two sets of goals for their students: the traditional curriculum and the Supposer curriculum. The group divided the goals of the traditional geometry curriculum into two parts. The first part encompassed the postulates and theorems of the course in the order in which they are introduced. Students demonstrated a knowledge of this part of the curriculum by being successful at problems that ask them to write simple proofs. Students were rarely asked to write complicated proofs of more than 10 steps or that require lemmas. The second part of the traditional curriculum, a second avenue for student demonstration of mastery of the course material, was a numerical part. Students demonstrated mastery by successfully solving problems that asked them to use knowledge of theorems and postulates to find missing measurements in figures when other measurements are presented. For example, in the drawing at right, O is the center of the circle. Students are asked to find length AC.

When describing the Supposer curriculum, teachers used the phrase "good explorers" to describe the expectations they had of their students. Later the group began to call the Supposer curriculum the "metacurriculum," because the goals of the Supposer curriculum were not related to the content—numerical or theoretical—of a geometry course, but were higher-order goals related to scientific exploration and the doing of mathematics.

During our third year, we decided it was important to become more articulate about the types of skills and beliefs that students needed and were lacking. As a result of group discussions (which included group exploration of problems and documentation of skills used and necessary beliefs), we created a list of types of inquiry skills and beliefs. We hoped that this list would provide a set of goals toward which we would design activities to help students become better inquirers. We decided on the categories in Figure 2.

Figure 2. Nine categories of inquiry skills and beliefs.

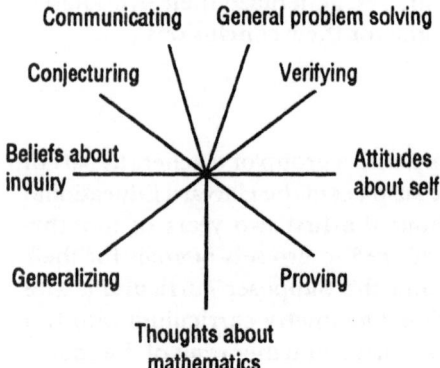

The group distinguished between verifying and proving. Verifying was defined as an activity, frequently involving measurement, carried out on a specific number of examples. Proving involves deductive reasoning about an infinite number of individual cases. The group also distingished between conjecturing and generalizing, a distinction which will be explained below. The group broke down each of these nine categories into simple skills or beliefs. These are given in Appendix I.

While this list of inquiry skills is clearly insufficient, it does delineate the wide range of skills and beliefs that students need. (For a theoretical discussion of the kinds of knowledge—resources, control, heurstics, and beliefs—students need, see Schoenfeld, 1985). Assessing each of these nine kinds of skills and beliefs is an enormous task. Therefore, in this paper we will focus on assessing three of the nine categories: students' verifying, conjecturing, and generalizing skills.

First, we will present an instrument that we have used in our research and which was designed to help us compare students' generalizations. While this instrument was helpful for evaluating the innovation, for comparing Supposer and non-Supposer students, it is not helpful for the kind of assessment that teachers do in their classrooms, that is, ongoing assessment to guide further instructional decisions. In order to do this kind of assessment, one needs a more careful description of the desired skills and an evaluation scheme that focuses on these skills and not on outcomes. After describing in greater detail the verifying, conjecturing, and generalizing skills that we wish to promote, we will present an adaptation of our research instrument that is designed for classroom use. However, before presenting any assessment instruments, we will clarify the meanings we assign to the words generalization and conjecture.

CONJECTURES IN GEOMETRY

We use the term conjecture to refer to statements whose truth value is not known at a given time. Although the statements have been explored and tested, there is, as yet, no reason to reject them (as opposed to hypotheses that haven't been tested yet). In geometry, conjectures have three key parts: the relationship described in the conjecture, the set of objects for which the relationship holds, and the quantifier, which determines the members of the set of objects for which the relationship holds. Conjectures are not always stated completely; sometimes one or more of these key parts is not explicitly stated, but is understood.

People create conjectures in different ways; calling a statement a conjecture implies no particular process of creation. A conjecture can result from belief, experience, attempts at explanation, deductive proofs,[1] or generalization. Generalizations are a particular kind of conjecture, conjectures created by using one of the following two generalization processes to reason from the specific to the general.

TWO GENERALIZATION PROCESSES IN MATHEMATICS

In mathematics, generalization processes do not produce definite, proven knowledge. Instead, they result in the creation of a special kind of conjecture—a generalization. Though it is difficult to determine how a particular generalization was made, we feel that it is valuable to distinguish two ways in which generalizations are created:

- *Induction* is a process for reaching generalizations by examining instances or examples. The generalizer examines an instance or a set of instances and identifies some of their properties. These examples are then identified as members of a larger set to which they belong, and the properties of the examples are then imputed to the larger set. Chi and Bassok (1989) argue that induction (generalization from examples) is based on the perception of similarity between examples.

- *Condition-simplifying generalization* (Holland et al., 1986) is a process that is carried out on a statement (in mathematics, either a conjecture or a proven statement). This process proceeds by the relaxation of conditions within the original statement to produce new statements. Studies investigating this generalization process suggest a connection between a person's ability to generalize in this way and their disposition towards constructing explanations for the original statement and their ability to do so (Chi and Bassok, 1989).

In geometry, statements usually include a diagram or some numerical information in addition to the written text. While the distinction between the two generalization processes described above is clear in theory, the presence of diagrams and numerical information causes this distinction to blur when applied to geometrical statements. For example, when the statement contains a numerical condition that is modified or relaxed, induction seems to be an appropriate, or natural, description of the process. In this view, the initial statement is one example, and any statement that substitutes a different numerical value is another example. A more general statement is reached by examining each of these specific cases. On the other hand, when a diagram is presented along with a statement, it is difficult to know whether a person involved in making a generalization is working from the example in the diagram or from the statement. To draw conclusions in such a case, one must infer what is taking place in the mind of the generalizer.

RESEARCH ON ASSESSMENT

We will begin our description of assessment instruments by describing a paper-and-pencil test designed by Yerushalmy (1986) that provides a structure for comparing outcomes (students' generalizations.) This test was designed for use in comparison studies. We were interested in assessing differences in competence at generalizing between Supposer and non-Supposer students. (See Yerushalmy, 1986, and Yerushalmy, Chazan, and Gordon, 1987, for results.) Given this goal, a paper-and-pencil test seemed appropriate, since it can be easily administered to a relatively large number of students. Also, as compared to an interview format, a pencil-and-paper test is less liable to change students' performance. During the course of an interview, considerable learning may occur. Finally, a written test forces students to formalize their thoughts in a way that does not necessarily happen in an oral interview.

This test also allowed us to examine other interesting questions: What kind of representations of geometric ideas do students choose to work with? Do they focus on a visual representation, or do they work with the numerical data that are given? Do students provide explanations for their generalizations? If so, what types of explanations do they provide? Do students present proofs as explanations for their generalizations? Since the test was given in a pre/post format, we were able to examine differences between students' performance at the beginning of a geometry course and after one year of studying geometry. Also, we were able to compare students' responses to problems that present numerical data and problems that give students a statement as a starting point. (See Yerushlamy, 1986, and Yerushalmy et al., 1987, for a discussion of these issues.)

While this test is by no means the last word on assessment of students' generalizations and though it focuses on outcomes—generalizations—and not on verification, conjecturing, and generalization skills, we still feel that it represents an advance. First of all, to our knowledge, this is the first test used with students in a classroom setting that focuses on students' ability to make generalizations when given a statement or a small amount of data and a limited amount of time. (See the description of the instructions given below.) The fact that students were able to accomplish the tasks set for them, that they produced generalization of different kinds, and that the test seemed to indicate meaningful differences between different treatment groups are all hopeful signs for future assessment of these types of skills. (See Yerushalmy, 1986, and Yerushalmy et al., 1987, for these results.) Also, based on this work, we have been able to design other types of assessment that might be useful for classroom teachers interested in assessing the level of their students' verification, conjecturing, and generalizing skills.

A CONJECTURE/GENERALIZATION TEST

We will present the version of the test that was used in a study carried out in 1985-86. (See Yerushalmy et al., 1987, for a description of the study.) In this study, the conjecture/generalization test, along with classroom observations, students' written work, an argument test, and teachers' and students' comments, were data sources for observing the types of generalizations students made in their geometry classes and for comparing Supposer and non-Supposer students.

A pretest and posttest were developed to assess students' ability to make generalizations concerning given data or a description of a geometric situation. The pretest and posttest consist of four and three problems respectively, each presenting a statement, a group of mathematical facts, or a mathematical idea from plane geometry, along with appropriate diagram(s). The geometric content of the problems was familiar. Thus, the problems on the pretest were about material covered in previous courses. The first two problems on both tests included numerical data. They were designed to provide insight into students' ability to generalize by induction based on given instances. Students could generate new data for themselves by drawing new instances, adding auxiliary lines, and following deductive lines of reasoning that made new information available. Problems 1 and 2 are from the pretest. The remaining problems (3 and 4) describe an idea abstractly. These problems were designed to provide insights into students' ability to carry out condition-simplifying generalizations.

Problem 1. The numbers on the diagrams below represent the measure of angles, lengths, and areas. For example: The length of AF is 3.85. The angle CAD is 41 degrees. The area of triangle CDG is 2.1.

List as many significant connected statements as you can make.

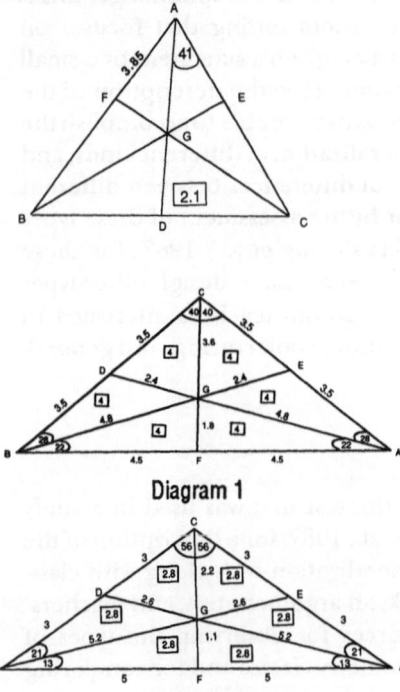

Diagram 1

Diagram 2

Problem 2. The right triangles on the grid below have 3, 6, and 8 points on their perimeter.

List as many significant connected statements as you can make.

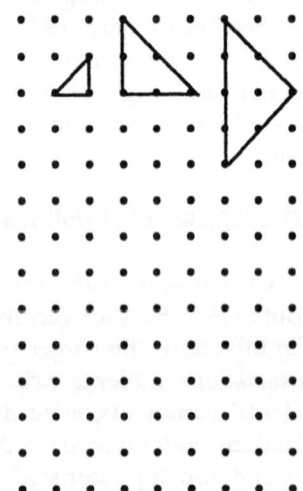

Problem 3. A line which passes through the center of a square and is parallel to two of its sides divides the area of the square into two equal areas.

List as many significant connected statements as you can make.

(Taken from Bell, 1976.)

Problem 4. P, Q, and R are points on the sides of triangle ABC.

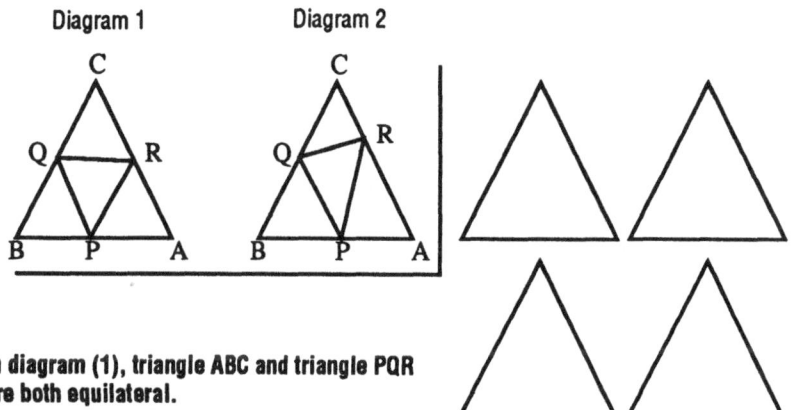

In diagram (1), triangle ABC and triangle PQR are both equilateral.

In diagram (2), triangle ABC is equilateral, triangle PQR is not.

List as many significant connected statements as you can make.

The instructions on these tests ask students to "list any significant statements" connected to the problem. The instructions are deliberately vague in order to ascertain what students consider "significant" and "connected." We asked teachers to refrain from explaining or elaborating on the instructions when they administered the tests. In framing the instructions, we wished to invite any plausible idea, not only generalizations, and as many statements as possible. There were no constraints such as demanding that the statements be true or be supported by arguments.

When these instructions were initially designed, a researcher in the area of thinking skills examined them and suggested that in his view the instructions were too difficult and that students would not be able to follow them (D. Perkins, personal communication, December 6, 1985). While the instructions are certainly laconic, students did not seem to have difficulty with them (Yerushalmy, 1986; Yerushalmy et al., 1987). In describing an instrument for classroom assessment, we will present expanded instructions designed to elicit specific inquiry skills.

Scoring the Test

We will present here a refined scoring scheme used by Yerushalmy and Maman (1988) in a study involving another version of the test presented above. (See Yerushalmy and Mamman, 1988, pp. 40-43, for a more detailed description of the scoring scheme.) The scoring scheme is predicated on the assumption that on each problem students are likely to make more than one statement.

Students' statements were scored using four central variables defined by Yerushalmy (1986). Of these variables, changes made to the original statement in order to create the generalization (CHANGES) is scored on a 0 or 1—exists or does not exist—scale. The level of the generalization (LEVEL), its originality (ORIGINALITY) (called plausibility by Yerushalmy, 1986), and its correctness (CORRECT) are rated on a 0-4 scale. These variables are related; we do not consider them to be distinct.

In order to assess CHANGES, for each problem on both tests, a list of attributes was compiled using Brown and Walter's (1983) analysis of similar problems in plane geometry. Each list was divided to three parts, using the Structure-Mapping Theory developed by Gentner (1983) and the Definition of Spontaneous Analogy by Clement (1983). The three parts are (i) geometric attributes, (ii) numerical attributes, and (iii) key (fixed) features. Since students working with the Supposer are trained to see geometric situations as involving an initial shape on which constructions are made, in assessing CHANGES, we added one category to the scheme suggested by Clement's and Gentner's work. We broke out the object of interest—the type of polygon given in the problem—from the category of fixed variables. Thus, we were interested in four types of changes that students might make to the problem situation. (Examples of the first three types of changes for the problems given above appear in Appendix E-2 of Yerushalmy et al., 1987.) The four types of changes are as follows:

- *Object of interest.* Replacement of the central geometrical object with a more general one; for example, any triangle instead of a right triangle.

- *Geometric relationships.* Replacement of one geometrical relationship by another; for example, movement of a point from inside a triangle to outside it.

- *Numerical variables.* Treating a numerical aspect of a situation as a variable.

- *Fixed variables.* A catch-all category for unexpected changes; for example, moving from two dimensions to three.

Students were given a 0 if their statement simply repeated the information given and did not change any aspect of the given information. They were given a 1 if they made any of the above changes.

The LEVEL variable examined how the students presented their data (both the given data and data that they generated) and their communication of their statement. Students' written work was examined and was assigned a LEVEL value from 0 to 4 according to the following scheme:

0: Students' do not write a statement. For the problems posed abstractly, their statement is less general than the given statement. For the problems posed with data, their statement simply repeats the data.

1-2: There is some discretion here in deciding between a 1 and a 2, depending on the type of change made and the number of changes examined. Students sometimes change some aspect of the problem, but do not make a more general statement. Students in this category may change the same aspect of the problem in several ways, but in their presentation of their data do not seem to connect these different changes systematically. They seem to address each change as a separate instance. For example, students may replace the midpoint in a construction with subdivision into four equal parts and then into six equal parts, but seem to consider each of the statements to be a completely separate idea.

3: Students in this case change an aspect of the problem systematically, but do not write a general conjecture that encompasses all of the cases. For example, students may systematically replace the midpoint with subdivision into 4, 6, and 8 parts. It is a matter of interpretation to distinguish between a level 2 and a level 3. In order to be scored as a 3, there must be some evidence, perhaps in data organization, of a systematic exploration.

4: There is a generalization that describes a general phenomenon, such as, even numbers..., in any quadrilateral..., and so on.

The LEVEL measure is the value assigned to the most general statement produced by the student on a given problem.

ORIGINALITY is a measure of the connection of the statement to the problem and to the school's curriculum. Thus, it is a function of students' classroom experience with geometry. Statements connected to the problem and not covered by the school curriculum receive a high score of 4, while those that are poorly connected to the problem or that are trivial because they were covered at length in class are rated at a low level of ORIGINALITY. Once again, since students offered many statements, ORIGINALITY is the value of the most original statement on a given problem.

CORRECT is a reflection of the percentage of student statements that are true for a given problem. Scores range from 0 for all false to 4 for all true.

VERIFYING, CONJECTURING, AND GENERALIZING SKILLS: A CLOSER LOOK

In order to assess an innovation, it is important to keep in mind the goals of the innovation. If one of the goals of the Supposer innovation is to have students become competent explorers of open-ended inquiry problems and if that process, by definition, includes having students be nimble conjecturers, it is important to have a better understanding of the verifying, conjecturing, and generalizing skills that successful students have. One way to understand which skills successful students have is to understand the capabilities of the Supposer and examine the kinds of difficulties some students have when exploring a problem. Below is a brief discussion of several kinds of difficulties students have.

The Supposer allows students to create sample geometrical objects easily and quickly and to generate data about these objects, but it does not evaluate the types of samples students create to test a hypothesis. Thus, students using the Supposer are prey to the kinds of sampling biases described in the literature (Nickerson et al., 1985). For example, the confirmation bias suggests that inexperienced students create limited samples, which only serve to confirm their incorrect or naive generalizations.

The Supposer also does not tell students how to analyze the data they have collected. For example, the Supposer cannot infer users' intentions and cannot alert users who ignore information provided by the Supposer that disconfirms their hypotheses. The Supposer cannot know when students will reject information because it contradicts a concept image (Vinner and Hershkowitz, 1980) that they have. For example, many novice inquirers seek control over the type of shape they are working with whenever the "random" members of the class conflict with their concept images. Some even choose to reconstruct a whole construction step by step on a new triangle without using the REPEAT option because they refuse to believe the results of the option. Only after seeing that they get the same result are they convinced that the data provided by the "uncontrolled" REPEAT option is correct.

The Supposer also does not teach students to order their sample in a particular sequence or add auxiliary lines to their figures. The Supposer does not suggest which numerical operations students should perform on their data and cannot correct mistakes in students numerical manipulations. For example, one student did not find a (correct) pattern which she had expected because she had unintentionally computed the ratios of two perimeters in two different ways. The first time her ratio was larger over smaller and the second time she compared smaller to larger.

Students working with the Supposer need their teachers' help to overcome the difficulties described above, yet it is helpful to have a positive description of the inquiry skills and beliefs that students should develop. Below, we describe the skills students should develop, not the difficulties they must overcome. We will integrate the skills that the ETC teachers thought important with dimensions suggested by research to outline the considerations that students, using the Supposer toward the goal of making general conjectures, have to take into account when generating hypotheses, deciding what data to collect, how to analyze it, and how to further their investigation.

While for the sake of description it is necessary to divide competent exploration into component skills, such a description does not capture the complex interactions between the skills which we are forced to describe separately.

Generating Hypotheses

Typically, the open-ended inquiry problems we give students to explore involve a geometrical construction. For example, we might ask students to connect points that are one third of the way in from each vertex of a square (Figure 3). (See Schwartz, 1989, for a description of student exploration of this problem.)

Students must decide what relationships in the construction are

Figure 3. Subdividing the sides of a square.

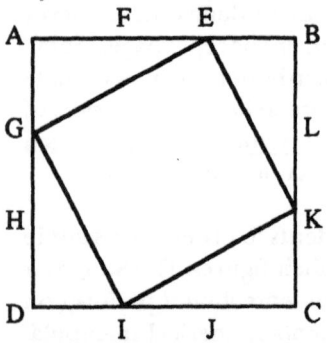

worth exploring; they must have some initial hypotheses to guide their data collection and some notion of which relationships might be of interest. Students should learn to look for the geometrical relationships explored in the curriculum, for example, congruence, similarity, parallelism, types of geometrical figures (squares, right triangles...). Students must also have strategies for coming up with initial directions when they are stuck and have no hypotheses. Useful strategies include repeating the construction on other figures to see if there are visual invariances that strike the eye, systematically varying some other aspect of the construction in search of interesting changes, and scanning of the measurement options (area, angle, and length) to collect data that might stimulate a hypothesis.

Creating Good Samples of Useful Data

Knowing what and when to measure. Once students have a hypothesis in mind, they must know how to test that hypothesis in a single case. They must use the definitions and theorems studied in class to know which measurements to make. Competent explorers know how to test a conclusion with the smallest number of measurements by investigating sufficient conditions. At the same time, a competent explorer also realizes when a particular measurement is unnecessary because it is directly entailed by the construction. Thus, in Figure 3, if the E is one third of the way from B to A, then AB = 3*EB by definition, and no measurement is necessary.

Figure 4. An extreme case.

Considering extreme cases. People often fail to generalize appropriately because they have only sampled stereotypic instances. It is important that students learn to try extreme cases in the set of objects they are conjecturing about. For example, if a student thinks that for this construction the resulting inner shape is a parallelogram, it is important to try an original quadrilateral that is not a parallelogram, a trapezoid, or a kite (Figure 4).

Collecting the "right" number of examples. "How should we determine that we have enough instances of a generalization?" (Holland et al., 1986, p. 232). This is a difficult question debated by many who write about induction. Given the dynamics of most mathematics classes, it is an especially hard question for students to understand. In most math classes, the amount of work required of the students in order to complete a certain task is clearly specified. Furthermore, traditional approaches to geometry do not make use of large quantities of information; one diagram is often considered sufficient.

We do not suggest a particular numerical answer to this question. Students should learn to make sure that their sample includes different types of shapes and covers a range of cases. Students should also learn to ask themselves if there is something special about the cases in their sample that might influence their conjecture. Thus, the particular number of examples examined is a function of students' knowledge of geometry and ability to convince themselves that the examples in their sample are indeed representative examples.

Analyzing the Data

Data display and organization. As students collect data, it is important that they organize it in a way that allows for easy analysis. Students should make charts and collect visual data by making good sketches. They can combine numerical and visual data by marking their measurements right on their diagrams. A final useful technique is the ordering of diagrams into a sequence based on a single characteristic.

Paying attention to negative data. The teachers we worked with made sure to remind their students regularly that conjectures are statements that are true for all members in the set of objects described in the given. (In high school geometry, there are few existential statements.) Students must learn to appreciate the power of a counterexample.

Manipulating numerical data. Being able to compare numbers is essential for looking for patterns. Students need to learn to use arithmetic operations to compare numbers; differences and ratios are especially important. It is also valuable to link geometrical objects and relationships with numerical operations, for example, linking the Pythagorean Theorem and the existence of a right triangle with squaring and addition. Students should also remember to look for patterns other than equality.

Manipulating visual information. Students need to learn to look at a diagram in different ways. For example, in looking at Figure 3, subdividing the sides of a square, students should be able to see it from inside out as made of four right triangles around a square or from outside in as a large square with a smaller square inside it. Students should also be willing to add auxiliary lines to diagrams. Adding auxiliary lines creates new geometrical objects and sometimes allows new relationships to become evident.

Evaluating Conjectures

As students develop and test a conjecture, it is also important that they use what Schoenfeld (1985) calls control. It is important to make sure that a result is not trivial. Is the conjecture something that has already been proven? Is it a direct consequence of something we already know? Once a conjecture seems supported, it is valuable to use the "what if not" strategy (Brown and Walter, 1983) to generate other avenues for exploration. Is the conjecture generalizable?

In geometry there are three aspects of statements that can easily lead to a generalization. These aspects are type of shape, number, and type of segment. Thus, when exploring the subdivision of the sides of a square (Figure 3), it is valuable to think of other quadrilaterals or other types of polygons. It is worthwhile to explore the numerical aspect of subdivision into two. Maybe there are interesting results which generalize from two to three or four. Finally, in conjectures that include median, for example, it is useful to explore the substitution of angle bisector or altitude for median (Brown and Walter, 1983).

CLASSROOM ASSESSMENT

There are many different types of assessment that are valuable and important in a classroom, yet as suggested by the California Assessment Program (Stenmark, 1989), classroom assessment should be assessment in the service of learning. In addition to assessment that examines outcomes and compares students' achievement, there is also classroom assessment, which helps teachers make instructional decisions based on examining the success of a particular instructional sequence in promoting certain student behaviors. If students do not exhibit the desired behaviors, then further instruction is necessary.

Turning now from research assessment instruments to opportunities for classroom assessment, we would like to sketch ways to assess the level of inquiry skills of students working with the Supposer. The first assessment tool that we will present allows a teacher to elicit some of the verifying, conjecturing, and generalizing skills we described above and to examine whether any are lacking. Later, missing skills can be discussed or explicitly modeled.

The instrument we will describe is a paper-and-pencil test that has five items. It is an adaptation of the research instrument described above, but with more focused and explicit instructions. Instead of a scoring scheme that assigns points for student responses, for each problem we will suggest a list of skills for which a teacher can check. Since in our approach to geometry students frequently work in pairs, it also makes sense to consider giving the test to pairs or as a take-home. Each paper should have written comments, but no numerical score. The most important information for a teacher is any pattern of missing skills and any unusual student behavior worth sharing with the whole class.

Since the literature on induction (especially Chi and Bassok, 1989) suggests that the level of students' generalization increases when they provide arguments for their generalizations, we suggest that the instructions ask students to provide explanations. However, we do not suggest that students write two-column proofs, because in a limited amount of time we prefer that students generate interesting connected ideas and argue for them rather than spending time writing detailed proofs. We also fear that if asked to write proofs students will not write complicated conjectures that they do not know how to prove.

Below (and at right) we will present the five problems on the test and for each problem, the important skills it assesses. This version of the

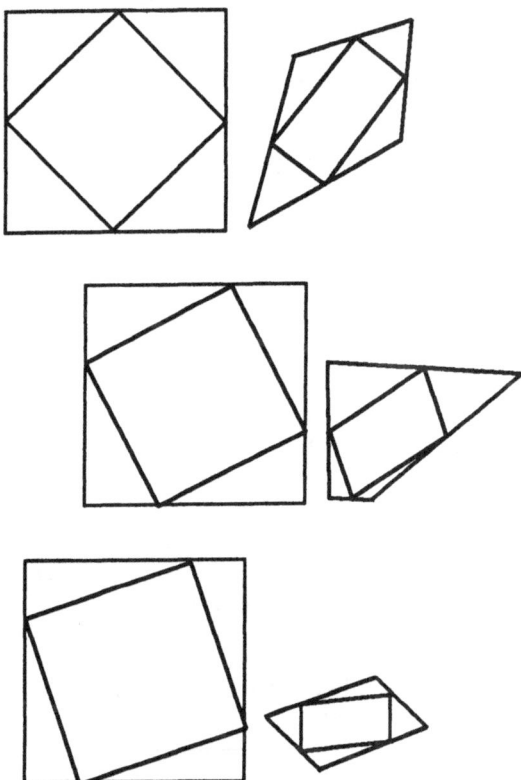

Test problem 1. Below are a series of diagrams. Write as many conjectures as you can that are related to the diagrams given below. For each conjecture indicate which diagrams are relevant to the conjecture and explain your conjectures.

In examining students' conjectures, look for the kinds of patterns students see in the visual data; whether they make conjectures for which counterexamples are present; whether they add to their data by making new figures; whether they add auxiliary lines to the figures; whether they put the diagrams in a sequence.

test is designed for students who have finished units on quadrilaterals and areas of polygons and who have begun to study similarity. Similar questions can be written for other diagrams and for other statements. In that sense, this test is but one example of a general assessment strategy. To create other tests, one would change the diagrams and statements, but not the instructions given with each question.

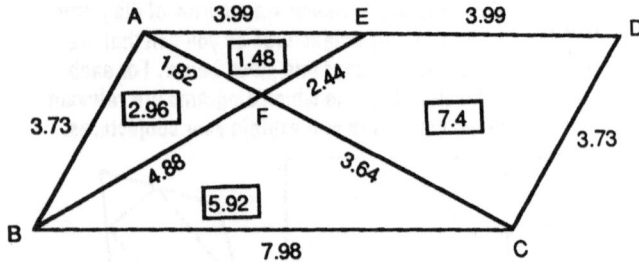

Test problem 2. ABCD is a parallelogram. AC is a diagonal of ABCD. E is the midpoint of side AB.

In the diagram, length measurements are provided next to each segment, and area measures are enclosed in a box.

What does this data tell you about this particular figure? *Does the data support the students' statements, or contradict them? Do the students manipulate the numerical data? Do students discuss patterns of inequality? Do they use both length and area measures? Are their statements trivial statements which can be deduced from the givens presented in the problem?*

Based on this one example and your knowledge of geometry, what general conjectures do you have? Explain your conjectures.

Do students examine different types of polygons? Do students explore figures with a different number of subdivisions to get point E? Do students examine any relationships involving angles?

Test problem 3. This diagram describes the following sentence:

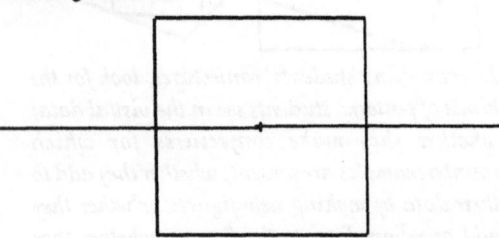

A line which goes through the center of a square and is parallel to two of its sides divides the square into two congruent rectangles. Make conjectures that are related to this sentence. Explain your conjectures and their relationship to the original statement.

Do students draw auxiliary lines? Do students change the type of polygon? the type of point? from a straight line to another type of line? Do they examine a relationship other than congruence?

Test problem 4. ABC is an acute triangle. BD and CE are altitudes in the triangle. The lengths of segments are listed next to each segment in the diagram. Angle measures are also provided.

Use this data to make conjectures about the relationships between the triangles in the diagram. Explain your conjectures.

Do students see all of the triangles that are in the diagram?

For each conjecture you have made, present the data which supports the conjecture.

Do student data support their conjectures? Are there contradictory data available that they are ignoring? Are their data sufficient? Do students manipulate the data to get new data? How do they organize their data?

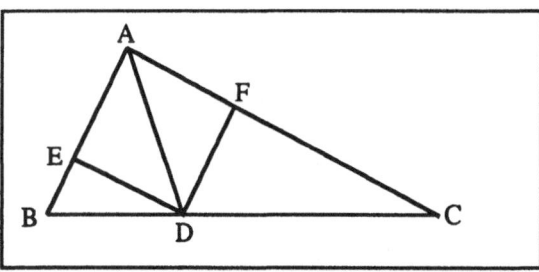

Test problem 5. The diagram below illustrates the following sentence: If AD is the angle bisector of BAC, then the length of the altitudes in triangle ABD (segment DE) and triangle ACD (segment DF) are the same.

Write a list of related geometrical questions that you would like to explore and explain the relationship of these questions to the original sentence. *What types of generalizations do students make?*

Explain how you would use the Supposer to explore each question you raised above. *Do students describe an appropriate construction? How many instances are they going to examine? What data will they collect? How well is their data sample constructed?*

As illustrated by these five test problems, one can gather information about students' performance in at least the following areas:

- *Inductive resoning skills:* data collection, sampling appropriate examples, relating conjectures to supporting data, and organization of information.

- *Relationship to diagrams:* willingness to add auxilary lines, organization of pictures in sequences, and seeing the same diagram in multiple ways.

- *Numerical manipulations.*

- *The number of connections* that students are able to make between different topics within the domain of geometry. Students inclination to look for connections.

- *Students' inclination to use the "what if not" strategy* and investigate related questions.

Knowledge of students' performance can suggest areas for concentrated instruction.

Evaluation of Students' Performance

There are other types of assessment used in schools. Teachers usually must evaluate their students and give them grades of different kinds. Below, we present a scheme for grading students' lab papers. It is designed for use in classes where students work in pairs using the Supposer to explore problems in a lab setting and then write up their explorations individually. This scheme examines many of the same issues as the test problems above. It can be used with a wide range of conjecturing problems, though we will illustrate it for only one:

The problem: Explore the figure formed by reflecting the intersection point of the altitudes in each side of a triangle and connecting the three image points.

The procedure:

- Construct an acute triangle ABC.

- Draw the three altitudes.

- Label G as their point of intersection.

- Reflect point G in each of the three sides of triangle ABC producing points H, I, J.

- Draw triangles DEF and HIJ.

- State your conjectures about the relationships among the points, elements, and triangles.

- Repeat this procedure for other types of triangles.

Assigning points

Figure 5 shows one scheme for assigning points to students' written work after they have worked in the lab on this problem. Teachers can choose to weight the relative value of these categories differently according to the needs of the class:

Figure 5. A matrix for assigning points.

	Plausible, but unsupported	Supported with data	Supported with arguments
Standard conjectures	x1	x2	x3
Special conjectures	x4	x5	x6

- Standard conjectures reflect students' knowledge of similarity. They include the similarity of DEF and HIJ, the 2:1 ratio of their sides, the 4:1 ratio of their areas, the parallelism of their corresponding sides, and the equality of their angles.

- Special conjectures might be about any of these: the type of triangle created for a certain type of starting triangle (for example, if ABC is equilateral, so is JHI and they are congruent); formulas for the relationships between the angles of JHI and DEF; the circle which circumscribes ABC; the altitudes of ABC (which are the angle bisectors of DEF and, if they are extended, of JIH);

- The reason for the unsupported, yet plausible, category is that some students may develop new conjectures while working at home and have no opportunity to collect data. Conjectures without supporting data should be recognized, but only if they are plausible. Otherwise, students might write any geometrical statements to receive credit.

- The teachers that we have worked with usually consider neatness and clarity of expression in their students work.

Assigning grades

Once points have been assigned teachers can use a variety of schemes for assigning grades. Below are two options that do not score on an absolute scale and that take into account that, with the above scheme for assigning points, there is no ceiling score. One option is to take the highest score and declare it to be a perfect paper (100 percent) and then compute percentages for every other score. This method reacts strongly to outliers. One student's score may be much higher than those of the rest of the class. Another method is to figure out the median score and assign it a median grade. Higher and lower scores are graded in relation to this median score.

We have found that it is important that teachers using the Supposers make clear to students that the work with the Supposer is an integral part of the course. One way to do so is to grade students' lab papers. Schemes like the one presented above help students' understand the types of expectations that their teachers have for them.

CONCLUSIONS

The goal of our approach to teaching geometry is to have students become competent explorers of inquiry problems and nimble conjecturers. Therefore, we are interested in assessing students inquiry skills. As Hawkins and Sheingold (1986) point out,

> While a move from one specific content to another need not alter measurement, a move toward emphasis on more general thinking and problem-solving skills or toward more abstract skills within a domain must change both standardized and less formal measurement techniques. Teachers will need to devise new ways of knowing how well their students are doing (p. 54).

Thus, if we really want people to try to teach geometry in the way that we have described, we need to work on this hard problem. We need to define the skills that we want to assess and then look at methods for assessing these skills. Our work must be practical, that is, useful in classrooms as they are now structured. Currently, this means paper-and-pencil assessment, yet it is extremely difficult to create a pencil-and-paper task that allows the teacher to follow the processes used by students. Furthermore, traditional assessment's approach to peer collaboration complicates our task (see Hawkins and Sheingold, 1986). Most of the high school teachers we have worked with now grade for individual achievement. At the same time, when using our approach to geometry they have students work extensively in pairs.

This paper represents our early efforts at assessing a small range of higher-order mathematical thinking skills. We focused mainly on conjecturing, verifying, and generalizing skills. We described two practical ways to assess the performance of students who have worked collaboratively. Both assessment instruments are designed to be administered with paper and pencil to individual students; one results in individual scores. We hope that these ideas will be helpful to those teaching other topics; for example, we believe that much of the analysis in this paper is relevant to suitably posed algebra problems.

APPENDIX I: DETAILED LIST OF INQUIRY SKILLS

The items in the following lists provide details for the nine categories shown in Figure 2 in the text. They indicate the kinds of behaviors, skills, questioning strategies, and beliefs that "good explorers" exhibit. There are areas where these lists overlap, and no one student will exhibit all of these strategies when solving a single problem.

These lists were developed by Harvard Educational Technology Center's Geometry Labsites group.

Conjecturing

Using knowledge about geometry—Checking the types of relationships discussed in class.

Looking for patterns other than equality.

Remembering that conjectures are "for all" statements.

Adding to the diagram—drawing auxiliary lines.

Has it been shown before?

Is it a direct consequence of a known relationship?

Is it generalizable?

Verifying

Using definitions to (i) know what to measure, (ii) know when measuring is unnecessary.

Using sufficient conditions as shortcuts.

Understanding the power of counterexamples.

Making one's own charts.

Marking measurements directly on drawings to keep organized data.

General Problem Solving

Getting organized to solve a problem.

Splitting a problem into parts.

Recording data in an organized way in a chart or on a diagram.

Working cooperatively on a problem.

Communicating

Working in pairs cooperatively.

Writing readable reports to summarize lab work.

Writing conjectures (in whatever form desired) that are intelligible to the reader.

Writing informal and formal proofs.

Proving

When does a statement need proof?

Isolating the "givens" from a drawing or construction procedure.

Determining the "to prove" from a conjecture.

Writing informal proofs by using markings on drawings for necessary conditions (givens in one color, derived in another color).

Learning to sequence conjectures so that if the first one is proven, it is then easy to prove the others.

Checking steps one is unsure of with Supposer measurements to gain confidence in truth of the step (doesn't help with reasons).

Drawing auxiliary lines.

Generalizing

Recognizing when a conjecture might be generalizable.

Using three aspects to generalize: (i) number (subdivisions, sides in polygon); (ii) type of line (e.g., altitude, median to angle bisector); (iii) type of polygon (different types of triangles or quadrilaterals).

Attitudes about self

I can create/discover/develop mathematics.

I can participate and talk in discussions of math problems.

Beliefs about inquiry

It's good to explore on your own outside the bounds of the problem.

A textbook is a valuable tool when exploring.

The final goal of an exploratory problem is to create conjectures and proofs. In such a problem, data are not enough, though they are important.

It is good to use deduction to avoid measuring things that you can know without measuring.

Measuring can't mathematically prove a "for all ..." statement.

A proof proves the statement for all of the drawings that satisfy the given.

Thoughts about mathematics

There are differences between definition, postulate, theorem, conjecture, and observation.

Choice is involved in making definitions and postulates. These choices determine what are theorems.

A proof proves the statement for all of the drawings that satisfy the given.

All Euclidean geometry has not been created/discovered/developed.

NOTE AND REFERENCES

Some people may find it strange to imagine that doing a proof can lead to conjectures, yet as one tries to explain a phenomenon, by deductive proof or otherwise, one sometimes comes to understand the key aspects of the phenomenon. In this way, one may understand that the phenomenon is more general than was thought, or one may find conditions that may be relaxed, resulting in new conjectures.

Bell, A. (1976). A study of pupils' proof explanations in mathematical situations. *Educational Studies in Mathematics, 7,* 23-40.

Brown, S., and Walter, M. (1983). *The art of problem posing.* Philadelphia: The Franklin Institute Press.

Chazan, D., and Houde, R. (1989). *How to use conjecturing and microcomputers to teach high school geometry.* Reston, VA: National Council of Teachers of Mathematics.

Chi, M., and Bassok, M. (1989). Learning from examples via self explanations. In L. Resnick (Ed.), *Knowing, learning, and instruction.* Hillsdale, NJ: Lawrence Erlbaum Associates.

Clement, J. (1983, April). Observed methods for generating analogies in scientific problem solving. Paper presented at the annual meeting of the American Educational Research Association. Montreal.

Gentner, D., and Gentner, D. R. (1983). Flowing water and teeming crowds: mental models of electricity. In D. Gentner and Stevens (Eds.), *Mental models.* Hillsdale, NJ: Lawrence Erlbaum Associates.

Hawkins, J., and Sheingold, K. (1986). Microcomputers and education. In Curbertson and Cunningham (Eds.), *The 1986 Yearbook of the NSSE.* Chicago: University of Chicago Press.

Holland, J., Holyoak, K., Nisbett, R., and Thagard, P. (1986). *Induction: Processes of inference, learning, and discovery.* Cambridge, MA: Massachusetts Institute of Technology.

Kidder, R. (1985, April 19). How high-schooler discovered new math theorem. *Christian Science Monitor.*

Kuhn, D., Amsel, E., and O'Loughlin, M. (1988). *The development of scientific thinking skills*. Boston, MA: Academic Press.

Nickerson, R., Perkins, D., and Smith, E. (1985). *The teaching of thinking*. Hillsdale, NJ: Lawrence Erlbaum Associates.

Schoenfeld, A.H. (1988). When good teaching leads to bad results: The disaster of "well taught" mathematics courses. In *Learning through instruction: The study of students' thinking during instruction in mathematics*. P. Peterson and T. Carpenter (Eds.). Special issue of *Educational Psychologist, 23*(2), 145-166.

Schoenfeld, A.H. (1985). *Mathematical problem solving*. Orlando, FL: Academic Press.

Schwartz, J., Yerushalmy, M., and Education Development Center. (1985). >*The Geometric Supposers* [Computer Software], Pleasantville, NY: Sunburst Communications.

Schwartz, J. (1989). Intellectual mirrors: A step in the direction of making schools knowledge making places. *Harvard Educational Review, 59*, >51-60.

Stenmark, J. (1989). *Assessment alternatives in mathematics: An overview of assessment techniques that promote learning*. Berkeley, CA: EQUALS and the California Mathematics Council.

Vinner, S., and Hershkowitz, R. (1980). Concept images and common cognitive paths in the development of some simple geometrical concepts. In R Karplus (Ed.), *The proceedings of the Fourth Conference of the International Group for the Psychology of Mathematics Education*. Berkeley, CA; pp. 174-184.

Yerushalmy, M., and Chazan, D. (1990). Overcoming visual obstacles with the aid of the Geometric Supposer. *Educational Studies in Mathematics, 21*, 199-219.

Yerushalmy, M., Chazan, D., and Gordon, M. (1988). Posing problems: One aspect of bringing inquiry into classrooms. Cambridge, MA: Educational Technology Center, Harvard Graduate School of Education. Technical Report 88-21. [In press for publication in *Instructional Science*.]

Yerushalmy, M., Chazan, D. and Gordon, M. (1987). Guided inquiry and technology: A year long study of children and teachers using the Geometric Supposer. (Technical Report 88-6). Cambridge, MA: Educational Technology Center, Harvard Graduate School of Education.

Yerushalmy, M., and Mamman, H. (1988). The Geometric Supposer as a basis for class discussions in geometry lessons. (Hebrew, Lab Report #9). Haifa: Laboratory for Research of Computers in Learning. Haifa University School of Education.

Yerushalmy, M. (1986). Induction and generalization: An experiment in teaching and learning high school geometry. Unpub. doctoral thesis, Harvard Graduate School of Education. Cambridge, MA.

PART II: New Items and Assessment Procedures

5 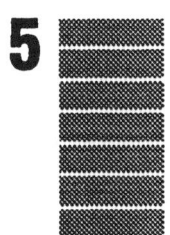 Balanced Assessment of Mathematical Performance

Alan Bell, Hugh Burkhardt, and Malcolm Swan

INTRODUCTION

The implementation of higher-order thinking in the school mathematics curriculum depends on the provision of appropriate assessment material. Teachers' natural and laudable desire to see students succeed at public examinations is bound to be reflected in their teaching. Short, closed, stereotyped examination questions are bound to encourage imitative rehearsal and practice on similar tasks in the classroom. (WYTIWYG or "What You Test Is What You Get"). Conversely, a range of high-quality tasks that assess a broader range of skills will convey messages about the nature of the desired learning activities more powerfully than any analytic description. It is hard for teachers to adopt new teaching practices, even those that offer innovative learning experiences focused on higher-level skills, if the teacher cannot see how the skills acquired will be recognized in their students.

What are higher-order skills? First, they are those general strategies and domain-specific tactics that govern the choices of lower-level technical skills and concepts used in a given activity. They enable a student to deploy mathematical knowledge and techniques effectively. They include the ability over a range of domains to generalize, represent, abstract, prove, check, generate questions, test a hypothesis, or practice a skill. They also include the ability to formulate a question in mathematical terms, or in terms appropriate for solving a problem, and to interpret a mathematical result in the context from which the problem originated. And they include the capacity to be aware of ones' state of knowledge and skill in a particular domain, and of ways of acquiring and retaining further knowledge. This is already an extensive list, which we shall flesh out and illustrate in the course of the following three chapters.

We cannot discuss innovative material for mathematics and its assessment without considering other ways in which we wish to improve currently conventional assessments. These include the following:

- *Practical relevance:* Too much current material offers a situation from real life, but then asks questions that have no practical significance.

- *The coherence or fragmentation of the task:* Many tasks lead students through a sequence of small, closed steps, entirely removing the decision load from the student. ("Solve equation A using method B," and so on). Few tasks invite students to select from their repertoire of techniques, carry through a chain of reasoning, or compare alternative methods—that is, show higher-order skills.

- *The range of possible responses:* To what extent can we set tasks that offer the opportunity for satisfying work to students of a wide range of ability and attainment? Traditionally, the level of response possible has been largely determined by the task rather than by the student.

- *The extent and value of the task:* Higher-order thinking is generally displayed more in extended tasks than in short ones. Tasks that occupy several weeks of school mathematics time are being used in the United Kingdom to assess such skills, and this has led to the realization that as more student (and teacher) time is taken up by assessment, it is important that these activities should themselves constitute valid and worthwhile learning experiences.

- *The mode of working on the task:* Traditionally, individual students have worked on written tasks in silence. Such unnatural conditions are imposed for the sake of "reliability," and this kind of assessment will probably always be with us. There is a great need, however, to explore how we might assess a student's ability to work cooperatively, perhaps using oral and practical forms of communication in a normal working atmosphere. Again, these aspects are currently being explored in Britain.

The aspects we seek to measure are key qualities that are sought by employers and that we need as effective citizens in the modern world. They represent higher standards than exist in most current assessment.

For the preceding kinds of reasons, we wish to focus not only on higher-order thinking but, more broadly, on the notion of balanced assessment of all those aspects of mathematical performance that are now widely recognized and described in documents such as the Mathematical Sciences Education Board reports (MSEB, 1989, 1990), the National Council of Teachers of Mathematics Standards (NCTM, 1989), the California Framework (California, 1985) and comparable guidelines in other tasks and countries (see, for example, Cockcroft, 1982; DES, 1989; NCC, 1989). Our key principles will therefore be

- curriculum balance, in which each assessment package consists of a set of tasks of varying length and style that, taken together, reflect the curriculum objectives in a balanced way, and

- curriculum validity, in which the assessment tasks themselves represent learning activities of high educational value so that the significant amount of time spent on them will represent a benefit rather than a loss to students' learning.

Such packages must also satisfy reasonable constraints of reliability and economy. While the development challenge is substantial, achievements so far suggest that these objectives can be met.

The preceding approach is in many ways opposite to "the psychometric ideal," where the assessment package takes very little time, measures only a tiny part of the student's range of performance, yet provides a full and reliable picture of his or her capabilities. This is not the place for a detailed critique of that approach; suffice it here to say that such an approach not only fails grossly to meet its own targets (correlation cannot carry so great a burden), but it sends out very unfortunate signals about the curriculum (see the excellent analysis in Ridgway, 1988, 1987). As we have indicated above, most customers of the psychometric approach are convinced that a narrow focus on the kind of tasks included in such tests is the best route to success in them, but the backwash effect on the curriculum can be disastrous. Effective performance in mathematics needs much more than this.

Thus, the kind of assessment proposed here is a prerequisite for widespread implementation of an effective curriculum. States and school districts that have appropriate objectives in mathematical education must have tests available that match those objectives for both formative feedback and accountability. Such assessments will also give an impetus previously lacking to developers of textbooks and other materials by stimulating demand for effective materials that meet the new standards.

It takes time to develop desirable materials. The issue of the development of effective support for teachers and students in the transitional period needs to be addressed at the same time, along with other issues related to the dynamics of curriculum change and the implementation process. In our chapter, "Moving the System," we focus on these later issues. Overall, the work we describe reflects an established approach in the United Kingdom and elsewhere, and it draws on a wide range of initiatives and experience there and in other countries, particularly the Netherlands, Australia, and parts of the United States. This area of work became a major part of the Shell Centre activity over a decade ago, when we suggested (Burkhardt, 1979, 1980) to the largest of the English examination boards that it should take more seriously the curriculum responsibilities arising from the influence of its examinations. Thus began a series of collaborative developments in assessment design and curriculum support that continues (Shell Centre 1984, 1986, 1987-89, 1989). This anecdote is useful in indicating the time scale of change achieved so far, which accords well with other experience, for example, in the Netherlands. These assessment challenges will not be solved in a year or two, but immediate progress can be made. So, it is important to make a start.

DESIGN PRINCIPLES FOR BALANCED ASSESSMENT

The principles that we believe should be applied to the design of balanced assessment of high curriculum validity, and which we have found to work well, are really quite simple. Their aim is to encourage the intended balance of mathematical activity and to observe the students' performance in it, assigning credit according to value judgments based on the aims of the curriculum. How is this done?

First, one must decide the range and balance of types of task that the "target group" of students should be able to do. Brainstorm and search until a reasonable set of attractive possible tasks are found. Then devise a form of presentation of each task that leads students to understand what is required and how to tackle the task. Finally, try it out, observing what happens, and revise the presentation, repeating this development cycle until the range of student activity matches that intended (or the task has to be abandoned). In the light of a sample of student responses (written or otherwise), devise a grading scheme for assigning credit, then check that this works well with those who will be doing the grading—external examiners, teachers, or students. Also, a monitoring or moderation procedure may need to be developed. Imagination plus realistic empirical development is the key.

In this process, notice how the nature of tasks is defined first, and the marking schemes are designed afterwards, to reflect the objectives sought. The creative talent needed to design such tasks and marking schemes must not be underestimated. This may seem obvious, but often, in current practice, we select only tasks that we know will be easy and cheap to mark using existing practices. Many of the best tasks have thus been filtered out.

We should like to stress the creative challenge of the whole process, of the design of beautiful assessment and associated support. So much that is offered in the mathematics curriculum is mundane. (If the English language curriculum were like this it would consist entirely of readings from dictionaries.) This need not be so; mathematics itself is not like that. It is possible to devise tasks that lift the spirit, that people will talk about afterwards because they are interesting. We think it is worth the effort to do so.

The Range of Tasks

We shall not review here the target curriculum that authentic assessment will serve; the descriptions in the documents referred to above represent a broad consensus, mirrored worldwide, as to what is now needed. Let us simply say that we recognize two broad types of task as representing applied and pure mathematics. The former is a situation or problem arising outside mathematics—perhaps an optimizing question, such as what is the best route for a postman's round in a given district, or perhaps a route is given and the question is what other routes are possible—where it is appropriate to construct and manipulate a mathematical model and to interpret the result in the original situation. The archetypal pure mathematical problem is to recognize relationships in a mathematical system and to attempt to generalize them, altering the constraints and observing the consequences. Our examples in this chapter will include both of these broad types, and we shall invite appreciation of our examples as displaying the well-recognized characteristics of authentic mathematical activity. In other words, we hope readers will study these examples and say, yes, this is an example of genuine pure or applied mathematical work at a level appropriate to the students to whom it is offered, that is, it displays characteristic modes of mathematical activity and deploys an appropriate range of mathematical concepts and skills.

According to our approach, there is a belief that assessment should emphasize positive achievement—*measuring what the students can do, rather than what they cannot* (Cockcroft, 1982). This is "inefficient" from a psychometric point of view but not from a curriculum viewpoint. The emphasis will be on problems with an "exponential ramp" of challenge, so that every student who tackles a problem can make productive progress while even the most able meet real challenges. Most tasks will involve the students in a broader range of mathematical activities than is common at present—investigating the problem domain, organizing a systematic attack on it, carrying it through with modifications in the light of experience, collecting and analyzing data, checking and reporting the results.

Scoring and Monitoring

The methods for assigning credit to student responses are integral parts of task design and development. General questions as to what aspects of performance should receive what credit require value judgments that are central curriculum issues. For example, in tackling a practical problem using mathematics, should credit be assigned for the overall success in solving the problem or only for the narrowly mathematical parts of the work? Methods for grading involve technical issues as well as those of principle—for example, how far should the grader take a holistic view of the student's attempt, and how far should it be analyzed under categories? The usual range of questions of validity and reliability enter.

The development of grading schemes (or assessment schedules or marking schemes or scoring rubrics—the terms proliferate) that enable teachers as well as outside scorers to measure assessment with adequate validity and reliability is a central part of the practical design of assessment, as are mechanisms for monitoring teacher assessment in a cost-effective way. (This process, if appropriately conceived and implemented, can contribute to teacher development in an important way.) There is much experience, worldwide, from which to learn.

Assessment Packages

Many forms of assessment packages have been tried, and more are worth investigating. We shall not attempt a comprehensive review; however, it may be useful to indicate something of the range of produced possibilities, which go well beyond the short test of multiple-choice items. Many of the design principles outlined above will be exemplified in chapters 6 and 7. Now, we turn to our first main theme, the range and balance of types of task.

TYPES OF MATHEMATICAL TASK

We aim to discuss this key issue through both exemplification and analysis. The latter is helpful, but we beg the reader not to take it too seriously. No classification scheme yet developed is adequate to capture the nature and variety of mathematical performance, for example; and a lot of harm has been done in the past through implementing both curricula and assessments that sought to develop performance in students through teaching the components of doing mathematics but without enough focus on putting these components all together.

We give much credence to balanced judgment of the face validity of a set of tasks as representing the kinds of things we want students to be able to do, and we urge readers to do so too. Accordingly we shall set out some examples of types of task that seem to us to have an essential place in the assessment of mathematics, emphasizing those that have been neglected in the past. To do this properly, one needs to present the following:

- *Several task exemplars.* The range of variation within the group of tasks is very important. If the tasks are very similar, a routine approach to teaching them, based on explanation and practice alone, will seem attractive to many teachers. Whereas, if tasks are more varied, this approach is obviously not enough, and an investigative learning environment becomes essential.

- *Sample student responses.* The responses of the students are central to judging a task; they also provide the platform for appropriate grading schemes.

- *Grading schemes.* These procedures describe how credit is assigned; that is, what aspects of performance are valued with what weight.

In order to cover the necessary range of tasks here, we shall be able to give only a partial picture; that is, we shall present some tasks with no more than commentary, leaving the reader's imagination to fill in the rest. However, we have chosen many examples from sources where the rest of the information is published, so the dedicated reader may find it. Such references are indicated by asterisks.

Dimensions of Task Description

While we have stressed the importance of viewing tasks holistically, we shall begin by writing down some of the dimensions that need to be covered in assembling a balanced set of tasks. These dimensions include the following:

- *Task length.* Doing and using mathematics involves tackling a great variety of types of tasks, from the quick mental calculation to the extended practical problem in which different parts of mathematics are used from time to time. We think it useful to distinguish *short tasks* (from a few seconds to, say, 15 minutes), *long tasks* (which may take 15 minutes to 2 hours) and *extended tasks* (which take many hours, often spread over several weeks).

- *Autonomy.* While the traditional mathematics curriculum is largely imitative, with students asked to tackle only tasks that are very similar to those they have been shown how to do, the need is for people who can use their skills and understanding with *flexibility* and *autonomy*, since in real life most problems do not present themselves in neat, standard form.

- *Unfamiliarity.* Some tasks will be entirely familiar and thus routine, but others need to be less so in order to develop the students' abilities to adapt and extend their mathematics. *Nonroutine problem solving* is thus closely connected with student autonomy. It is important to recognize that such problems have a high *strategic load* in finding a route through the problem, and that the *technical load* must be correspondingly lighter if the overall difficulty is to be the same. Equally, students can only use autonomously those skills that they have thoroughly absorbed and linked to other aspects of their understanding. In practice we have found there is roughly a four-year gap between the level of autonomous technical performance of students and that which they show in imitative exercises of a familiar kind. This critical factor is often missed.

- *Practicality.* Some tasks will involve the use of mathematics in practical applications, while others will be purely mathematical. Each supports the other.

- *Context.* Applications of mathematics cover a very wide range of practical contexts and assessment tasks should cover some of that spectrum, particularly those that relate to the experience and interests of the students.

- *Mathematical content.* This hardly needs stressing, but the tasks should sample the whole range of strategies, concepts and skills in the target curriculum.

In the rest of this chapter, we exemplify and discuss these characteristics for a range of short and long tasks. (Extended tasks are discussed in the next chapter.)

IMPROVING SHORT TASKS

We start in the most familiar zone, with tasks that (i) take from a few seconds up to 15 minutes or so, (ii) are focused on particular areas of mathematical skill and conceptual understanding, and (iii) are meant to be straightforward. We call these *short tasks*.

The principal way of testing student performance in the technical skills of mathematics should be through their ability to recognize the need for such skills, and their ability to select and deploy them effectively in worthwhile tasks. However, there remains a need for the external assessment system to embody some curriculum guidance derived from the collective wisdom of longer and wider experience. For example, although knowledge of table facts is constantly needed in classroom mathematical work and in examination tasks, many students still do not possess these facts at the level of fluency necessary for efficient work. It is not only higher-order skills that need to be encouraged by appropriate inclusion in assessments. Thus, we see a value in retaining, at least for a transitional period, some short tasks that test technique directly, giving them a modest total weight (perhaps 20 percent). But even these can be greatly improved, particularly in developing flexibility and practical relevance beyond the standard types of "fill-in-the-blank" problems such as

$9 \times 5 = \underline{}$, or simplify $x^3 y^5 / x^2 y$.

For example, Examples 1 and 2 provide ample practice, and in Example 2, a little thinking leads to some amusing features accessible to children at the same stage (from *The Power Series*, Shell Centre/UCSMP, 1991). Notice that Example 2 requires a calculator for posing the problem but not, for most children, for its solution. Of course, part of the credit in each case must be given for the mathematical insights, and part for reliable technical manipulation. Furthermore, these possibilities must be clearly indicated to the student. Such problems (from Tyler, 1984; Shell, 1984) furnish particularly good examples of the exponential ramp—everyone can make quite a lot of progress but few students, or adults, will discover all its possibilities.

Example 1. Answer 5.

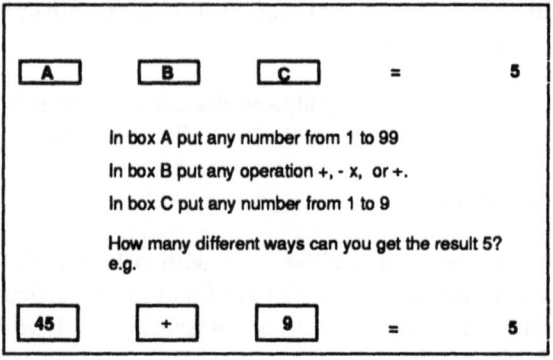

Example 2. Target.

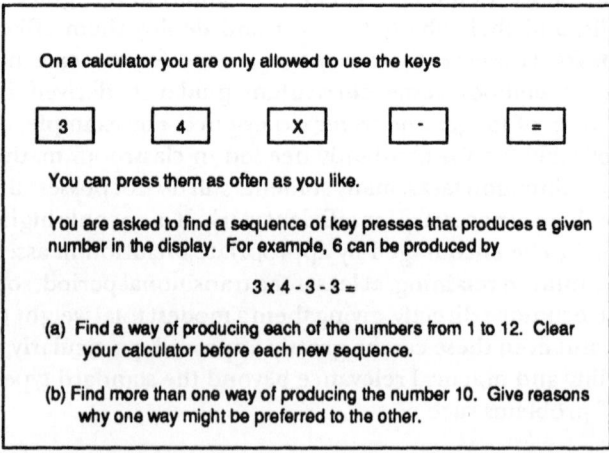

We have suggested that the degree of practical realism of questions should be improved. But if this is done, the reality must clearly be more than cosmetic. For example, in the question illustrated in Example 3 from a 1988 GCSE (General Certificate of Secondary Education) examination for 16-year-old students (Shell Centre/SEC, 1989), the ironing board does not make the question any more real; it is still a formal exercise in trigonometry. Nonetheless, there is a good real question lurking here, namely, where one should place the "stops" under the top surface so that the board can be conveniently used by people of various heights. Of course, like all real problems, this brings in other factors, but if mathematics is to be any use to the students, they must be able to integrate it in this way. When looking at a real question, one should ask "Why?" or "What use is the answer?"

Worse than in the preceding case, some other questions require misconceptions to yield a "right answer." For example, consider the 1988 GCSE question shown in Example 4. Example 4 is meant to demand scale-reading and simple extrapolation based on a linear model. In fact, however, the temperature will pause at 0 C while the ice melts. Also, freezers should operate below -18 C (0 F) This question amounts to disinformation. Sensible questions can usefully be set in a practical context that runs over several tasks; for example, the tasks in Examples 5 and 6 use a series of questions to tackle a coherent theme.

Example 3.

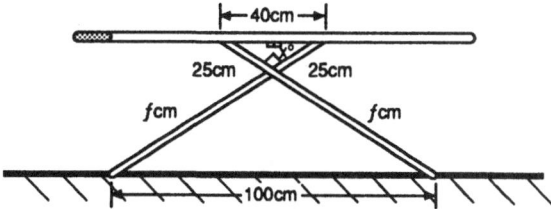

The diagram shows the side view of an ironing board. The two legs cross at $x°$ and are equal in length.
(a) Use the information in the diagram to calculate angle $x°$.
Give your answer to the nearest degree.
(b) Calculate the value of f.

Example 4.

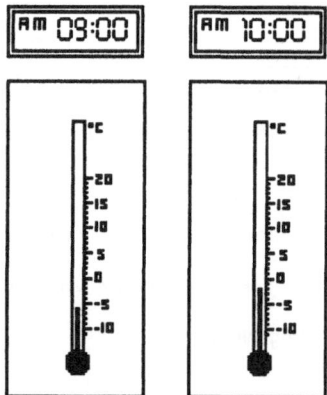

A freezer is switched off at 0900 in order to defrost it. The diagrams show the temperatures in the freezer at 0900 and one hour later at 1000.

(a) What is the temperature in the freezer when it is switched off?
(b) By how much does the temperature rise in the hour between 0900 and 1000?
(c) The temperature rises by the same amount in the next hour. What is the temperature at 1100?

Example 5. Before and after birth.

Task B: Growth.

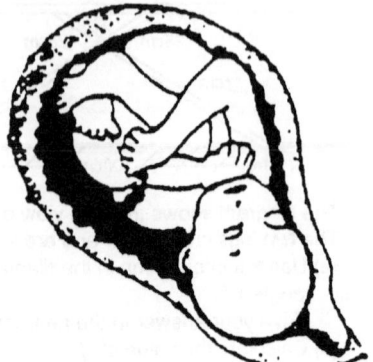

1. Before and after birth
   The table below shows how a typical baby grows during the first months inside its mother's womb.

Month number	2	3	4	5
Length in centimeters	4	9	16	25

   (a) Describe any patterns you can see in the numbers in the table.
   _____
   _____
   _____

   If this pattern continues, what length would the baby be at month 9?
   _____

   (b) Luckily for the mother, this pattern doesn't continue!
   For months 6 to 9, the approximate length of the baby is given by the formula

   (Length in cm) = 4 × (Month number) + 6

   Complete the table, using this formula:

Month number	6	7	8	9 (birth)
Length in centimeters				

   (c) The baby is born at 9 months.
   During the first 6 months after birth, she grows 18 cm.
   During the second 6 months, she grows a further 10 cm.
   On the next page, draw a smooth curve to show how the baby grows from 0 months to 21 months.

   This task comes from the Mathematics through Problem Solving examination (NEA, 1990*) referred to in chapter 6.

Example 6. College entrance.

Sample open-ended questions (and response)
from CAP Grade 12 Test, 1987-1989

James knows that half of the students from his school are accepted at the public university nearby. Also, half are accepted at the local private college. James thinks that this adds up to 100 percent, so he will surely be accepted at one or the other institution. Explain why James may be wrong. If possible, use a diagram in your explanation.

one student's response:

| people accepted by the public university ◐ |
| people accepted by both ➔ ● |
| people accepted by the private college ⊕ |
| people not accepted ○ |

James' school's acceptance outcome

AS MUCH AS 50%
MIGHT NOT
BE ACCEPTED

Imagine you are talking to a student in your class on the telephone and want the student to draw some figures. The other student cannot see the figures. Write a set of directions so that the other student can draw the figures exactly as shown below.

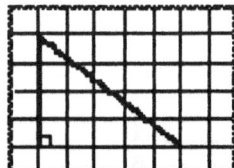 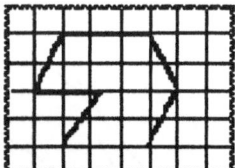

John has four place settings of dishes, with each place setting being a plate, a cup, and a saucer. He has a place setting in each of four colors: green, yellow, blue, and red. John wants to know the probability of a cup, saucer, and plate being the same color if he chooses the dishes randomly while setting the table.

Explain to John how to determine the probability of a cup, saucer, and plate being the same color. Use a diagram or a chart in your explanation.

(This is a page from *Assessment Alternatives in Mathematics*, a booklet from the California Mathematics Council and EQUALS.)

These questions are essentially technical tasks in a practical context. They should not be confused with questions that are practical in purpose, where the emphasis is on understanding a real situation through the languages of mathematics. For example, Example 7 is a 15-minute task that has been widely discussed (from Shell 1986). Should we give credit in mathematics for the quality of the commentary, as a commentary, or even for an explanation? We should, if learning and performance are to be advanced. The marking scheme for this question is designed to give credit for the effective display of some of the following skills: (i) interpreting mathematical representation using words or pictures, (ii) translating words or pictures into mathematical representations, (iii) translating between mathematical representations, (iv) describing functional relationships using words or pictures, (v) combining information presented in various ways and drawing inferences where appropriate, (vi) using mathematical representations to solve problems arising from realistic situations, and (vii) describing or explaining the methods used and the results obtained.

Script C and Script E, shown in Example 8, illustrate two student responses. Table 1 summarizes the features for which credit is given in this particular problem and the marks awarded to six student responses, including the two shown. A more extended discussion is given in the reference, which also shows how such discussions can be used in a type of in-service training session for teachers that is both popular and effective. Example 9 came from a student's own initiative.

Example 7. The hurdles race.

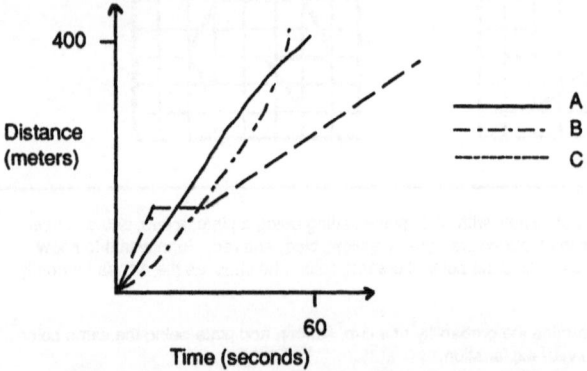

The rough sketch graph shown above describes what happens when 3 athletes, A, B, and C, enter a 400- meter hurdles race.

Imagine that you are the race commentator. Describe what is happening as carefully as you can. You do not need to measure anything accurately.

Table 1. Suggested marking scheme applied to scripts based on Example 7.

		Script					
		A E	B F		C	D	
1 mark for each of these	At start, C takes lead		1	1		1	1
	After a while, C stops	1	1	1	1	1	
	Near end, B overtakes A		1	1	1	1	
	B Wins		1	1	1	1	
2 marks for 4 of these, or 1 mark for 2 of these	A and B Pass C		✓	✓			
	C starts running again		✓		✓	✓	
	C runs at slower pace					✓	
	A slows down or B speeds up	✓	✓	✓	✓	✓	
	A is second or C is last		✓	✓	✓	✓	
Quality of commentary		0	2	1	2	0	1
TOTAL		1	8	6	7	3	4

Example 8. Sample scripts.

Script C (Simon).

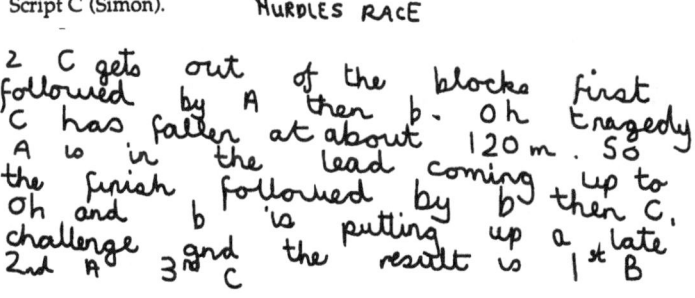

Script E (Jackie).

Athlete A on the first 100 m is in second place when he has past the 100 m mark the time is about 10 seconds His speed stays about the same through the next 100 m and as he passes 300m mark the time is about 50 seconds
He finishes the race in about 1 minute 10 seconds

Athlete B on the first 100m is slower on the first 100 m than Athlete A his time ofter 100m is about 20 seconds. His speed stays about the same through the next 100m and as he passes the 300m mark the time is about 60 seconds.
He finishes the race in about 1 minute 5 seconds so he quickened up near the end.

Athlete C is quicker than Athlete A, B in the first 100 m at about the 150 metre mark he goes steps gradually but quickens up again on the last 200 m but he finishes the race in about 1 minute 40 seconds

Example 9. Feelings.

These graphs show how a girl's feelings varied during a typical day.

Her timetable for the day was as follows:

7:00 am	Woke up	1:30 pm	Games
8:00 am	Went to school	2:45 pm	Break
9:00 am	Assembly	3:00 pm	French
9:30 am	Science	4:00 pm	Went home
10:30 am	Break	6:00 pm	Did homework
11:00 am	Math	7:00 pm	Went to 10-pin bowling
12:00 pm	Lunchtime	10:30 pm	Went to bed

(a) Try to explain the shape of each graph, as fully as possible.

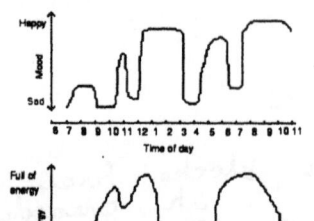

(b) How many meals did she eat? Which meal was the biggest? Did she eat at breaktimes? How long did she spend eating lunch? Which lesson did she enjoy the most? When was she "tired and depressed?" Why was this? When was she "hungry but happy?" Why was this?

Make up some more questions like these, and give them to your neighbor to solve.

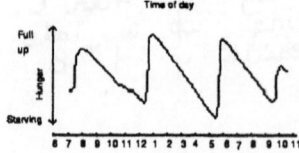

(c) Sketch graphs to show how *your* feelings change during the day. See if your neighbor can interpret them correctly.

© Shell Centre for Mathematical Education, University of Nottingham, 1985

Short Tasks to Test Some General Mathematical Strategies

Some aspects of mathematical strategy lend themselves quite well to assessment by short tasks (Bell et al., 1978), and such tasks have been used extensively in research studies and occasionally in public examinations. We shall give some items related to proof and to symbolization.

We can distinguish three major dimensions of development in the use of proof strategies. These are (i) the degree of regularity or rationality expected by the pupil (some are unsurprised by stark inconsistencies); (ii) the explanatory quality of the proof response (an awareness that a proof or explanation must go beyond the restatement of a result in general terms, that it must connect it with existing knowledge and must avoid circularities); and (iii) the level of sophistication of the proof techniques or logical transformations.

The problem "Add and Take" (Example 10) tests the ability to distinguish two proposed arguments, one the checking of a number of cases, the second a general argument applying to all cases. This item has been used with various groups of students aged 11 to 15. Vagueness was the keynote of many answers to this item, such as, "Brenda has explained it better," or "Brenda's reason is based solely on facts." Some students argued for whichever of the alternatives they found most simple — a subjective view of proof. The following are two typical answers, one fair, one poor:

> Because Jane has assumed that just because two numbers (1 and 9) work out as 20, then all other numbers under 10 must work out as twenty whereas Brenda has given a detailed explanation.

> Jane has explained it most easily than Brenda has.

These are typical responses; very few students were sensitive to the invalidity of assuming what one is trying to prove.

Items that asked students to explain why certain well-known principles were true elicited good responses from very few. For example, in an item called "Adding a Nought," students were asked whether the principle exemplified by 243 x 10 = 2430 was true for all whole numbers; they were asked also to explain or justify their answers. Most simply gave a few more examples of the use of the principle. Questioning the limits of the truth of the principle was not on the agenda.

Recognition of circularity of argument in a geometrical proof is tested in Example 11. One good answer to the question said, "Step 2 is wrong because if he can safely assume this, there is no need to do the proof in the first place. Step 3 is wrong because everything falls apart at Step 2."

Example 10. Add and take.

Choose any whole number less than 10. choose [1]
Add the number to 10. ans. [11]
Take the first number from 10. ans. [9]
Add the last two numbers. ans. [20]

Do the same beginning with 9. choose [9]
[ ]
[ ]
[ ]

Show that the answer is again 20.

JANE says,	'Begin with 1, answer is 20. Begin with 9, answer is 20. So begin with any number between 1 and 9, answer will be 20. The answer is always 20'.
BRENDA says,	'You have 10+ number. You add 10- number. You add and take the same number so you will always be left with 2 tens. The answer is always 20'.

(1) Who do you think has given the better reason for the answer always being 20?   JANE / BRENDA

(2) Use the space below to explain why. (If you think they are equally good explain why.)

Example 11. Angles.

Barry wants to prove that any two adjacent angles on a straight line make up 180°.

He draws this diagram

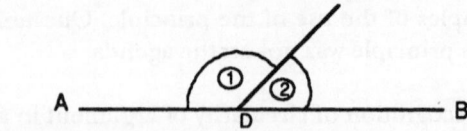

and aims to show that ① + ② = 180°.

Barry's proof goes like this.

First he extends CD in a straight line to E to make his diagram look like this.

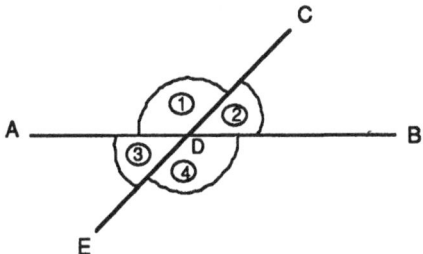

Step 1 ② = ③
Step 2 ① + ③ = 180° since $\overline{CE}$ is a straight line.
Step 3 Combining steps 1 and 2, ① + ② = 180°.
Step 4 Hence two adjacent angles on a straight line add to 180°

For each of steps 1 to 4 explain why it is right or wrong.

Step 1 is right because they are opposite angles on a straight line

Step 2 is wrong because if he can safely assume this, there is no need to do the proof in the first place.

Step 3 is wrong because everything falls apart at Stage 2.

Step 4 is wrong because of the reason for Step 3

Sensitivity to definitions is tested in Example 12, while the items in Example 13 are intended to test ability to work with symbolic representations. "Turning the arrows" (Example 13) involves the first familiarization with the movements P and Q, a half and a quarter turn respectively. Then, by requesting lengthy combinations of these, the pupil is led to formulate rules by which such sequences may be reduced. These rules are (i) the order in which the symbols appear is irrelevant, then (ii) P^2 is an identity movement and so is Q^4. This is a typical process in which the thrust towards generalization leads to the abstraction of relationships. Here, the expression of the generalizations is also asked for. One might reasonably assume that some familiarity with the process of interrelating the symbol system and a geometrical situation, and with the process of reducing words using algebraic rules, might lead to an improvement in the process.

Example 12. Quadrilaterals.

In this question the meaning of quadrilateral is as given in the following statement.

Definition: A quadrilateral is what you get if you take 4 points A, B, C, D in a plane and join them with the straight lines AB, BC, CD, DA.

Horace says, "the angles in a quadrilateral at A, B, C, D always add up to 360°.

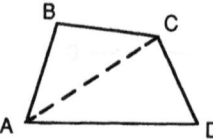

Here is my proof.
Every quadrilateral can be made into 2 triangles by joining a diagonal.
Each triangle has 180° and
2 x 180° = 360°."

Warwick says, "Look at my quadrilateral.

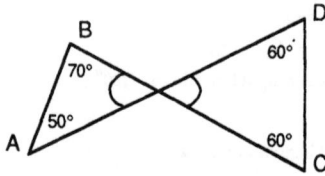

50° + 70° + 60° + 60° = 240°
My angles make up 240°.
Horace is wrong."

(1) Has Warwick drawn a quadrilateral?   Yes/No

(2) Is the first sentence that Horace says correct?   Yes/No
Explain why or why not.

(3) Explain what is right or wrong about Horace's proof.

Example 13. Turning the arrows.

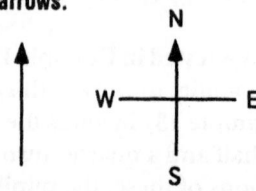

Two changes P and Q can be made to this arrow head

P turns it to point in the opposite direction.
Q turns it through a quarter turn, clockwise.

It always starts by pointing north.

1. After doing PQ (P then Q) which way does it point?   ....

2. After doing P^2 (P twice) which way does it point?   ....

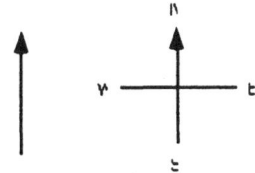

Two changes P and Q can be made to this arrow head

P turns it to point in the opposite direction.
Q turns it through a quarter turn, clockwise.

It always starts by pointing north.

1. After doing PQ (P then Q) which way does it point? ....

2. After doing P^2 (P twice) which way does it point? ....

"Roofs" (Example 14) is another item where the relationship between a geometrical figure and a symbolic code is exploited. Here the question asks what conditions on the numbers constituting the code are necessary to ensure that a roof can be drawn. The conditions are A + B = C and D = B. Students may argue the truth of these either empirically (by looking at a number of usable codes) or structurally or deductively (by arguing from the features of the diagram). Thus the different aspects of proof come into this item too.

Example 14. Roofs.

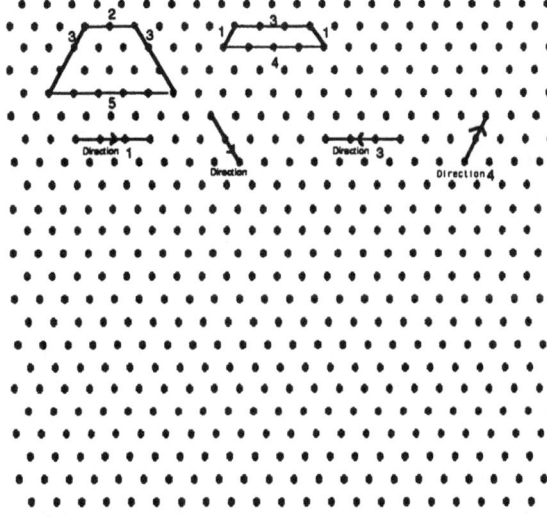

Roofs can be drawn in different shapes and sizes, using the dots provided. The first one drawn at far left is a 2 3 5 2.

The second is a 3 1 4 1.

(The first number tells you how many units to draw in direction 1, the second in the direction 2, the third in direction 3, and the fourth in direction 4.)

1. Draw a 2242 and a 4151.

2. Try to draw a 3 2 5 1 and 1 4 3 4. Explain what happens.

A number of the above items were developed as part of a project to evaluate achievements in process aspects of mathematics in a curriculum for 11- to 13-year-olds in which problem investigation was the main way of developing knowledge of the syllabus content (Bell, Rooke, and Wigley, 1978, 1979). Subsequently, some interview-type assessments were developed, complete with scripts specifying prompts and hints to be offered if needed (Fowler, 1983). Such questions were used by the Freudenthal Institute Assessment of Performance Unit in its national surveys, 1978-82. Some twenty teachers were recruited and trained, and spent two weeks interviewing, in different schools, about six students per day, each for 45 minutes, covering three topics. The valuable experience gained of the feasibility and consistency (high) of their procedure is reported by Foxman et al. (1989).

NONROUTINE PROBLEM-SOLVING TASKS

The development of mathematical performance demands flexibility and adaptability, as well as reliable technical performance and the ability to communicate what has been tried, and what found. While the tasks set out so far have demanded some flexibility, they have been essentially straightforward, with a fairly obvious approach to finding what is asked. We now want to discuss tasks where the strategic demand—finding a successful approach to and route through the problem—is a major aspect of the task. Again, we emphasize the importance of keeping the total cognitive load in line with the student's abilities, and the "four-year gap" (between autonomous and imitative performance) that this seems to imply for the technical level.

We consider first a mode of assessment in which the attempt is made to test the various components of the generalization process through a short investigation in a traditional examination setting. One early development was a joint project between one of the major British examining boards, the Northern Universities Joint Matriculation Board (JMB) and the Shell Centre (Burkhardt, 1980). Each year one new type of question was included in the JMB examination for abler 16-year-olds (the GCE O-level). The Shell Centre provided the questions and (equally or more important) a substantial module of teaching material with guidance for teachers to support preparation of students for this question (Shell Centre, 1984). The teaching materials, designed to cover three or four weeks of mathematics time, are based around well established problem-solving strategies. They offer substantial initial support for students (and for the teacher), with decreasing support as experience develops. The set of strategies (cf. Polya, 1945) used was (i) try some simple cases, (ii) find a helpful diagram, (iii) organize systematically, (iv) make a table, (v) spot patterns, (vi) find a general rule,

(vii) explain why it works, (viii) check regularly. The questions were in the form of longer (20-25 minute) problems requiring an investigative approach. The grading schemes were based loosely on the preceding strategies, giving credit for (i) understanding the problem, (ii) organizing an attack, (iii) carrying it through successfully, and (iv) explaining or justifying the solution.

In Example 15 (from Shell Centre, 1984), students, who have not seen this particular problem before, take a variety of approaches. Some break the "tower" into four legs and a center, some take horizontal slices, finding number patterns and generalizing them into verbal rules with more or less success. Few, in fact, express their verbal rule in algebra, or sum their arithmetic progression, even though they are able to do so if these are presented as separate technical tasks. This is an illustration of the four-year gap between autonomous and imitative performance that we have already noted—the technical level of this question, apart from the last part, is at elementary school level, but the whole task is a suitable challenge for able students at age 16. It was also interesting to note how few students (less than 5 percent) take a geometric approach, breaking off two opposite "legs" and putting them upside down on the others to form a rectangle.

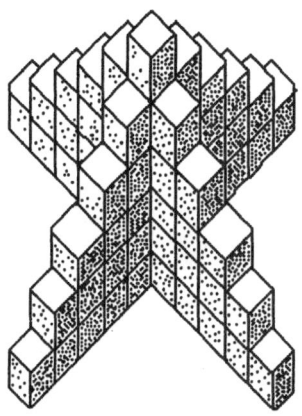

Example 15. Skeleton tower.

a. How many cubes are needed to build this tower?

b. How many cubes are needed to build a tower like this, but 12 cubes high?

c. Explain how you worked out your answer to part (b).

d. How would you calculate the number of cubes needed for a tower n cubes high?

Example 16, "Stepping Stones" is another question that tests ability to generalize. The elements of the solution to this question consist of (i) explaining how the girl will stop only on even-numbered stones, (ii) identifying values of n that entail stepping eventually on every stone, that is, those divisible neither by 2 nor 7, (iii) generalizing this to numbers that have no divisors in common with the number of stones in the ring—with explanation. Thus, the first part involves trying a given simple case, displaying understanding of the problem; the second requires some systematic organization of trials of other numbers to cover all significant cases, and the third involves making, stating, and explaining a generalization. (In many such questions, a little algebra is appropriate at this final stage, though surprisingly few students can translate the rules they have found verbally into algebra (Shell Centre, 1986).

Example 16. Stepping stones.

A ring of 'stepping stones' has 14 stones in it, as shown in the diagram.

A girl hops round the ring, stopping to change feet every time she has made 3 hops. She notices that when she has been round the ring three times, she has stopped to change feet on each one of the 14 stones.

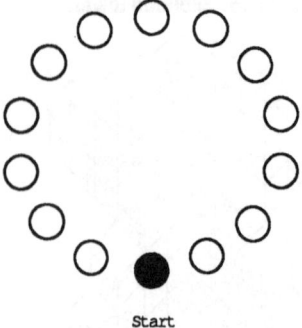

Start

a. The girl now hops round the ring, stopping to change feet every time she has made 4 hops. Explain why in this case she will not stop on each one of the 14 stones no matter how long she continues hopping round the ring.

b. The girl stops to change feet every time she has made n hops. For which values of n will she stop on each one of the 14 stones to change feet?

c. Find a general rule for the values of n when the ring contains more (or less) than 14 stones.

Such questions have some but not all of the characteristics of a mathematical investigation—generalizing from examples, and stating and explaining the generalization are required, but with little formulation or extension of questions and little choice of where next to go in the inquiry. The steps to follow are prescribed. We may regard these as relatively closed investigations testing a fairly well defined set of strategies. Of course, this still represents a substantial extension of the range of attainments normally sampled in written examinations. Since the design and use of marking schemes for such questions presents different demands from the more traditional questions, the module of support material contains examples of a number of such questions with mark schemes and students' scripts (Shell Centre, 1984) as well as the classroom materials and some other in-service support for teachers new to this kind of work.

In the next chapter, on extended tasks, we shall consider how assessment can be made of those further higher-order skills and strategies that are brought into play in extended pieces of work, particularly those in which the openness of the situation leads to responses differing widely in content and style. Such assessment is often nonspecific in some respects, and some approaches entail an extensive moderation system, which we shall also discuss.

REFERENCES

References marked with an * contain detailed examples of tasks, student work and grading schemes.

Bell, A., Rooke, D., and Wigley, A. (1978). *Journey into maths*. Nottingham: Shell Centre.

Burkhardt, H. (1979). *The influence of the examination boards on the school curriculum.* Nottingham: Shell Centre.

Burkhardt, H. (1980). *Board examinations and the school curriculum.* Nottingham: Shell Centre.

California. (1985). *Mathematics framework for California Public Schools K-12.* Sacramento: California State Department of Education.

Cockcroft, W. (1982). *Mathematics counts.* Report of the Committee of Inquiry into the Teaching of Mathematics in Schools. London: HMSO.

DES. (1989). *Mathematics and the national curriculum.* Statutory orders from the Department of Education and Science and the Welsh Office. London: HMSO.

EQUALS. (1989) *Assessment alternatives in mathematics.* Berkeley, CA: EQUALS.

Fowler, N. (1983). Oral tests of process aspects of mathematics: Development and evaluation. M. Phil. thesis, University of Nottingham.

Foxman, D., Ruddock, G., and Thorpe, J. (1989) *Graduated tests in mathematics.* Windsor: NFER-Nelson.

Mathematical Sciences Education Board. (1989). *Everybody counts: A report to the nation on the future of mathematics education.* NRC. Washington, DC: National Academy Press.

Mathematical Sciences Education Board. (1990). *Reshaping school mathematics: A philosophy and framework for curriculum.* National Research Council. Washington DC: National Academy Press.

National Council of Teachers of Mathematics. (1989). *Curriculum and evaluation standards for school mathematics.* Reston, VA: NCTM.

Northern Examining Association. (1990). *Mathematics through problem solving,* A GCSE syllabus. Manchester: NEA.

Polya, G. (1945). *How to solve it* (2nd ed.; 2nd printing 1973). Princeton: Princeton University Press.

Ridgway, J. (1987). *Review of mathematics tests.* Windsor: NFER-Nelson.

*Ridgway, J. (1988). *Assessing mathematical attainment.* Windsor: NFER-Nelson.

Schools Examinations Council. (1985). *National criteria for the GCSE: General criteria and mathematics criteria.* London: SEC.

*Shell Centre. (1984). *Problems with patterns and numbers: An examination module for secondary schools.* Manchester: Joint Matriculation Board.

Shell Centre. (1986). *The language of functions and graphs: An examination module for secondary schools.* Manchester: Joint Matriculation Board.

Shell Centre/JMB. (1987-89). *Numeracy through problem solving* (five modules: Design a Board Game, Produce a Quiz Show, Plan A Trip, Be a Paper Engineer, Be a Shrewd Chooser, from the Shell Centre for Mathematical Education and the Joint Matriculation Board). Harlow: Longman.

Shell Centre/MEG. (1989). *Extended tasks for GCSE mathematics* (a set of modules to support school based assessment from the Shell Centre for Mathematical Education and the Midland Examining Group). Nottingham: Shell Centre.

Shell Centre/SEC. (1989). *Practical work in GCSE mathematics.* London: Secondary Examinations and Assessment Council.

*Shell Centre/UCSMP. (1991). *The power series* (six modules for the single micro classroom, Exploring Graphs, Exploring Numbers, Strategic Problem Solving, Patterns and Functions, More about Graphs, Graphs and Functions, for the University of Chicago School Mathematics Project). Nottingham: Shell Centre.

Tyler, K., and Burkhardt, H. (1984). *Calculator maths 1 - 7.* Glasgow: Blackie.

6 Assessment of Extended Tasks

Alan Bell, Hugh Burkhardt, and Malcolm Swan

INTRODUCTION

In this chapter, we discuss the assessment of extended tasks, by which we mean mathematical activities covering many hours, usually spread over several weeks. Typically, this may range from three to fifteen hours of class time, often with additional private study outside. This type of task and its assessment has become a major focus of attention in England over the past few years. Such work originated from the desire of some teachers to increase the level of positive involvement on the part of their students in mathematical activity. Discussion led by the teacher was aimed at enabling the student to get started on the investigation of a problem that students had, as far as possible, formulated for themselves and adopted as their own. Ideally, each student would be exploring a different problem, although in practice, several might have arisen from a discussion of the same basic situation. Two examples of work initiated in this way are given in the early part of this chapter. Originally, most such investigations were located in pure mathematics rather than in applications, but in current examination schemes, it is normal to require candidates to submit work from several different fields. For example, one typical scheme (Shell Centre/MEG, 1989) requires work from practical geometry, statistics, everyday applications of number, and pure mathematical investigation. Since the tasks are not specified, such work cannot be assessed by a content-specific scoring scheme. Profile assessments are required, involving global evaluation of the work under a number of headings. These will be discussed below.

In Britain, all General Certificate of Secondary Education (GCSE) examinations (for age 16) must include the assessment of extended tasks of the preceding kind, but the schemes of the different examining groups differ considerably in the amount of freedom allowed. For some teachers this provision is the culmination of a long campaign to have students' individual, autonomously chosen work recognized in the examination system. For the majority, it is a new, substantial, and probably not entirely welcome demand. To satisfy the need for support for innovations in assessment, a number of publications have emerged recently. These offer possible "starters" for investigations and, at best, some careful general guidance about how to initiate investigation of a suitable field of activity and to lead students to identify a particular problem of their own to pursue. Some extracts from our own publication, *Extended Tasks for GCSE Mathematics* (Shell/MEG, 1989) developed with one examining board, will be given below. This forms the second section of this chapter.

Another way in which extended tasks may arise is in the course of work on a substantial class practical activity as, for example, in our *Numeracy through Problem Solving* modules (Shell Centre/JMB, 1987-1989). In the third and longest part of this chapter, we shall describe the approaches illustrated, in particular, from "Plan a Trip," "Be a Paper Engineer," and "Be a Shrewd Chooser." (Other modules are "Design a Board Game" and "Produce a Quiz Show.") The pattern of classroom activity prescribed provides additional support to teachers, but there is a wide scope for individuality of response in the actual games designed, objects made, or products assessed, and in the design methods adopted. The assessments for these modules are well defined, covering a variety of modes; they take place in the course of the work on the module, and afterwards, in a manner which will be described below.

Two Examples of Open Assignments

These examples (one of an "applied" problem explored by a 17-year-old boy, the other of a "pure" problem investigated by a 13-year-old girl) have been chosen because they show quite well the strategic choices being made by the student—about what questions to ask, what next steps to take—and the explanation of the course of the investigation to a reader. The authenticity and autonomy of the work shine through.

In "Filter Paper" (Example 1), a 17-year-old boy investigates a question concerning the standard way of folding filter papers in the chemistry laboratory. He asks whether a cone containing greater volume could be obtained from a different method of folding. He concludes that 80 percent more volume could be obtained, and he offers a brief suggestion about why this is not used in practice. The crux of the task is the modeling stage—the identification of the relations among the vertical height, slant height, and base radius of the possible cones as determined by the possible folding, and the choice of a suitable independent variable (A to B in the script). Following this, there is a consideration of the shape of the resulting graph (B to C), and the application of the standard method of finding a maximum point by differentiation (C to D). Dealing with the algebra involving several quantities is a significant task which is handled well. The last main section (D to E) considers the normal method of folding and calculates the volume given for a paper of radius l. Finally there is the summary of results (E to F; for further discussion, see Association of Teachers of Mathematics, 1978*).

Example 1. Filter paper.

What is the cone of greatest volume that can be made from a piece of filter paper of any given radius?

A

"A piece of filter paper is a perfect circle, of radius, let us say, L

FILTER PAPER

f this piece of filter paper is folded into a cone, L becomes the slant height of the cone and the centre of the circle becomes the apex of the cone.

The cones which can be made from that sheet of filter paper can vary a lot in dimension, but the slant height will always be constant and equal to the radius of the paper.

For a typical cone,

    h = perpendicular height
    r = radius of base cone
    L = slant height

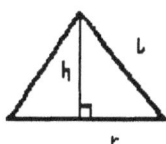

Now by Pythagoras, who stated that in a right angled triangle (as in the case above) the square of the hypotenuse was equal to the sum of the squares on the other two sides:

$$L^2 = h^2 + r^2$$

whence $r^2 = L^2 - h^2$

and $h^2 = L^2 - r^2$

The formula for the volume of a cone = $\frac{1}{3}\pi r^2 h$

where r = radius of base of cone
h = perpendicular height of cone

r and h are variables in the equation, but for a constant slant height, r may be expressed in terms of h and vice versa so that we may substitute in volume = $\frac{1}{3}\pi r^2 h$ and hence end up with one variable.

Thus substituting $r^2 = L^2 - h^2$ in Vol = $\frac{1}{3}\pi r^2 h$

we get $\frac{1}{3}\pi (L^2 - h^2) h$ X

Assessment of Extended Tasks

i.e. vol $V = \frac{\pi}{3} h L^2 - \frac{\pi}{3} h^3$

B

If a graph of $V = \frac{\pi}{3} h L^2 - \frac{\pi}{3} h^3$ we drawn, i.e. a graph showing how volume varies with the perpendicular height of the cone, (L always remains constant for that piece of filter paper) a cubic graph would be obtained since there is a term in h^3 present in the equation. This graph will be of the form

$$y = ax^3 + bx^2 + cx + d$$

we have $V = -\frac{\pi}{3}h^3 + \frac{\pi}{3}L^2 h$

i.e. $y = ax^3 + cx$

Where the graph $V = \frac{\pi}{3}h^3 + \frac{\pi}{3}L^2 h$ crosses the x axis

$V - 0$: letting $\frac{\pi}{3} = a$, and $\frac{\pi}{3}L^2 = b$

we get:
$0 = -ah^3 + bh$
dividing by h we obtain
$0 = ah^2 + b$
$\therefore -b = -ah^2$
$\therefore \frac{-b}{-a} = h^2$

multiplying both sides by $\frac{-1}{-1}$ we obtain

$$\frac{b}{a} = h^2$$

$$\therefore h^2 = \pm\sqrt{\frac{b}{a}}$$

which means the h has real roots.
Also in $V = -\frac{\pi}{3}h^3 + \frac{\pi}{3}L^2 h$, when $h = 0$, $V = 0$
$\therefore$ h has three real roots, 0 being the third

On this graph of $V = -\frac{\pi}{3}h^3 + \frac{\pi}{3}L^2 h$

There are thus • turning points on the graph, hence a maximum and a minimum point.

C

Where there is a maximum point, here the V co-ordinate is the maximum volume attainable from the filter paper and the h co-ordinate the corresponding height of that cone

$\therefore$ differentiate $V = \frac{\pi}{3} h L^2 - \frac{\pi}{3} h^3$ with respect to h remembering that L is a constant for that piece of paper.

remembering that l is a constant for that piece of paper, $\frac{dV}{dh} = \frac{\pi}{3}L^2 - \frac{3\pi h^2}{3}$

$$= \frac{\pi}{3}(L^2 - 3h^2)$$

Thus there are turning points when

$$\frac{\pi}{3}(L^2 - 3h^2) = 0$$

multiplying both sides by $\frac{3}{\pi}$ we obtain

$L^2 - 3h^2 = 0$
$\therefore L^2 = 3h^2$
$\therefore \frac{L^2}{3} = h^2$
$\therefore h = \pm\sqrt{\frac{L^2}{3}}$

ii.e. $h = \overset{+}{\sqrt{\frac{L^2}{3}}}$ or $\overset{-}{\sqrt{\frac{L^2}{3}}}$
   A   B

to find out whether A is a maximum point or not we find $\frac{d^2 V^2}{d^2 h^2}$ of $\frac{\pi}{3}L^2 h - \frac{\pi}{3}h^3$

$$\frac{dV}{dh} = \frac{\pi}{3}L^2 - \pi h^2$$

$$\therefore \frac{d^2 V}{dh^2} = -2\pi h \text{ when } h = \overset{+}{\sqrt{\frac{L^2}{3}}}$$

thus $= -2\pi, \frac{L^2}{3}$

since $\frac{d^2 V}{dh^2}$ is a negative quantity,

when $h = \overset{+}{\sqrt{\frac{L^2}{3}}}$ there is the maximum point

hence the maximum volume

to check, substitute h for $\overset{-}{\sqrt{\frac{L^2}{3}}}$

$\therefore \frac{d^2 V}{dh^2} = -2\pi h \quad -2\pi h - \sqrt{\frac{L^2}{3}}$

$= +2\pi\sqrt{\frac{L^2}{3}}$

$\therefore$ when $h = -\sqrt{\frac{L^2}{3}}$ there is the maximum volume, obtained from the cone of this height

D

Now compare the volume of the cone formed by folding laboratory filter paper in the normal way, to the volume of the cone that could be made by folding it in such a way that the height of the cone was $\sqrt{\frac{L^2}{3}}$, where L equals the radius of the filter paper.

In the laboratory, the filter paper is taken and folded in half, it is then folded in half again so that a quarter segment of a circle is produced. If this radius of the filter paper is L, as we have supposed all along, the circumference = $2\pi L$

∴ the length of the quarter segment of the circle = $\frac{1}{4} 2\pi L$ = $\frac{\pi L}{2}$

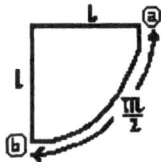

Along (a), (b) there are 4 folds of paper, and when the cone is produce from this paper, 3 folds are made to form one side, the 4th fold, the other thus:

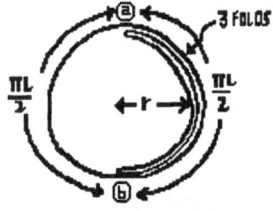

This is bottom view of the cone, i.e. from the other diagram looking down on the lines along (a) (b) and separating 3 lines (folds of paper) to the right as in the diagram, and 1 to the left.

Thus the perimeter of this circle at the base of the cone

$$= \frac{\pi L}{2} + \frac{\pi L}{2} = \frac{2\pi L}{2}$$
$$= \pi L$$

Let the radius of the base of the cone = r
circumference of this circle = $2\pi r$
but the circumference also = πL

∴ $2\pi r = \pi L$
∴ $r = \frac{\pi L}{\pi 2}$
$= \frac{L}{2}$

Now if the radius of the base of the cone = $\frac{L}{2}$, and the slant height is L, pythagoras's theorem may be applied to find the height of the cone, h.

$$\left(\frac{L}{2}\right)^2 + h^2 = L^2$$

∴ $h^2 = \frac{4L^2}{4} - \frac{L^2}{4}$

∴ $h^2 = \frac{3L^2}{4}$

∴ $h = \frac{L}{2} \sqrt{3}$

Now volume of the cone is given by
$$V = \frac{1}{3} \pi r^2 h$$

∴ Volume (given that $r = \frac{L}{2}$, $h = \frac{L}{2}\sqrt{3}$)

$$= \frac{1}{3} \pi \left(\frac{L}{2}\right)^2 \cdot \frac{L}{2} \sqrt{3}$$

$$= \frac{\pi L^3}{8} \cdot \frac{\sqrt{3}}{3}$$

∴ Vol $= \frac{\pi L^3}{8 \cdot \sqrt{3}}$

$\sqrt{3} = 1.732$

∴ Vol in terms of π and L

$$= \frac{\pi L^3}{13.856} \text{ cu units}$$

If the paper is folded so that the height = $\sqrt{\frac{L^2}{3}}$

Referring back to page 2 of the investigation from equation X

$$\text{Vol} = \frac{1}{3}\pi(L^2-h^2)h$$

substituting $\sqrt{\frac{L^2}{3}}$ we get

Vol = ...

$$= \frac{\pi L^3}{7.794} \text{ cu units}$$

since $\frac{\pi L^3}{13.856} < \frac{\pi L^3}{7.794}$

E

Thus filter papers in the laboratory, as regards obtaining the maximum volume from them when folded into a cone, if they were folded so that the height of the cone was $\sqrt{\frac{L^2}{3}}$, could be made to hold

(13.856 - 7.794), 7.794) × 100% more = 80% more.

However, the difficulty lies in folding the paper so that the height = $\sqrt{\frac{L^2}{3}}$, where L is the radius of the paper, that is why, for simplicity's sake, the paper is folded in the lab how it is.

F

This example brings out several points. There is not only the initial modeling and final interpretation, but also a continuing need to move between the actual problem and the mathematical processing, keeping both aspects in mind. Then there is the selection and use of appropriate knowledge and technique at appropriate points—Pythagoras' theorem, volume formula for a cone, cubic form recognized through a multiletter expression with independent variable h, not x, the shape of a cubic graph, and the crossing of the axis, the process of differentiating to find turning points, presentation of results in percentage form, and in particular there is the decision to make a temporary substitution to simplify the algebra. *Each of these could form the subject of a short test item, but it is clear that the ability to recognize their relevance and to deploy them in this way is a different, more demanding, and essential accomplishment.*

Our next problem is to describe and to evaluate this accomplishment. One approach, for example, uses the description of mathematical processes as comprising abstracting, representing, generalizing, and proving. For instance, the preceding example involves much representation, some abstracting (if one includes in this recognizing relevant concepts and methods), and some generalizing (if one so interprets the initial question, "How could this situation be different?"). A more natural description would be of an interplay of representing (modeling) and processing.

To quantify our evaluations we have to ask whether it makes sense to describe a person as good or bad at generalizing or abstracting or representing *in general* (or at least over some specified domains), and whether such capacities can be improved simply through relevant experience, or by more specific naming and practice of the processes. The principle of devising such grading schemes is clear: founded on principle, *a grading scheme must give results of good face validity in practical use over a representative sample of scripts*. We shall return to these questions later. Meanwhile, it important to have one or two further and different examples before us.

The "Remainder Problem," described in Example 2, was investigated by a 13-year-old girl. This work shows a number of characteristic features of mathematical investigation. It starts with a particular problem (4r2, 5r1) of modest interest which becomes more interesting as it is generalized. The concept of least common multiple is almost, but not quite, created in the course of the solution; and there is the arrival at an algorithm (for the starting numbers) when a direct explicit procedure is not available.

Example 2. The remainder problem.

It always works in the same pattern.

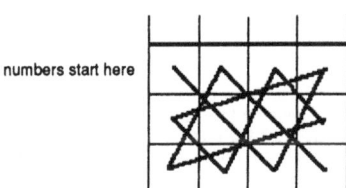

This kind of table works for any numbers:

	11r0	11r1	11r2	11r3	11r4	11r5	11r6	11r7	11r8	11r9	11r10
4r0	0	12	24	36	4	16	28	40	8	20	32
4r1	33	1	13	25	37	5	17	29	41	9	21
4r2	22	34	2	14	26	38	6	18	30	42	10
4r3	11	23	35	3	15	27	39	7	19	31	43

As we know how the table works we can predict what the first number is going to be by working out from the table but not actually writing all the numbers in. If the numbers are large it would take a long time.

We also found a few cases that didn't work at all;

e.g.   5r0   3r2   4r3
       10r1   6r4   8r1

But didn't really have time to go into this.

<u>Conclusion</u>

I couldn't find a really efficient way of predicting what the first number is going to be, only by the method of the tables. What I found about the way the numbers go up has been

Assessment of Extended Tasks

mentioned earlier. When I tried using more than 2 counts the results seemed to be the same and fitted in with my patterns or rules about two counts.

One of the things I didn't cover was when it was impossible to find any whole numbers above 0 which would fit. This happened when there was two number of which one was a multiple of the other with a completely different remainder. As there are no numbers there is not much you can do with them.

I found out that in some cases you can predict what the last digit is going to be. In the last example, 2r1 means it must be an odd number and 5r0 means to divided by 5 exactly, the numbers must all end in 5 or 0. As 0 isn't an odd number, the numbers must all end in 5.

This doesn't happen every time. But I did find that there was a pattern in the tables of first numbers.

	3r0	3r1	3r1
5r0	0	10	5
5r1	6	1	11
5r2	12	7	2
5r3	3	13	8
5r4	9	4	14

	4r0	4r1	4r2	4r3
5r0	0	5	10	15
5r1	16	1	6	11
5r2	12	17	2	7
5r3	8	13	18	3
5r4	4	9	14	19

The numbers go from 0 to (in this case) 19 in order in a pattern. The numbers go from 0 diagonally down from the top corner.

	2r0	2r1
3r0	0	
3r1		1
3r2		☐

Then you want the next number down at ☐ but as it goes off the table you have to look horizontally across for the 2. Then the 3 goes off the table so you look vertically upwards from what the 3 should be. The 4 also goes off the edge of the table so you must look horizontally across find it and then as usually diagonally down till you reach the bottom and the table is finished.

	2r0	2r1
3r0	0	3
3r1		1 ← 4
3r2	2	

	2r0	2r1
3r0	0	3
3r1	4	1
3r2	2	5

I thought that when all the first numbers are prime numbers the numbers go up in those numbers multiplied together, but when those number are not all prime numbers, the numbers go up in half this multiple. But another case cropped up and put me off the trail:

3r1
6r4   10, 16, 22, 28, 34, 40, 46

This time the numbers should go up in a third of 3 × 6

$3 \times 6 = 18$ $\frac{1}{3}$ of 18 = 6

To test this rule we tried some more examples:

4r1   4r2   5r2
<u>8r5</u>   <u>8r6</u>   <u>10r7</u>
13,21,29   14,22,38   17,27,37

The rule seemed to be correct. If the larger number is a multiple of the smaller number, then the numbers always go up in the multiples of the two numbers divided by the smallest number, e.g.

     3r1
              13,22,31,40,49,58
     9r4

Nine is a multiple of three

    3 x 9 = 27   27 ÷ 3 = 9

To multiply two numbers together and divided by the smallest always leaves you with the largest number.

If the numbers have no connection, they go up in the multiples of the two numbers and if one number is a multiple of the second, the numbers go up in the largest number.

Next I tried using three or four counts. Here are a few examples of them:

    3r2
    4r1   53,113, 173, 233 .....
    5r3

3, 4 and 5 have no connection so the numbers go up in 60s - 3 x 4 x 5

    2r0
    3r1
    4r2   58, 118, 178, 238 .....
    5r3

As 4 is a multiple of 2 the numbers go up in 2 x 3 x 4 x 5 ÷ 2
In this next example, the number have no connection so go up in 2 x 3 x 5 x 7 = 210

    2r1
    3r2
    7r4   95, 305, 515 .....
    5r0

"The original problem was:

    "When a boy counted his sweets in fours, he had two left over; when he counted them in
    fives, he had one left over. How many sweets did he have?"

I thought about this problem and wrote it out like this:

    4r2
              6, 26, 46, 66, 86 ...
    5r1.

The first number of sweets he could have had was 6 but after that there were many more numbers which continued to go up in 20s. I noticed that

$$5 \times 4 = 20$$

and the numbers went up in 20s. To check this rule I tried a couple more examples:

    3r2   8r4
        gives 2, 14, 26, 38, 50, 62   gives 28, 100, 172, ...
    4r2   9r1

    3 x 4 = 12 the numbers go up in 12s   8 x 9 = 72   It goes up in 72s.

We looked at the problems:

1. How do you find what the starting number is going to be?
2. How many do they go up in each time?
3. What happens with three or more counts?

First I tried to find what the starting number is going to be. But at first it didn't seem a very easy problem.

There were exceptionally cases when 4r2 and 5r1 would add up to 6 the first number, but this rarely worked.

For a while I left this and went on to the problem of what the number went up in. I had found that the first numbers, when multiplied together made the amount the numbers went up in. But then I found these two cases

                         4r2,
gave 22, 34, 46
                         6r4

                         4r3,
gave 19, 31, 43,
                         6r1

This made us think that the numbers always go up either by the multiples of the two numbers or half that amount.

When this example came up:

                2r0
                          6, 10, 14, 18, 22,
                4r2

Problems of grading such work depend somewhat on whether all students in the group have been given the same starter, or have selected or developed the question for themselves. The more natural situation with work of this kind, in or outside school, is that the question has provoked the person's curiosity, making for a self-motivated creative activity. This is an important aspect to preserve; indeed, it is essential if we want students to share the experience of genuine scholarly activity, of the pursuit of some new aspect of knowledge for its own sake—or rather for their own sake. In these circumstances one has to construct a grading scheme that is general rather than specific to the problem. A number of such profile schemes have been developed in recent years for the assessment in the 16+ public examination (the GCSE) of individual assignments, like those discussed here. Two such schemes are shown on the following pages, one of them in the version adapted for students' use. Such schemes are now in extensive use. They provide quite well for the assessment of work on a wide variety of topics, and they make it possible to display to the students the set of general criteria of quality on which their work will be assessed. With these schemes an essential part of the procedure is a moderation process, by which samples of work graded by different people are checked by others.

"Starters" and General Assessment Schemes

We now give two examples of practically oriented starters, together with some assessment schemes of the broad kind described above. They come from a set of 10 books designed to support school-based assessment of individual work (Shell Centre/MEG, 1989). These books provide less classroom support than other Shell Centre materials referred to earlier, but they suggest many examples of tasks (with brief teaching notes) and provide substantial guidance on the assessment (illustrated by student work at different levels with a commentary on each problem).

The "Celebration" (Example 3) is an example of a "starter" that the student might receive for one such task (in fact, this runs on for two further pages of specific suggestions of things to investigate in the early stage). "Orienteering" (Example 4) is another specific suggestion from another extended task, on maps (A is a hill top). The examples from the board's grading schemes (Figures 1-3) give an idea of the type and level of guidance provided for teachers by this board. The illustrative examples of graded student work in the books have been found to provide helpful, perhaps essential, support to their inter-

Example 3. The celebration.

As you complete this task you will be involved in planning something like a birthday celebration for a group of friends. You can go about this in any way that you like.

You do not have to consider a birthday celebration, it can be anything that you would like to organise, or you feel would be useful to you in the future. You may choose to go in any direction that you wish. Basically you can plan or organise anything that interests you, for whatever reason and in any way.

If possible, try to link it to something that you are going to have to do in real life, whether it is in the near future or just a possibility in the long term. You will be able to gain a lot from this experience, even if it is only a dummy run at organising something. You may be lucky enough to have the chance of putting your plan into action.

Don't forget to record all of your ideas and decisions as you go along. These need to be discussed in your final report. Your report should outline your problem and how you tackled it. It should include any information you collected, and decisions you made, comparisons of alternatives, calculations made, and many more things.

© Shell Center for Mathematical Education/Midland Examining Group 1989.

pretation. (They are unfortunately too long to include even small extracts here.) The last example (Figure 4) shows a similar approach that was developed by the California Assessment Program (EQUALS, 1989).

Example 4. Orienteering.

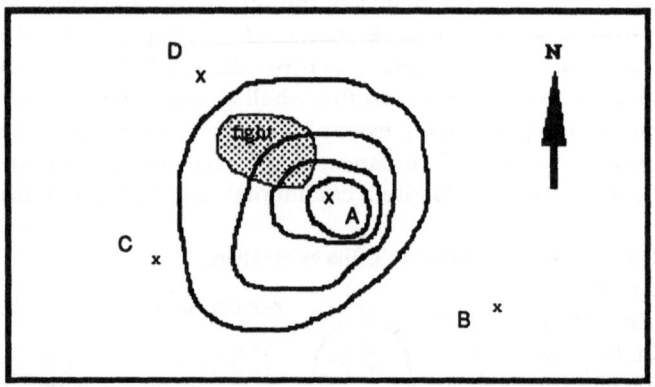

ORIENTEERING

Orienteering is sometimes described as *the thinking sport*. Orienteers have to pass through each check point, and complete the course in the *shortest time* possible. It is a sport which is growing in popularity, because it demands both physical and mental skills.

Map reading is very importnat. Orienteering maps show contour lines, and orienteers need to learn how to imagine what the land is like from looking at these contours.

Orienteering maps also show what is growing on the land and whether you can run, walk, or have to fight it.

* Try to describe the journeys of three orienteers who travel to A from B, C, and D.

* Describe the route you would take to travel from B to D.

* You may find it interesting to investigate a local orienteering course and to describe your experiences.

© Shell Center for Mathematical Education/Midland Examining Group 1989.

Figure 1. Scheme of assessment for course work component.

CLASSIFICATION	MAXIMUM MARKS	GUIDANCE FOR MARKING
OVERALL DESIGN AND STRATEGY	4	• Do you have a definite problem that you are looking at? • Have you been following your own ideas? • Have you asked and answered your own questions? • Have you tackled your problems in a suitable way?

MATHEMATICAL CONTENT	4	• Have you used a suitable range of mathematics? • Have you developed the mathematics as you have gone through the work?
ACCURACY	4	• Is your work accurate? • This can be calculations, drawings, graphs, practical work, collecting data, and conclusions.
CLARITY OF ARGUMENT AND PRESENTATION	4	• Have you explained in your report what you were doing at all times? • Have you explained the link from one stage to another? • Have you an introduction and conclusion? • Have you used mathematical tables, graphs, diagrams, language, symbols, etc.?

Figure 2. Summary profile.

MATHEMATICS SUMMARY PROFILE

PUPIL'S NAME.......................... FORM.............. GROUP.............. DATE...................
MODULE TITLE..

ASSESSED COURSEWORK

(a) *Planning*

Foundation: requires assistance to write a plan to tackle the first stage of the task and to choose appropriate equipment.

Intermediate: can plan the first stage of the task, recognizes further stages but requires some assistance to tackle these. Can select useful equipment and information.

Higher: can write a plan, which is capable of modification, to deal with all stages of the task.

(b) *Carrying out the task*

Foundation: can follow the plan carrying out some measurements and calculations accurately. Can recognize familiar patterns, sometimes with assistance.

Intermediate: can make reasonably accurate measurements and calculations and write rules from the patterns observed with some help.

Higher: can make accurate measurements and calculations and write algebraic rules from patterns observed. Can predict results, check their accuracy and amend if necessary.

(c) *Communication*

Foundation: can write about the work done using tables, graphs, diagrams and calculations where appropriate with some observations. Can talk about the work.

Assessment of Extended Tasks

Intermediate: can summarize the work in a logical sequence using techniques including tables, graphs, diagrams and calculations. Can explain the method used and results. Can discuss ideas used in the task.

Higher: can give a detailed explanation of the reasoning behind each stage of the task and results using a wide range of mathematical techniques. Can discuss fluently the ideas used in the task and possible implications.

PUPIL'S COMMENT: to comment on my most satisfying piece of work, any difficulties experienced during the module and suggestions for improvement.

TEACHER'S COMMENT (to be discussed with pupil. May include comments about working with others, ability to complete the tasks, homework and problem solving).

Figure 3. Internal assessment sheet.

TASK: WHY ARE WE WAITING? GCSE INTERNAL ASSESSMENT SHEET

CLASSIFICATION	MAXIMUM MARKS	GUIDANCE FOR MARKING
IDENTIFICATION OF TASK AND SELECTION OF STRATEGY	5	5 - Shows clear understanding of the principles involved by defining and classifying a board range of queueing systems. Clearly identifying the areas for investigation and the appropriate questions to be answered. 3 - Shows understanding of the task by defining and classifying a selection of queuing systems and showing some appreciation of the effect of changing appointment intervals. Identifies sufficient areas for investigation and deals with some of the appropriate questions to be answered. 1 - Shows poor understanding of the task: few examples of queueing systems and shows little or no appreciation of the effects of changing appointment intervals. Identifies insufficient areas for investigation and does not deal with the appropriate issues and questions to be answered.
IMPLEMENTATION AND COMMUNICATION	10	10 - Generates and processes data accurately. Applies sound reasoning in interpreting data and recommends a viable system. 6 - Generates and processes data with few errors. Recommends a viable system but with incomplete reasoning or explanation. 2 - Generates inaccurate or incomplete data and makes no recommendation or provides a recommendation which is not supported by sound reasoning or explanation.
INTERPETATION AND COMMUNICATION	5	5 - Selects an appropriate and clear method of recording results, with effective use of mathematical language and notation, diagrams, lists and tables. States results achieved and supports results and recommendations by clear explantions and reasoning. 3 - Selects an appropriate method of recording results, with adequate use of mathematical language and notation, diagrams, lists and tables. States results achieved and supports results and recommendaitons by some explanation and reasoning. 1 - Makes limited or no use of appropriate methods of recording results and draws few or no valid conclusions.

Figure 4. EQUALS Scoring rubric.

General Scoring Rubric for Open-ended Questions
Used for Grade 12 CAP questions

Please Note: For each individual open-ended question, a rubric should be created to reflect the specific important elements of that problem. This general rubric is included only to give examples of the kinds of factors to be considered.

Recommendations: Sort papers first into three stacks: Good responses (5 or 6 points), Adequate responses (3 or 4 points), and Inadequate responses (1 or 0 points). Each of those three stacks then can be re-sorted into two stacks and marked with point values.

Demonstrated Competence

Exemplary Response ... Rating = 6
Gives a complete response with a clear, coherent, unambiguous, and elegant explanation; includes a clear and simplified diagram; communicates effectively to the identified audience; shows understanding of the open-ended problem's mathematical ideas and processes; identifies all the important elements of the problem; may include examples and counterexamples; presents strong supporting arguments.

Competent Response ... Rating = 5
Gives a fairly complete response with reasonably clear explanations; may include an appropriate diagram; communicates effectively to the identified audience; shows understanding of the problem's mathematical ideas and processes; identifies the most important elements of the problems; presents solid supporting arguments.

Satisfactory Response

Minor Flaws but Satisfactory ... Rating = 4
Completes the problem satisfactorily, but the explanation may be muddled; argumentation may be incomplete; diagram may be inappropriate or unclear; understands the underlying mathematical ideas; uses mathematical ideas effectively.

Serious Flaws But Nearly Satisfactory ... Rating = 3
Begins the problem appropriately but may fail to complete or may omit significant parts of the problem; may fail to show full understanding of mathematical ideas and processes; may make major computational errors; may misuse or fail to use mathematical terms; response may reflect an inappropriate strategy for solving the problem.

Inadequate Response

Begins, But Fails to Complete Problem ... Rating = 2
Explanation is not understandable; diagram may be unclear; shows no understanding of the problem situation; may make major computational errors.

Unable to Begin Effectively ... Rating = 1
Words do not refelct the problem; drawings misrepresent the problem situation; copies parts of the problem but without attempting a solution; fails to indicate which information is appropriate to problem.

No Attempt ... Rating = 0

NUMERACY THROUGH PROBLEM SOLVING

This project development provides an example of a more tightly controlled and varied assessment scheme with similar objectives. The project has as its main aim the teaching and learning of numeracy, that is, the ability to deploy mathematics and other skills in tackling problems of concern or situations of interest in everyday life. It supports two important recommendations in the Cockcroft Report *Mathematics Counts* (Cockcroft, 1982):

Most important of all is the need to have sufficient confidence to make effective use of whatever mathematical skill and understanding is possessed, whether this be little or much (paragraph 34).

Our concern is that those who set out to make their pupils "numerate" should pay attention to the wider aspects of numeracy and not be content merely to develop the skills of computation (paragraph 39).

The resulting scheme is modular and includes both curriculum and assessment materials, although the former can be used separately. Each module has a range of practical targets that require the students, usually working in groups, to tackle a problem "for real." This means that the students themselves are responsible for planning, designing, organizing, and choosing within the module theme and have to live with the consequences of their decisions. The teacher's role, meanwhile, becomes that of advisor or counsellor, chairperson, encourager, and clarifier.

Five modules have been developed:

- "Design a Board Game," in which each group of students designs and produces a board game that can be played and evaluated by other members of the class.

- "Produce a Quiz Show," in which students, working in groups, devise, schedule, run, and evaluate their own classroom quizzes.

- "Plan a Trip," in which students plan and undertake one or more class trips, and possibly some small group trips.

- "Be a Paper Engineer," in which students design, make, and evaluate three-dimensional paper products, such as pop-up cards, envelopes, and gift boxes.

- "Be a Shrewd Chooser," in which students research and provide expert consumer advice for clients in their class.

The modules are published by Longman Group (Shell Centre/JMB 1987-89); details are obtainable from the Shell Centre.

Many contexts, designed to cover the areas of planning and organizing, designing and making, and choosing, were considered and tried in the early stages of development to see which led to the best balance of classroom activities and learned skills. Those that were chosen all have a practical outcome, are interesting and relevant to the students' present circumstances, and require the use of a wide range of skills while not making unreasonable demands on classroom or school organization. We found that practical situations with an element of fantasy, stimulating the students' imagination, seem to work best; for example, in Produce a Quiz Show, students tend to see themselves as television performers and producers. This corresponds with our observation that people best develop the strategic skills we seek in the course of solving problems that are realistic, stimulating, and within their capabilities (Binns et al. 1987, 1989).

The learning activities typically fall into four stages.

Stage 1. Understanding the problem: An exploration and critical review of existing examples to help students get to know the variables and possibilities, to identify strengths and weaknesses, and to suggest improvements.

Stage 2. Making a rough plan: Generating ideas (brainstorming), sorting them out and making a rough plan for their own solution to the problem.

Stage 3. Carrying out the plan: Detailed planning and design, followed by implementation.

Stage 4. Evaluating the outcome.

The strategic skills generally needed in these processes, with some indication of the criteria for success and for quality, are

- understanding the problem, both general ideas and details;

- following instructions precisely;

- distinguishing between essential constraints and desirable features;

- generating and listing viable possibilities (brainstorming);

- developing a rough plan, including reviewing the prepared suggestions, reaching and recording agreed decisions; maintaining a broad level of description and avoiding excessive detail, identifying needed information and materials, making estimates of quantity and cost, and describing, testing and evaluating the plan;

- identifying faults;

- correcting faults;

- making the final plan, product, and/or detailed instructions with comprehensiveness, accuracy, clarity, and quality; and

- testing and evaluating the plan or product comprehensively.

In addition to a specific tactical realization of this set of general strategies in the domain of the problem, each module brings into play some knowledge and skill specific to the topic. In the assessment of students' attainment, one has to take account of these specifics and to recognize that performance is strongly context-dependent. It is in fact a question for research (which we plan to carry out) to determine what aspects of strategy acquisition carry through from one module to another; informal indications from the development/research work suggest there is significant transfer. We shall illustrate the scheme by outlining the assessment procedures for "Plan a Trip," adding brief references to other modules.

The "Plan a Trip" Module

Most school trips are organized and run by teachers, and the educational objectives are met by activities that take place at the destinations. The students therefore learn little about the processes that go into the original planning. In this module, however, the students take on the responsibility for planning, organizing, and going on a trip during school time. The destination is largely irrelevant, as the main objective is to increase the students' ability to make, implement, and learn from their own decisions and thus feel more confident when moving beyond the immediate surroundings of their own home, school, or town. The planning skills developed may also be utilized within other contexts. As outlined above, the planning process is considered in four stages.

Stage 1. Looking at trips. In a card game simulation, groups undertake and record imaginary trips, encounter problems and errors of judgment, then seek to correct them by better planning.

Stage 2. Making rough plans. Groups share ideas of possible places to go and produce a leaflet explaining these ideas. The class then work together to reach a decision on the best destination and look at possible means of transport.

Stage 3. Making detailed plans. The class lists the tasks to be done and shares them out. Students working in groups undertake the preparatory tasks that need to be done before the trip can take place.

Stage 4. Going on a trip and evaluating it. The trip now takes place, and afterward, the students reflect on what happened.

The classroom materials consist of student booklets which guide them through the various stages, masters for photocopying "structured stationery" of various levels to help the planning process, the "card game" packs and a teacher's guide covering objectives, the classroom activity, developing the mathematics involved, and assessment.

Assessing the "Plan a Trip" Module

The assessment procedures are designed to verify that a student can, in the context of planning a class trip, meet the following criteria: (i) evaluate a plan and identify faults in it (including expenditure of money and time); (ii) generate lists of alternatives; (iii) devise a satisfactory rough plan (including sensible costings and time schedules); (iv) use and describe a decision-making process, obtaining and interpreting information from a telephone directory, aural and written timetables, a street map, and a route map; (v) identify omissions in given information; (vi) place given jobs in a logical order; (vii) complete a clear and comprehensive final plan; (viii) take an active part in the planning process; and (ix) evaluate the plan that has been implemented.

Several methods of assessment were considered, including the following:

Method 1. Teachers observing and/or holding discussions with individual students while they work, and recording evidence of achievement with a checklist of some kind. This method is often unsatisfactory for the following reasons:

- Each student's performance is affected by the nature of the group he or she is in. A student may take an active role in one group, but be overshadowed by a dominant personality in another.

- When groups are working cooperatively, group members tend to adopt complementary roles. This means that the teacher is unlikely to see every student carrying out every task.

- Teachers often find it difficult to reconcile/combine the roles of assessor, helper, and supervisor in the limited time available.

Method 2. Written assessment tasks, administered during or after the work. This has some disadvantages:

- Some skills cannot be assessed in this way; and

- Many students have difficulty in expressing themselves fully in writing, leading to responses that fail to reflect their true abilities.

Method 3. Students carrying out self-assessments, describing those aspects of the work in which they have made particular contributions. While invaluable insights may be gained using this mode of assessment, its subjective nature makes it unsuitable for our purpose.

We finally decided on written tasks supplemented with teacher-observation. During or at the end of each module stage, we offer a number of short "coursework" tasks designed to suit 90-95 percent of all 14- to 16-year-old students. These are administered by the teacher in as supportive a way as possible. For students who have successfully completed the coursework component of a module, there is an opportunity to demonstrate that they have retained the skills they have learned and can transfer them to other situations within the same context. This is done in the external examination component. These are considered in more detail below.

The coursework assessment component

At convenient points in the work, a group of students is split up and its individual members are invited to complete a number of short tasks. These tasks closely reflect the general planning processes the student has recently followed. Students with learning difficulties may need help from their teacher in reading the tasks or recording their answers, but they are usually able to tackle the tasks successfully. Example 5 is designed to assess criterion (iv), for example, and is administered at the end of Stage 2, in which the students have themselves been considering the best destinations for a class trip.

Example 5. The vote.

Six people are planning a trip. Six different places have been suggested.

a) The ice rink
b) Zoo
c) Bowling alley
d) A castle
e) Snooker hall
f) Swimming pool

In order to choose between these places, they decide to vote.
Each person is given a list, and they write down their preferences.

This is what they write:

```
Sanjay
a) Ice Rink    6th choice
b) Zoo         1st choice
c) Bowling     3rd choice
d) Castle      2nd choice
e) Snooker     4th choice
f) Swimming    5th choice

John
a) Ice Rink    1st Choice
b) Zoo         2nd Choice
c) Bowling     5th Choice
d) Castle      4th Choice
e) Snooker     3rd Choice
f) Swimming    6th Choice

Claire
a) ice Rink    6th choice
b) Zoo         5th choice
c) Bowling     1st choice
d) Castle      2nd choice
e) Snooker     4th choice
f) Swimming    3rd choice

Mike
a) Ice Rink    4th choice
b) Zoo         6th choice
c) Bowling     5th choice
d) Castle      3rd choice
e) Snooker     1st choice
f) Swimming    2nd choice

Elaine
a) Ice Rink    6th choice
b) Zoo         3rd choice
c) Bowling     2nd choice
d) Castle      1st choice
e) Snooker     4th choice
f) Swimming    5th choice

Jenny
a) Ice Rink    6th choice
b) Zoo         5th choice
c) Bowling     4th choice
d) Castle      2nd choice
e) Snooker     3rd choice
f) Swimming    1st choice
```

Which place would be the best choice for their trip? _____

Explain how you get your answer:

_____
_____
_____
_____

In Stage 3, groups in the class carry out a variety of planning tasks as they make detailed preparations for their trip. Here, it is natural to find that each group is working on different planning tasks. While one group is, for example, using the telephone to explore the possibility of hiring a coach, another may be finding out costs and times of trains and buses. Thus, the sharing of tasks prevents each student from demonstrating every skill. The assessment tasks help to overcome this problem. In this stage, a "circus" approach is used with individual students tackling a variety of tasks in their own time in any order. These tasks include the interpretation of street maps, bus route maps, timetables (including a prerecorded telephone "talking timetable"), identifying omissions in a "letter to parents," and the logical sequencing of jobs (criteria (iv) to (vi); see Example 6).

Example 6. Sorting out jobs.

> You will need a set of 8 leisure centre job cards.

A class have chosen to go to a Leisure Centre for the morning.
Some want to go swimming, while others want to play table tennis, squash, badminton or snooker.
There are 3 leisure centres within a bus ride from their school.

They have written all the jobs that need to be done on cards.

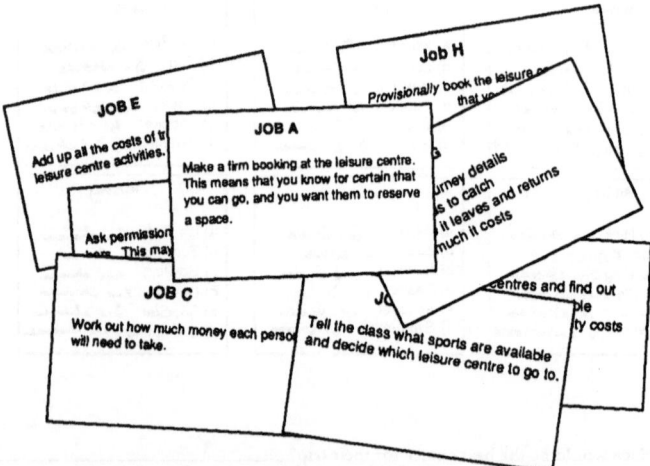

- Try to sort your job cards into order of priority.
  (Which job needs to be done first? second? third? and so on.)

- When you have done this, write down the jobs in order in the boxes below.
  (The first one has been done for you.)

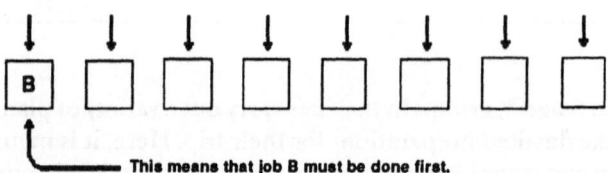

During the course of their work, students are also asked to complete "Brainstorming," "Rough plan," "Final plan," and "Evaluation" sheets that refer to their own trip. As well as providing an impetus for group discussions, these tasks are used to assess individuals' understanding of the work of their groups.

On the basis of good performance on the coursework tasks (including taking an active part in the planning of the trip), students are awarded a Basic Level pass. Nearly all students (over 90 percent) are capable of passing at this level.

The external examination component

There are two examination papers for each module, set and marked externally to the school, one at Standard Level and one at Extension Level. About 80 percent of students should be able to pass at Standard Level (where the examination tasks are fairly closely related to the module context). Extension Level demands that students show the ability to transfer skills to more complex, less closely related tasks; 40 percent of students should be able to pass at this level.

Examples 5 and 6 show part of a Standard Level examination paper for the Plan a Trip Module. Much of it requires students to discover and correct mistakes in someone else's plan for a trip. In devising such papers, we try to meet the following criteria:

- The mathematical skills (that is, timetable reading, money calculations, or geometrical skills) must be tested in real and relevant contexts.

- The tasks must be coherent and unfragmented.

- The data used must be reproduced as they appear in everyday life. (Numbers are not cleaned up, and redundant information is included.)

- The tasks should, if at all possible, be a valid and enjoyable educational experience in their own right.

In addition, the marking procedures must be reliable and efficient in use, and this sometimes means that a few compromises have to be made in achieving these aims. The setting of such examinations involves much development, and piloting, and thus the expense of a few highly skilled person-weeks, but we feel it is necessary if real problem solving is to become a part of every classroom.

In the remainder of this chapter we focus on the two modules, "Be a Paper Engineer" and "Be a Shrewd Chooser." In particular, these two modules illustrate how the teachers are provided with a secure structure within which to work, while the students are still allowed to retain "ownership" of the fairly open problems being tackled.

The "Be a Paper Engineer" Module

Geometry, at the secondary level, is almost always presented as a static, two-dimensional subject (for a splendid exception, see de Lange, this volume). On the few occasions when we leave this "Flatland," it is usually only to construct a few polyhedra or to perform some sterile, abstract, technical exercises. Students are rarely given the opportunity to explore a rich, three-dimensional environment. There are, however, some indications that this situation is improving.

Recently, in England, national criteria have specified that all examinations leading to a GCSE award in mathematics must assess practical work. Some examination boards are now encouraging students to present extended pieces of coursework that involve a "practical geometry" task such as designing a cardboard box to hold five tennis balls. (Even these tasks sometimes give the feeling that mathematics is being "dragged in" in an artificial way. For example, students are asked by their teachers to calculate the volume of air in the box after the tennis balls have been introduced.)

In the "Be a Paper Engineer" module, we offer students the opportunity to design and make a product from paper or thin card and then produce a kit containing full instructions so that someone else can recreate it. They may then, perhaps, set up a small business enterprise based on the products. The design process is again arranged into four stages:

Stage 1. Looking at examples. Before designing an original product, it is sensible to look at a few examples of existing products to stimulate ideas and become familiar with the techniques involved. In the module package, we provide a collection of thirty-two pop-up cards and gift boxes that may be photocopied and given to groups of students. After making these, students are asked to find more examples from home,

Example 7. Two examples from Stage 1.

8. A Barbecue Invitation

and then reflect on structural differences using a classification game. Example 7 illustrates the kind of products provided.

Stage 2. Exploring techniques. In this stage students explore and develop, in more depth, some of the techniques that have already been introduced. They may, for example, discover what happens when they change the positions and angles of both cuts and fold lines when making popup cards, and devise theorems that must hold true if the products are to function properly. Students are expected to keep full written records of all their discoveries, including failures (Example 8).

Stage 3. Making an original prototype. In groups, students now brainstorm ideas for their own products, then work individually to prepare rough prototypes. This process involves a combination of the techniques and theorems developed in stage 2 and trial and improvement, where ideas are successively refined until satisfactory results are achieved (Example 9).

Stage 4. Going into production. Students now attempt to draw accurate templates for their products, accompanied by full instructions that enable other people to recreate them. These kits are now photocopied and tested by other students in the class.

Example 8. Extracts from the Student's Booklet for Stage 2.

Stage 2
Exploring techniques

In this stage you will work mostly on your own using some of the Exploring techniques sheets.

You will try some

- investigations, which ask you explore what happens when you fold, cut or stick things in different ways,
- challenges, which show you pictures of finished articles that you can try to make.

You will also

- keep a record of everything you discover for use later on, when you come to design your own item.
(Page 14 describes how to do this in more detail.)

KEEPING A RECORD

Record all you do in an exercise book or in a folder. This will help you to remember what you've found out.
For example:

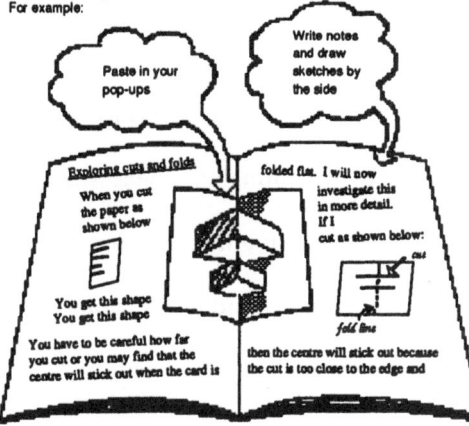

Keep everything you make -- even the things that go wrong.

Make a note of
- what you tried to do,
- what happened,
- what you learnt.

You'll need all this information for Stage 3.

Assessment of Extended Tasks

Example 9. Extracts from the Student's Booklet for Stage 3.

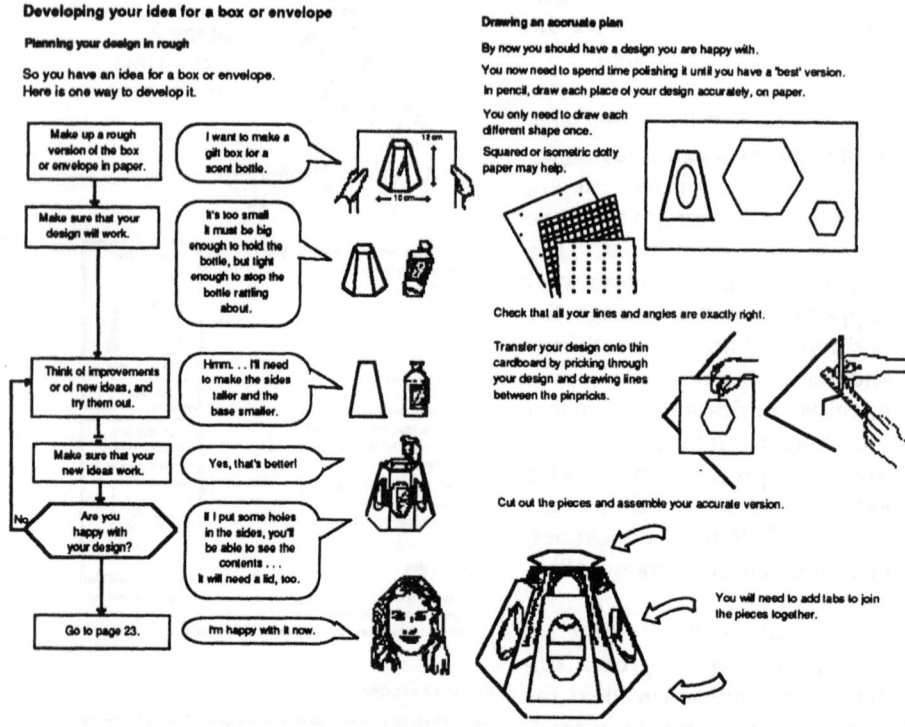

During these four stages, students call on a variety of mathematical and technical skills. These are likely to include

- understanding and using angles and symmetry;
- estimating and measuring lengths and angles;
- following instructions presented in words and diagrams;
- making and testing conjectures, explaining and proving;
- visualizing;
- creating three-dimensional objects from two-dimensional representations;
- drawing two-dimensional representations of three-dimensional objects;

- designing, making, and using levers and linkages;

- writing clear, concise, and complete instructions using diagrams or photographs where appropriate; and

- perspective drawing.

The level at which such skills are deployed depends both on the nature of the product being made and on the ability of the student. Clearly, however, there is a challenge here for students at every level of ability.

The "Be a Shrewd Chooser" Module

Students frequently face and make consumer decisions. Such decisions are often made on impulse, with little appreciation of the many factors that could be taken into account to help the student make a better choice. In mathematics lessons we often simplify the situation to comparisons of prices per unit weight, a model that rarely applies to real situations. In this module students reflect on how people really make consumer decisions, and they produce consumer reports to inform better choices. The material is again presented in four stages.

Example 10. Extract from the Student's Booklet for Stage 1.

The Shrewd Chooser Radio Show
Part 3: Looking at a consumer report

Three students have written a consumer report called 'Drinking Orange!'
You will need to refer to it as you listen to the tape.

In the radio programme the students describe how they
· visited shops and looked in the public library.
· carried out a classroom survey.
· carried out some classroom 'taste' experiments.

After hearing about each section of the report, you will be asked to
· describe exactly what the students did.

Stage 1. Learning from experience. Students listen to a "radio show," recorded on audiotape, which contains a number of interviews with people who have just purchased different items, and an interview with two students who have produced a report on choosing orange drinks. This is supported by a written copy of the report which students can discuss critically. These activities enable students to consider the factors and methods involved in decision making and the processes and difficulties involved in writing a consumer report (Example 10).

Stage 2. Preparing the research. In this stage students begin to plan their consumer research based on items of their own choice. This choice has to be restricted to ensure that a rich variety of classroom activities can take place, using the following criteria:

- People in the class must have some experience of choosing the item.

- The item must be cheap enough for samples to be brought to school.

- It must be possible to carry out tests/experiments on the item to measure its quality.

Suitable items are confectionery, breakfast cereals or other foodstuffs, soft drinks, batteries, writing instruments and other small, frequently purchased items. The students then have to list their research aims and methods and prepare any tables and questionnaires that will be required for data collection.

Stage 3. Carrying out the research. Students now carry out the research they have planned. These are mainly surveys and experiments to discover, for example, whether more expensive products perform or taste any better than cheaper products. Students also need to choose appropriate ways of presenting data, to draw conclusions from their data, and to prepare a report. These activities give students the experience of using important statistical concepts which may later be applied to other situations, such as sampling techniques, graphical representation, graphical interpretation, and measures of central tendency. Students may have experienced all of these but are unlikely to have been asked when and where to use them.

Stage 4. Presenting and evaluating the reports. In this final stage all the written reports are circulated around the other groups in the class, and any group wishing to make an oral presentation does so. The reports are then evaluated by the class, and each group is given an opportunity to improve its own report taking these comments into account.

During this module, students are likely to call on a variety of mathematical and technical skills, including

- devising questionnaires and conducting interviews;

- designing and carrying out experiments;

- analyzing and presenting data in various ways;

- selecting and using sources of information;

- interpreting data and presenting clear and reasoned recommendations; and

- handling money, other everyday measures, percentages, statistics, graphical representation, ratio, and proportion in real-life contexts.

SOME ASSESSMENT ISSUES

The scheme has been designed to assess strategic skills (for example, the ability to plan and design) as well as the more traditional technical skills (for example, the ability to measure accurately). The procedures are designed to assess whether or not a student can satisfy a number of criteria. In Figure 5 we list some of these criteria as they appear in the two modules under discussion. (Gillespie et al., 1989, give details about criteria and other assessment issues relating to the Plan a Trip module.)

Figure 5. Assessment criteria for two modules.

The module criteria - students are given the opportunity to show that they can...

	in *Be a Paper Engineer*		in *Be a Shrewd Chooser*
i	follow instructions	i	identify important factors and methods involved in decision-making
ii	cut, fold and glue accurately to assemble a 3-dimensional product		
		ii	obtain and interpret information from oral interviews
iii	make a 3-dimensional object from a 2-dimensional representation		
		iii	obtain and interpret information from tables and graphs
iv	recognize structural features of a design		
v	draw a 2-dimensional representation of a 3-dimensional product	iv	identify possible research aims
		v	select appropriate research methods
vi	give a reasoned explanation for design features	vi	devise suitable methods for the collection and organization of data
vii	identify and correct design faults		
		vii	present a summary of research data in a clear, organized way
viii	develop an existing idea for a paper product		
		viii	draw sensible conclusions from a collection of research data
ix	generate possibilities for a design with original features		
		ix	take an active part in compiling their own reports
x	draw a design to an acceptable degree of accuracy		
		x	evaluate a report and suggest improvements to it.
xi	construct a prototype with original features		
xii	devise instructions to enable someone else to make the product		

Such criteria assist in providing a useful profile of relative strengths and weaknesses, but they have little absolute meaning without specifying the context, the frequency of success, the amount of help that was given, the distance of transfer from the student's previous experience and the mode of response (written or oral). As the students work through the modules in this series, they are likely to demonstrate similar strategic skills in a variety of different contexts. This enables the teacher to make more general statements about the students' progress.

Example 11. A basic-level task from Be a Shrewd Chooser.

Tea bags

Two students have done a small shopping survey to find out the different kinds of tea sold at a supermarket.

They have made notes on scraps of paper.

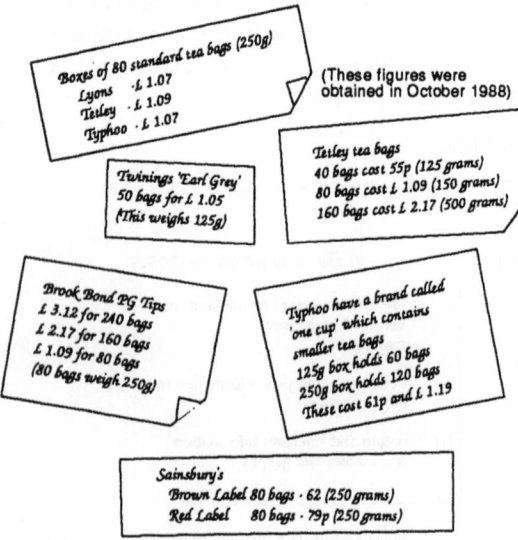

Make a table showing all this information as clearly as you can. Which way of buying tea is most economical?

What other factors would you take into account when buying tea?

Each module is accompanied by a collection of short, basic-level assessment tasks, some of which are completed in the normal course of the work and some of which are administered at the end of the appropriate stage in the module. These are intended to be accessible to the vast majority of students. They are supplemented by two written examination papers at Standard and Extension levels, which about 80 percent and 40 percent of students respectively should be able to pass (although these are not norm referenced). One (or both) of these papers is sat during the term following that in which the student completed the module. On the basis of these assessments, a student may be awarded a short Statement of Achievement for each module. This lists the criteria that a student has satisfied. If a student has been successful in three or more modules then he or she may be awarded a Certificate in Numeracy through Problem Solving which gathers together in a more generalizable form the entire collection of criteria that have been satisfied. This assessment scheme has proved popular with the schools that have been involved.

In Examples 11 and 12 we give examples of a basic-level task from "Be a Shrewd Chooser" and part of a standard level examination paper from "Be a Paper Engineer."

Example 12. A basic-level task from Be a Paper Engineer.

A Pop-up castle

The picture shows a Castle pop-up card which Ann and Steve are making.

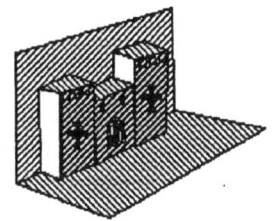

When the card is closed,

- the castle must not stick out beyond the edges of the card,
- the card must fit into an 8 1/2 cm by 16 1/2 cm envelope.

1. Using a copy of the design, cut out and make the card. Now complete the design for the card shown below as accurately as you can. (Use ____ for outlines, ... for hill folds, - - - - for valley folds.)

2. Here are four instructions (A, B, C, D) to help someone else make the card from the design:

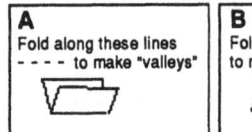 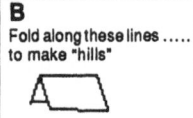 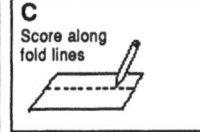 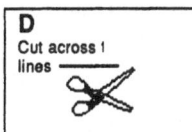

In what order should the instructions be?

3. Ann and Steve decide to change their design by making a bridge which leads to the door. Add this bridge to the card you have already made.

Postscript

The importance of group work: In one respect our thinking has moved on, spurred by the importance of the ability of people to work effectively in groups. We took the decision not to include assessment of group achievements (such as the Board Game design, for example) because of the evidence that a given individual performs differently according to the group he or she is in. We recognize that this omits a key element in the assessment, which we tried to cover in other ways, for example, through the ability to transfer to related tasks in the external examination component. We should now choose to give credit for group achievements to all the students involved in the group. Varying group membership reduces chance elements.

A final comment: The main approach that the Shell Centre has adopted in the developments described in these two chapters has been based on successive small-but-profound annual changes over several years. Experience seems to confirm its soundness. As well as the new kinds of task, it involves developing support that enables typical teachers to face and absorb the required extensions of their teaching style, effectively and enjoyably.

REFERENCES

References marked with an * contain detailed examples of tasks, student work, and grading schemes.

*Association of Teachers of Mathematics. (1987). *Mathematics for sixth formers*. Derby: ATM.

Binns, B.S., Burkhardt, H., Gillespie, J., and Swan, M.B. (1989). *Mathematical modelling in the school classroom*. ICTMA 3.

Binns, B.S., Burkhardt, H., and Gillespie, J. (1987). *Bottom-up numeracy*. ICTMA 2.

Cockcroft, W.H. (1982). *Mathematics counts*. London: HMSO.

EQUALS. (1989). *Assessment alternatives in mathematics*. Berkeley, CA: EQUALS.

Gillespie, J., Binns B., Burkhardt H., and Swan, M.B. (1989). *Assessment of mathematical modelling*. ICTMA 3.

*Shell Centre for Mathematical Education, Joint Matriculation Board. (1987-89). *Numeracy through problem solving: Design a Board Game, Produce a Quiz Show, Plan a Trip, Be a Paper Engineer, Be a Shrewd Chooser*. Harlow: Longman

Shell Centre/MEG [Midland Examining Group]. (1989). *Extended tasks for GCSE mathematics*. Shell Centre, University Park, Nottingham NG7 2RD UK.

7 Moving the System: The Contributions of Assessment

Alan Bell, Hugh Burkhardt, and Malcolm Swan

INTRODUCTION

In the two previous chapters, we discussed the principles and examples that are typical of some recent efforts to design external assessment that reflects curriculum objectives in mathematical education in a balanced way. Here, we want to review the processes of implementation and the roles of assessment in the dynamics of educational change, illustrated mainly from experience in the United Kingdom.

We have already noted the heuristic observation that the implemented curriculum is strongly influenced, perhaps even dominated, by the nature of any assessment procedures whose results directly affect the students and teachers involved. This is, of course, neither surprising nor accidental. Since public assessment represents an official measure of the achievement of students and (although there are other obvious important factors) the performance of their teachers, it is a brave teacher who will devote much time to aspects of mathematical performance for which no direct "credit" accrues. Indeed, some would regard it as irresponsible to do so, although there are groups of dedicated teachers who regularly include other elements of the curriculum that are not tested because of their perception of the educational benefits that these elements offer. Politicians have long recognized, and used, assessment as a powerful lever for putting pressure on those who work in the education system.

In such a climate, the hope that "teachers will not teach for the test," frequently expressed by both politicians and educationalists in England, seems futile. Indeed, as we have made clear, we regard it as an abdication of responsibility by test designers and those they serve, who should ensure that any assessment fairly reflects the objectives of the curriculum. Then teaching for the test leads to a balanced curriculum.

In Britain, the traditional posture of the examining authorities has contained two key elements: "We are the servants of the teaching profession," and "We don't test *that* but, of course, all good teachers do it." Most teachers do not see it that way; they see external examinations as something over which they have no influence and for which they must prepare their students as a matter of priority.

We should not exaggerate the sense of conflict that arises, since the examinations and the standard curriculum have been locked together symbiotically. When the education system is not in a state of change, there is no sense of cause and effect. In a climate of educational change, however, the situation did and does act as a strong brake on such change. Individual teachers are discouraged from trying to improve the curriculum in their classroom. Projects that propose change have their impact reduced. Indeed, it is a feature of all the British projects that have had substantial impact on secondary education (where assessment looms large) that new alternative examination syllabuses were introduced to match the new curriculum as part of the project.

Many creative innovations in mathematics teaching have been developed by teachers who recognized that the traditional offering to students was in some way seriously deficient. Their efforts at improvements, however, had to overcome the handicaps not only of a conservative environment in general, but of the particular and strong influence of the system of external assessment.

In Britain, the strongest component of this influence has been the public external examination at the school leaving ages of 16 and 18. There have been successful attempts on the part of development groups to gain official approval for elective modifications to these examinations so as to give credit for attainments in a broader and richer range of mathematical activities than the traditional short tasks and applications of techniques. These modifications have generally consisted of some change in the written questions (which, in Britain, are nearly always of an open response rather than a multichoice type), or of introducing a component of assessed "coursework" into the examination scheme to accommodate the innovatory material.

There have also been some experiments with more specific assessment of a wide range of general strategies (Bell, 1979). However, such schemes attracted only a small minority of schools until the advent of the General Certificate of Secondary Education (GCSE), first examined in 1988. The criteria for the GCSE (SEC 1985) require the assessment both of submitted assignments for all students and of practical and oral aspects of mathematics, as well as many innovations in the examination of other subjects. More recently still, the National Curriculum (DES 1989, NCC 1989) includes using, applying, and investigating mathematics among its attainment targets. This National Curriculum is to be tested at ages 7, 11, 14, and 16, partly by teacher assessment and partly by the nationally administered Standard Assessment Tasks. These tests will be used not only to assess individuals but also to monitor school performance. The current debate centers on the need for these tasks to be broadly based, particularly when so much depends on the results.

The pattern of influences in the United States is similar, though it is only in recent years that public assessment seems to have been recognized for its curriculum influence. Even now, in many states where assessments are used only as overall monitoring devices, with no specific reporting on individual pupils or teachers, the influence on the curriculum is not so great. However, the trend is in the same direction as in Britain, with more tests associated with increased pressure for accountability. The kind of improvements in the quality of the assessment that we discussed earlier are thus becoming increasingly urgent.

It is ironic that some U.S. states with excellent curriculum guidelines are still content that their students' progress be measured by tests that do not reflect those guidelines in any balanced way. One factor is surely that alternative tests are not yet generally available as choices for schools, school districts, or states. Our earlier discussion indicates why and how that situation should soon be remedied. Now we address the practical "engineering" issues of implementation, many of which also raise issues of principle.

HANDLING BROAD-SPECTRUM ASSESSMENT

First, we discuss the remarkably different processes that are used for the design and delivery of assessment in various countries. On the one hand, we have machine-graded multiple-choice tests, with tasks taking a minute or two at most for students to answer. On the other, there are now assessment packages containing a considerable variety of types of tasks (taking from a few seconds to many hours spread over several weeks) and of modes of response by the students (oral, written, and constructed as well as simply chosen). These latter types of complex tasks are largely graded by teachers (who may also, with their students, have devised some of the tasks), but sometimes they are also subject to a fairly complex monitoring procedures to ensure comparability of standards.

It is not surprising that those who are used to the former model should have doubts about the latter — on all sorts of grounds, including cost and fairness. Here we want to look at these issues, and the gains and losses of various approaches. As a preliminary global assurance on losses, we simply report that it is customary in Britain to deliver assessment of the second—broad-spectrum—kind in a way that commands public confidence and at a cost that is similar to that charged in the United States for multichoice tests. We shall look at the elements one by one.

The Design of Tasks

The range of practice here is astonishing. Public examination questions in Britain are generally devised by chief examiners and reviewed by a monitoring committee of teachers. The examiner is paid a modest fee, corresponding to the few evening's effort required to write a draft examination paper. In the time available, and in the absence of any official opportunity to pilot test (although we know that some people do so unofficially!), it is not surprising that papers are closely similar to those of previous years. Any attempts by the chief examiner to introduce novel questions are further restricted by the monitoring committee, whose members tend to be anxious to ensure for their clients (teachers and students) that the questions are fair—which includes being familiar.

How does this compare with the process of producing "objective" multichoice tests? Uniformity of the tasks is a common factor; indeed, the psychometric machinery of piloting and analysis is designed to ensure that new tasks added to a bank are as close to those that they replace as is possible without being identical. The introduction of new types of tasks should be easier with these resources; however, task types that do not fit into the narrow mould of the machinery are uncomfortable as they often require different modes of response and are too complex to fit the simplistic definitions of "reliability" that are used. So they are rejected.

We have already stated our view that (i) this conception of "reliability" is an artifact, statistically sophisticated but educationally naive, and (ii) the valid assessment of mathematical performance in a balanced way is a fairly rough-and-ready matter whose genuine reliability is fairly low, as far as details of performance by individuals is concerned.

The credibility of the whole assessment enterprise is rescued by the Central Limit Theorem of statistics—that is, the reduction of fluctuations by the averaging process across tasks, across subjects, and across individuals. Even in the narrower technical sense of reliability, test/retest variation is substantial for individuals between one occasion and another, even on very simple, few-step technical tasks (see, for example, Woolf, 1986; Foxman, 1989). This situation is potentially satisfactory, however, since we obtain broad indications of a student's overall performance of reasonable reliability, together with more detailed information that, while it is only indicative, can be useful to students and to teachers in formative or diagnostic ways.

We take validity more seriously. What is the point of assessing individuals except on those things we actually want them to be able to do? Correlation cannot carry the load of extrapolating measurements from one kind of performance to another. *We believe it is essential from both a measurement and a curriculum influence point of view that the assessment be balanced across the curriculum objectives, not simply in mathematical content but in the types of task assessed.*

Balance is not easy to achieve. From our own experience, a substantial design effort seems essential to produce a balanced set of tasks of high educational quality that span the curriculum objectives. While this coverage is a matter for detailed analysis, we give great weight to the <I>face validity<$> of the set of tasks as reflecting what we want students to be able to do in a balanced way, and for allowing them the opportunity to show their abilities. The pilot testing of such tasks, with detailed study of a significant sample of student responses (up to perhaps about 100 students), is a key factor in establishing the feasibility of the activities; another key factor involves the detailed structuring of the tasks to ensure that students will be presented challenges with an appropriate range of difficulties. The devising of appropriate grading schemes for assessing the range of responses to be expected also rests on the piloting. Even in open tasks, the great majority of tracks through the problem tend to be found in this way. The responses of the rare individual students who take entirely novel approaches can be referred to the chief examiner, even when hundreds of thousands of students are involved in all.

Both normative and criterion-referenced judgments enter into this process. The norms need to reflect, to a greater or lesser extent, the levels of difficulty and the degree to which students have been prepared for novel types of tasks. As we have noted, this approach seems to work well in large-scale assessment; we believe that the losses arising from trying to force tasks into a tight psychometric mould far outweigh the gains.

We note in passing that the new assessment packages for the English National Curriculum at ages 7, 11, and 14 are being designed not by individual chief examiners but by groups of specialists under substantially funded contracts from the authorities. However, the set of constraints under which these groups are operating has been steadily tightened so that it can no longer be regarded as an example of the mode of operation we have just advocated. Indeed, there are serious doubts about whether the total set of constraints allows any solution at all, let alone one of high educational quality. It will be interesting to see which constraints are ignored and what emerges.

Grading by People or Machines

Machine-graded multiple-choice tests have a number of advantages. No skill is needed in the grading process, and their "reliability" can be carefully established by piloting. Machine grading is therefore often preferred, particularly in the United States. Since no judgments on individual students are involved, such tests are often regarded as fair. We shall return to the question of fairness below, but it is worth pointing out that judgments are made in advance of doing the tasks and that the multiple-choice format requires that the judgment made on the students' performance is of the crudest possible type—except on the narrowest tasks, a student with a different approach may well be undervalued.

Grading by people, on the basis of more or less tight grading schemes, permits a choice of the level of subtlety with which the students' response is analyzed. A balance must be struck between adequate intergrader consistency and a sufficiently sensitive analysis of each student's response. Where many markers are involved, training and standardization procedures are also an issue.

Teams of as many as 100 assistant examiners are used by British examination boards (where grading is called "marking"). Typically, these assistant examiners are provided with a written marking scheme prepared by the chief examiner, and are asked to mark a preliminary batch of 25 carefully chosen student scripts. Then there is a standardizing meeting, where scripts are discussed in detail, differences between examiners are discussed, and any points of confusion are sorted out by the chief examiner and noted for future checking. Finally, the main batch of scripts is marked. This typically involves marking of about 600 scripts by each examiner, occupying two weeks of vacation time, for which examiners are paid. Mark-remark consistency varies from subject to subject, but in traditional mathematics examinations, deviations are usually about 1 percent, or one tenth of a grade width. In other subjects with less tightly structured questions, including essay writing for example, substantially larger variations are common and are accepted. Thus, mathematics could broaden its range of task types.

Increasingly, teachers are being asked to grade the assessments of their own students as part of the public examining process. The same issues of training and standardization apply, but rather different methods have to be used (see below). The direct cost of examining is then substantially smaller, since the assessment function is assumed by teachers as part of their normal work. One should ask how much time teachers should spend assessing their students, as opposed to teaching them. We believe that diagnosis is a critical component of teaching, and appropriate feedback is an essential and helpful element of this; therefore, some increase in the time spent on standardized assessment is desirable. If substantial amounts of time are to be used for assessment, however, we again stress that assessments should have high curriculum validity. That is, *the assessment activities should be high-quality learning activities in their own right.* This implies that assessment should lead curriculum change, not hold it back.

Standardization

The central problem of test design is, as always, the conflict between validity and reliability. The attainments we are most concerned with are those that govern larger-scale activities. Here, reliability is measured by the allocation of a large number of independent marks, although a holistic view of the students' response is important. The way in which a compromise is effected between these demands is a focal point of interest in these test innovations.

A variety of approaches is used to try to ensure that uniform standards are maintained from student to student, class to class, and school to school. All are imperfect and some are quite costly. There are two rather different approaches, either or both of which can be used. They are sometimes summarized as *personal moderation* and *statistical moderation.*

Personal moderation methods involve review by designated individuals of samples of work from different groups of students. It may be thought of as a partial regrading exercise. This method has high intrinsic validity, but is also expensive unless the sample is kept very small (when its effectiveness may be called into question). A variation on this is *consensus moderation,* in which a group of schools meet to compare each other's gradings of their students, mediated by a moderator from the examining board. This too is expensive, but the approach has been found to be a valuable mode of professional development for teachers, with the assessment tasks providing a common focus for discussion of questions of curriculum, of teaching, and of student performance, in ways we have illustrated earlier.

Statistical moderation uses students' performance on a subset of the assessment tasks to standardize the overall distribution of grades. Clearly, consistency of grading of this subset must be achieved by external marking or personal moderation of some kind. If the standardizing instrument can be graded inexpensively, statistical moderation is correspondingly cheap. However, such an approach suffers from a major disadvantage: the performance of students on the standardizing part of the assessment normally determines the average level of a group, so this is a strong indication to teachers to focus the curriculum efforts of the class or the school on the part of the curriculum that is tested by the standardizing instrument (and to neglect other aspects that are equally or more important educationally). For example, if a short multiple-choice test of basic skills in arithmetic is used to standardize an assessment that also includes other kinds of mathematical activity, including extended problem-solving work, it will not be surprising if teachers focus most of their effort on preparing their students for the short, narrow test. Differences in performance on the other parts can only move children around within a group whose average level will be fixed by the tests.

Statistical moderation is acceptable, we believe, if and only if the moderating instrument is itself balanced across curriculum objectives. This is roughly what was proposed for the National Curriculum in England by the Government's Task Group on Assessment and Testing (TGAT, 1988). A few extended standard assessment tasks (SAT) are to be used to moderate teacher assessment carried out on a continuing basis. These SATs are currently being designed under a set of constraints that seem to preclude their balance in the sense we have described. In particular, the new requirement that performance on each task be reported under detailed behavioral objectives, largely focused on content, neglects the higher levels of demand imposed by the integration of such elements that is needed for more extended or less routine tasks.

We have so far talked about the standardization of grading. The standardization of tasks is also an issue. How far can one allow (or indeed encourage) teachers and their students to choose the specific tasks they tackle, within more general descriptions of the types of task to be used? On the one hand, the difficulty of similar-sounding tasks can vary greatly depending on factors such as the familiarity of the context, the complexity of the situation, the strategic, tactical, and technical demands of different approaches to it, and the climate of support in which the student works. This all suggests that, for fairness, the tasks themselves should be defined externally. On the other hand, the process of defining interesting questions is an important part of mathematical activity, and there is much to be said for allowing students to tackle some tasks that they themselves find in areas of their own specific interest.

This situation has arisen in Britain most recently and specifically in the context of coursework for the GCSE examination at age 16, already mentioned. Typically, the examining boards require several pieces of coursework, each of which takes about two weeks of mathematics time. In their attitude toward this issue, boards presently cover the whole spectrum. Some define precisely the tasks that students must study, perhaps allowing some choice, while others leave the choice of specific tasks within some general guidelines entirely to the schools. Those boards that include both possibilities find that nearly all their schools use the standard tasks they provide, rather than devising their own. However, this may be a function of sophistication; there is some evidence that after a year or two, some teachers want the freedom to devise their own tasks.

Standardization of tasks is an area where continuing empirical investigation is needed. Our inclination is to believe that some tasks should continue to be set by the boards; these might be used as part of a standardizing instrument. However, we see great advantages in a number of ways in asking teachers and their students to devise some of the tasks themselves.

REPORTING RESULTS

The extent, the form, and the methods of reporting results are almost as sensitive an issue as the assessment process itself. How far should results on the performance of individual students, of particular classes, and of particular schools and school districts be a matter of public knowledge? These are questions of political and public policy with strong potential effects on individuals and communities. On the one hand, an atmosphere of freedom of information and a search for increased accountability suggests the widest possible dissemination. On the other hand, much of the information may be regarded as personal, almost akin to medical records, and to be primarily the concern of the students involved, their parents, and the teachers who are responsible for forwarding their educational and social development.

In the United Kingdom, a government with a strong belief in market forces is insisting on the public dissemination of results, reluctantly excepting only the performance of individual students at age 7. The government sees this as information that should be available to parents to guide their choice of schools. The Task Group on Assessment and Testing (TGAT, 1988) recommended that such results should be distributed only in a form that makes allowance for other factors that have a strong influence on performance, such as the level of the children on entry into the school and factors of economic, social, or educational background. This issue needs careful thought and experiment in each society.

Processing

We shall not attempt to describe the detailed administrative arrangements by which public examinations are set, marked, and reported in various countries. Suffice it to say that these are important, and that all the systems we have discussed succeed in assessing the great majority of children at various ages in their various school subjects in a way that commands general, but not uncritical, public support.

Fairness

Fairness encompasses a set of issues that arouse strong feelings in discussions of assessment. Fairness has many aspects, however, and most systems tend to take very seriously the need to ensure fairness in some respects, while ignoring others. For example, fairness suggests that assessment should be even-handed in the following instances:

- Between different students on the same occasion. This is always a concern; however, in traditional mathematics, mark/remark variation is generally small.

- Between different students on different occasions. Research on test/retest consistency for given individuals on similar tests shows there is no room for complacency here; it receives some attention, but variations are broad. Where different boards set parallel examinations, comparability is equally elusive.

- Between different aspects of performance. Typical traditional tests cover only a part of a satisfactory contemporary curriculum. In mathematics, it is a part whose importance is shrinking year by year through the impact of technology. This is now becoming a matter of wider concern. Hence, this book.

- Between students taught by different teachers. Unfairness arising from the effects of good or bad teaching is universally ignored, using the argument that the tests are assessing performance not ability; however, the results of public assessment are used as predictors of future performance in job selection, entry to further education, and other areas where promise is as important as current performance.

- Between students from different kinds of home background. This too is ignored using excuses similar to those described above; however, results are used not only for the selection of students but for the selection of schools.

Fairness is an issue that above all needs both sensitivity and good sense.

Costs

Cost-effectiveness is an important criterion in the design of assessment, as in most other fields. How far assessment costs are seen as a serious problem again depends upon perspective. In the United Kingdom, fees for public examinations, covering nearly all children, are a tiny proportion (about 0.3 percent) of total educational expenditure. Still, this is a substantial sum that is, for example, about ten times as much as the total expenditure on systematic research and development. Without attempting to resolve this issue, it seems to us that the quality of feedback to all involved is such an important feature for the performance of any system that a significant proportion of time, effort, and money can appropriately be devoted to it in education too. As far as assessment itself is concerned, *it is not assessing but policing the assessment that costs money.*

It is true that a certain amount of investment must be made in the design of sets of tasks and of ways of administering them, evaluating them, and reporting the results, but these costs can be spread over many students. The major expense, because it must be provided individually for each student, is in the time used in the work on the tasks—particularly in the grading of the students' responses. This can be undertaken by students and teachers themselves as long as they are trusted; it is then a proper use of learning and teaching time, with no additional costs. Some training is needed but, again, this is not a per student cost. *It is the use of external markers or moderators, or other monitoring devices, that requires additional expenditure.*

We do not question the need in some societies and social situations for some such monitoring, but it is good to be aware that others do not share that need. For example, teachers in higher education in the United States and Britain (and elsewhere as well as—school teachers in Germany, for example) are commonly relied upon to provide assessment of their students and former students without significant external monitoring. There is room for experiment, and for progress, here.

Assessment Packages

We have left any discussion of the way assessment may be packaged until near the end, because many of the main issues we have discussed apply to many forms of packaging that satisfy the principles we set out at the beginning. Of course, the practicalities of use in schools and beyond are intimately tied up with the form and mode of administration of the assessment package. Equally, they raise social and educational issues of the kind we have touched on. We shall content ourselves here with describing some possibilities for consideration, adding the assertion that a variety of solutions that all meet reasonable constraints is usually available.

Many forms of assessment packages have been tried, and more are worth investigating. It may be useful to indicate something of the range of possibilities, beyond the short test of multiple-choice items. These include the timed written examination and various extended practice tasks.

The timed written examination has a great variety of forms and possibilities. In the United States, such examinations are familiar at the higher levels in high school and at college level. We have said something of the extensions to this format in English public examinations, involving other modes of presentation or response as well as types of task and, particularly, the addition of coursework to bring extended tasks into the package.

The assessment month is a nonstandard possibility worth consideration. That is, an extended practical task is used that runs over several weeks of mathematics time (ten to fifteen hours), punctuated by one-hour sessions each week containing shorter tasks, a one-hour pure mathematical investigation, a half-hour analysis of some data leading to recommendations for action, several fifteen-minute practical problems or interpretations of mathematical presentations of real events, and some ten-second to five-minute exercises on technical skills in mathematics that need to be fluently available to the student for such purposes. Such a package (Burkhardt, 1991), dominated by vivid and interesting tasks, can represent a high-spot in the curriculum. Curriculum support, including materials for the extended task, would be valuable, particularly in the transitional phase for those many teachers with little experience of such classroom activities. (It is worth noting that such activities normally have a strong cross-curricular flavor, though it is important to focus on the mathematical aspects of performance, firmly but not exclusively.)

The assessment week might be an appropriate way of mounting a similar range of activities in an elementary school, filling the mornings for a special week in each semester. In a secondary school, it would perhaps represent a less wide ranging package, with the longest tasks taking no more than an hour. Even a mathematics assessment day could embrace a far broader range of tasks than is assessed at present, though it would not have the same curricular impact as a longer package. Exemplars are under development.

IMPLEMENTING LARGE-SCALE CHANGE

How can all these aspects be assembled into well-engineered packages, including assessment, that effectively promote and support the intended curriculum changes? We do not have space here for a detailed discussion of the dynamics of large-scale curriculum change. We would point out, however, the importance of systematic study and experiments in this area (see, for example, Burkhardt, Fraser, and Ridgway, 1989).

We believe that *there are no established methods of planned curriculum change that are known reliably to yield outcomes in reasonable accord with their intentions,* at least not for the kind of changes that we are currently trying to advance in mathematical education. Both politicians and educators in many countries behave as though there are straightforward methods of effecting planned curriculum change—that once the difficult questions as to what should be done are resolved, achieving it is relatively straightforward. We would suggest that this is the reverse of the truth, and that the *how* of planned curriculum change is an unsolved problem that needs separate attention.

This need is beginning to be recognized in a few places. In mathematical education for example, methods of implementing curriculum change is a major theme for the 1992 International Congress on Mathematical Education, in Quebec in 1992. One of us (Burkhardt) is organizing the working group on this theme. This is the first time the topic has had this status, although it was a sub-theme at the 1988 Congress in Budapest. It is to be hoped that it will bring together whatever systematic work is being done to find better methods and that it will act as a stimulus for further systematic work in the future. (This sudden appearance of an obviously important but neglected field is not new. Alan Schoenfeld has remarked [Burkhardt et al., 1988] that problem solving was slated to appear in the 1980 Congress only under the general heading "unusual aspects of the curriculum." By the 1984 Congress, it was a major theme and remains one of the main focuses of interest in the field.) Here we can only take a brief look at the factors involved, the methods used, and ways that they might be improved, concentrating particularly on the role of assessment in influencing curriculum change.

The educational system is complex. Although it basically involves teachers teaching children in classrooms in schools in school districts in society, there is a great variety of types of players, with different roles, needs, wishes, and expectations. Teachers and their students are at the heart of it all, but politicians and parents, education leaders at national, state, district, and school level, textbook publishers and authors, and designers and examiners (the purveyors of assessment), all play their parts. If the system is to move in the direction that is intended, all must work together.

Common sense and retrospective analyses of past innovations (see, for example, Fullan, 1980) suggest that change is only likely to happen if all those involved are subject to adequately designed patterns of pressure and support. A few may move in the desired direction without pressure, but amid all the other stresses of professional and personal life, they are not likely to be many. Equally, support that reliably enables those involved to implement the required changes is essential if one is to avoid the key problems of *low take-up, dilution, and corruption of the change*. On the whole, governments are better at providing pressure than support, which tends to be more costly.

Assessment is commonly used as a key element in the pattern of pressure. We have seen how that pressure often inhibits changes that are supposed to be implemented because the assessment does not reflect the target curriculum. The essence of our contributions to this volume lie in suggesting how assessments may be designed so that their introduction will promote desirable change, rather than undermining it. There is some evidence that this can be done. For example, the special examination syllabuses associated with the successful British projects of the 1960s in many fields were at least a factor in enabling those projects to achieve widespread impact.

We ourselves explored a different model, working within the main mathematics syllabus of a major English examining board. The board agreed to introduce a sequence of small but profound changes, each amounting to about 5 percent of the syllabus or about three weeks' work. Such changes were necessary to ensure that examination papers reflected already declared assessment objectives in a more balanced way. We designed exemplars of the new types of questions, classroom materials that would enable teachers to prepare their students for them, and materials for do-it-yourself, in-service training within the school. These were packaged in boxes and made available for sale to schools, along with the announcement of the syllabus change.

A number of factors have prevented as full an evaluation of the impact of these materials as one would wish—in particular, the abolition of the examination involved after two such changes, on the introduction of the GCSE! However, the new questions were taken by a significant proportion of the students, even though they were optional for the first two (and only) years. The sales of the boxes of materials, "Problems with Patterns and Numbers" and "The Language of Functions and Graphs" (Shell Centre, 1984, 1986), now correspond to about half the number of secondary schools in England and Wales. In addition, significant numbers have been sold around the world, including in the United States. All those involved agreed that the adoption of such novel questions, and the level of student performance on them, would not have been comparable without a carefully designed pattern of pressure and support.

Note that this is a gradual change model, contrasting with approaches that seek to introduce a whole new ideal curriculum at once (and that are so often corrupted). We have found that *the pace of change is a critical variable*. What proportion of their lessons can teachers be reasonably asked to change substantially each year? Five percent seems about right. This pace may seem slow, but it compares well with what has actually been achieved in the past.

The Numeracy through Problem Solving enterprise was rather different—the support was similar, but the pressure was less direct. A new assessment scheme was offered, but schools did not have to take it seriously. The pattern of response has been interesting here. The scheme has achieved significant penetration so far (about 20 percent of secondary schools in Britain take some part in it), but it is mostly used with low achieving students. In this area, teachers of mathematics are always searching for materials that work well, and the activities supported by these five boxes (Shell Centre/JMB 1987-89) regularly transform the students' attitude toward and persistence in mathematics. For above-average students, for whom the scheme is equally important and successful, the adoption is much lower, probably because teachers perceive the priority there as achieving good grades in the GCSE. (The modules are used for assessed coursework in some schools.) It will be interesting to follow the recent introduction of a GCSE option based upon Numeracy through Problem Solving.

Other models of change in England involve assessment in a central role. For example, the Graded Assessment in Mathematics project (GAIM, 1989) has developed an approach based on teacher's assessing the students "on the hoof" in the course of teaching. This assessment is based on specific criteria at fifteen levels related to every aspect of the curriculum. Not surprisingly, it places considerable demands on teachers, who must absorb the classification scheme of mathematical performance involved and accept the present limitations of such schemes. In practice, the designers have developed sets of tasks for teachers to use, to help them come to grips with the criteria, so the approach has many of the features we have advocated.

Meanwhile, the main examination boards in England are seeking to develop their examinations to meet the broader objectives. Coursework has provided the most substantial change for most teachers. The effects of matching the GCSE to the National Curriculum, and the knock-on effects on the later examinations at age 18, have yet to be worked through.

Parallel developments in some other countries are described in other chapters of this book, and we would like to note a few of particular promise. The situation in the Netherlands, where the national assessment system is closely and sensibly linked to a long-range program of curriculum development (at least equal to any in the world), has produced substantial change over twenty years, rooted in basic research and systematic curriculum development (see the chapter by de Lange in this volume). The Australian states, each with its own education system, are pioneering high-quality assessment as part of a program of change; their approaches seem to have much in common with the constructive chaos of the British scene. In the United States, a number of promising initiatives involve major states, so there is hope of change in the fairly near future.

The third important general feature of an implementation method, complementing balanced pressure and support, is that the balance should be dynamic. That is, it should be designed to rapidly discover what are its successes and failures and to take action to modify itself accordingly. This is rarely attempted. Educational innovations are normally presented in an atmosphere that implies that they are bound to work well, and it is generally regarded as seditious to question that assumption. Models that have all three aspects we call "Dynamic Pressure Support Models"; they deserve particular attention in all current and future reform programs. Clearly, there is much systematic work to be done, and much to learn, for a decade or two at least.

REFERENCES

Bell, A. (1979). The learning of process aspects of mathematics, *Educational Studies in Mathematics, 10*.

Burkhardt, H. (1991). *A model for a standard assessment task in mathematics for age 14*. Nottingham: Shell Centre.

Burkhardt, H., Groves, S., Schoenfeld, A., and Stacey, K. (1988). *Problem solving: A world view*. Nottingham: Shell Centre.

Burkhardt, H., Fraser, R., and Ridgway, J. (1989). The dynamics of curriculum change. In I. Wirszup and R. Streit (Eds.), *Developments in school mathematics education around the world*. Proceedings of the Second UCSMP International Conference, Reston VA: National Council of Teachers of Mathematics.

DES. (1989). Mathematics and the National Curriculum. Statutory Orders from the Department of Education and Science and the Welsh Office, London: HMSO.

Foxman, D., Ruddock, G., and Thorpe, J. (1989). *Graduated tests in mathematics*. Windsor: NFER-Nelson.

Fullan, M. (1980). *The meaning of educational change*. New York: Teacher's College Press.

Graded Assessment in Mathematics. (1989). *GAIM development pack graded assessment in mathematics*. London: Macmillan.

National Curriculum Council (1989). *Mathematics: Nonstatutory guidance*. York: NCC.

Schools Examinations Council. (1985). *National criteria for the GCSE: General criteria and mathematics criteria*. London: SEC.

Shell Centre/JMB (1987-89). *Numeracy through problem solving: Five modules* (Design a Board Game, Produce a Quiz Show, Plan A Trip, Be a Paper Engineer, Be a Shrewd Chooser). Harlow: Shell Centre for Mathematical Education and the Joint Matriculation Board. Longman.

Shell Centre. (1984). *Problems with patterns and numbers, an examination module for secondary schools*. Manchester: Joint Matriculation Board.

Shell Centre. (1986). *The language of functions and graphs, an examination module for secondary schools*. Manchester: Joint Matriculation Board.

TGAT. (1988). *A report of the Task Group on Assessment and Testing for the National Curriculum*. London: HMSO.

Woolf, A., and Silver, R. (1986). *Workbased learning: Trainee assessment by supervisors*. Research and Development Monograph 33, Training Agency. London: HMSO.

8 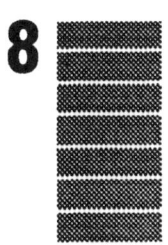 Assessing Mathematical Skills, Understanding, and Thinking

Jan de Lange

INTRODUCTION

Many mathematicians, among them Ahlfors, Bers, Birkhoff, Courant, Coxeter, Kline, Morse, Pollak, and Polya, signed a 1962 statement from which the following quotes were taken (Ahlfors, 1962):

> To know mathematics means to be able to do mathematics: i.e., to use mathematical language with some fluency, to do problems, to criticize arguments, to find proofs, and, what may be the most important activity, to recognize a mathematical concept in, or to extract it from, a given concrete situation. Therefore, to introduce new concepts without sufficient background of concrete facts, to introduce unifying concepts where there is no experience to unify, or to harp on the introduced concepts without concrete applications which would challenge the students, is worse than useless. Premature formalization may lead to sterility; premature introduction of abstractions meets resistance especially from critical minds who, before accepting an abstraction, wish to know why it is relevant and how it could be used...Extracting the appropriate concept from a concrete situation, generalizing from observed cases, inductive arguments, arguments by analogy, and intuitive grounds for an emerging conjecture, are mathematical modes of thinking...The best way to guide the mental development of individuals is to let them retrace the mental development of the race (the genetic principle). (p. 8.)

This statement was a strong reaction to the excessive abstraction and emphasis on "content at the expense of pedagogy" (p. 195) that were invading the school mathematics curriculum. In 1959 at the Royaumont conference, Dieudonne gave his now infamous address, "New Thinking in School Mathematics," which resulted in the rise of modern school mathematics based on set theory. Thom (1973) later concluded, "The emphasis placed by modernists (structuralists) on axiomatics is not only a pedagogical abbreviation—which is obvious enough—but also a truly mathematical one" (p. 225).

One of the three Dutch participants at the Royaumont conference in 1959 was Vredenduin. After experimenting with the concept of structure in his geometry books for secondary education, he concluded many years later in an interview with Goffree (1985): "It was beautiful edifice, but I do not think there was one student who shared that opinion" (p. 235).

CONCEPTUAL MATHEMATIZATION: FROM CONCRETE TO ABSTRACT

Freudenthal (1973), being a mathematician influenced by Brouwer's constructive or intuitionistic view of mathematics, introduced the slogan, "Mathematics as a human activity." He argued that (i) mathematics should never be presented to students as a ready-made product, (ii) the opposite of ready-made, dehumanized mathematics is human mathematics *in statu nascendi*, and (iii) the student should reinvent mathematics. What Freudenthal calls reinvention is often described as discovery or rediscovery.

Freudenthal's view is that the learner is entitled to recapitulate in a fashion the learning process of mankind. This means that instruction should not start with the formal system—which is, in fact, a final product—nor with embodiments, nor with structural games. Instead the phenomena by which the concepts appear in reality should be the source of concept formation. The 1962 statement called this process "extracting the appropriate concept from a concrete situation," or as de Lange (1987) states it, "conceptual mathematization."

To put the preceding views a little more precisely, the real-world situation (or problem) is first explored intuitively, for the purpose of mathematizing it. This means organizing and structuring the problem, trying to identify the mathematical aspects of the problem, and discovering regularities and relations. This initial exploration (with a strong intuitive component) should lead to the development, discovery, or (re)invention of mathematical concepts.

From our years of observations in classrooms, it is clear that by depending on factors such as interactions between students, interactions between students and teachers, the social environment of the student, and the student's ability to formalize and abstract, students will sooner or later extract the mathematical concepts from the real situation. This is what we call conceptual mathematization.

After formalizing and abstracting these concepts, students can use them by applying them to new problems. This leads to a reinforcement of the concepts that have been constructed, and to a readjustment of the perceived real world. In this way, the learning process has an iterative character, and the learning cycle may be modeled as shown in Figure 1.

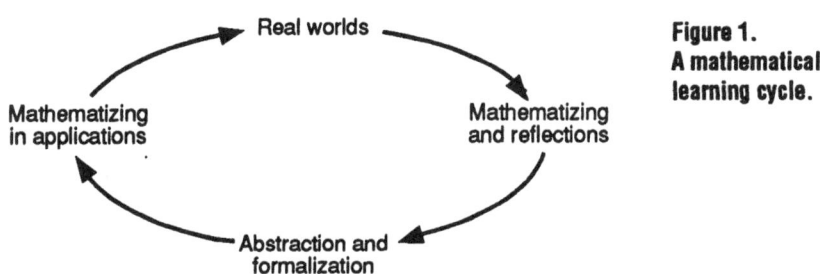

Figure 1. A mathematical learning cycle.

This model shows a remarkable similarity to the Experiential Learning Model of Lewin (1951). Two aspects of Lewin's model (Figure 2) are noteworthy. First it relies on concrete experience to validate and test abstract concepts. We may call this part of the problem-solving process *applied* mathematization. Secondly, the feedback principle plays an important role in the process. Lewin used the concept of feedback to describe a social learning and problem-solving process that generates valid information to assess deviations from desired goals.

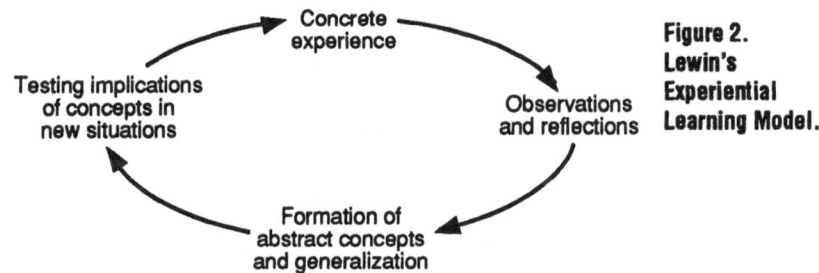

Figure 2. Lewin's Experiential Learning Model.

In a more recent study Kolb (1984) adapted this Lewinian model and compared it with Dewey's Model of Experimental Learning and with Piaget's Model of Learning and Cognitive Development. In Kolb's opinion, all the models suggest the idea that learning is, by its very nature, a tension and conflict-filled process. Learners need four different kinds of experiences to be successful: concrete experience, reflective observation, abstract conceptualization, and active experimentation.

Free Productions

Components similar to those in Lewin's model should be found in the learning of mathematics, and the weakest link in the cycle of mathematics education seems to be the active experimentation stage. This is why, in the Netherlands, we currently give students real-world, open-ended problems to solve. Nonetheless, current practices can be improved. For example, one way to achieve this would be to have students make more productions—not only mental contributions. That is, students should be asked to produce more concrete things.

Treffers (1986) stresses the fact that by making free productions, students are forced to reflect on the paths that they themselves have taken in their learning processes—while, at the same time, anticipating their continuation. In this way, free productions can form an essential part of assessment (as we will see later). For example, to design exercises that can be used in a test, or to design a test for other students in the classroom, we may ask students to write an essay, to do an experiment, to collect data, and to draw conclusions.

Interactive Learning

As mentioned before, interaction between students (and between students and teachers) is essential. We agree with Balacheff (1985) when he states:

> Pairwork is not only a source of explanation but also a source of confrontation with others. This adds greatly to the dynamics of the activity. Contradictions coming from the partner, due to the fact that they are explained, are more likely to be perceived than contradictions confronting the solitary learner, derived only from the facts. They are also harder to refute than in a conflict resulting from the individual, or from temporary hesitations between two opposing points of view that the solitary learner experiences when confronted with a problem. (p. 181)

Doise and Mugny (1984) speak of a socio-cognitive conflict. Their research shows clearly that inter-individual encounters lead to cognitive progress when socio-cognitive conflicts occur during the interaction. The social and cognitive poles are inseparable here, because it must be a matter of conflict between social partners about the ways to resolve the task. In mathematics education, students are usually operating on different cognitive levels and there are large differences in cultural background. This means interactions among students are excellent starting points for discussions resulting in socio-cognitive conflicts. It is for the teacher to make the best use of this situation, but the conditions can also be shaped by researchers and curriculum developers.

Integrated Learning Strands

Mathematics is integrated with the real world(s). This principle is one of the key factors of mathematics education as it has been developed by IOWO (Institute for the Development of Mathematics Education) and later OW&OC (the Dutch research group on mathematics education). Also, the integration of mathematical strands is essential because applying mathematics is very difficult if various subjects are taught separately, neglecting the cross-connections (Klamkin, 1968). In applications, one usually needs more than algebra alone or geometry alone. In real applications, students often must compare different models and integrate them. This implies integration on yet a third level. For example, the "rat problem" (below) taken from a college-level biology book, shows that algebra, probability, linear algebra, and calculus all function within one simple problem that can be solved by drawing a schema or by visualizing it. (Notice that we consider solutions that consist of a beautiful visualization just as good as any other solution that may have a mathematical formula in it.) Because it makes our point clear, we give parts of the solution of this problem:

Students read the following statement:

> ...It might be interesting to estimate the number of offspring produced by one pair of rats under ideal conditions. The average number of young produced at a birth is six; three out of those six are females. The period of gestation is twenty-one days; lactation also lasts twenty-one days. However, a female may already conceive again during lactation, she may even conceive again on the very day she has dropped her young. To simplify matters, let the number of days between one litter and the next be forty. If then the female drops six young on the first day of January, she will be able to produce another six forty days later. The females from the first litter will be able to produce offspring themselves after a hundred and twenty days. Assuming

> there will always be three females in every litter of six, the total number of rats will be 1808 by the next first of January, the original included....

Students were then asked, "Is the conclusion that there will be 1808 rats at the end of the year correct?" Solutions by a student and a teacher are shown in Figures 3 and 4.

Figure 3. One student's schematic solution to the "rat problem."

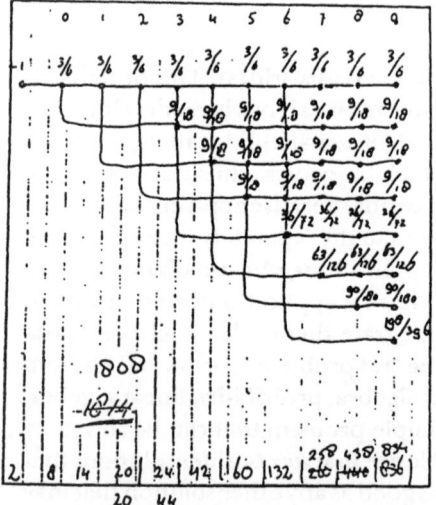

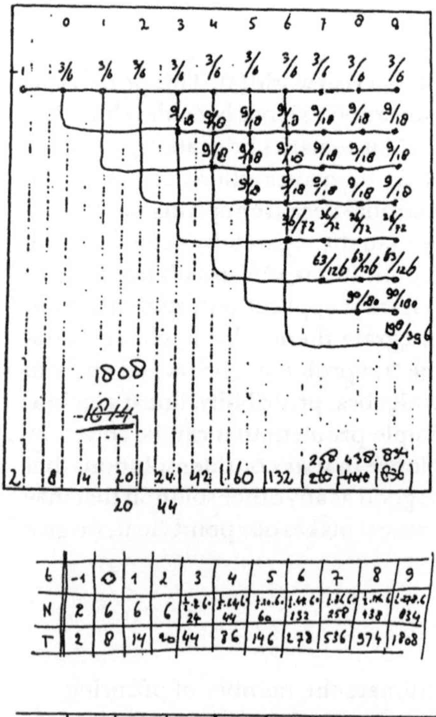

During an in-service teacher training course, only 20 percent of the teachers were able to solve the preceding problem within half an hour. On the other hand, nonmathematics majors (16 years old) from some of our experimental schools did very well on the problem. Results depend, of course, on the conditions. In the classroom, with a limited amount of time, both students and teachers tend to find it very difficult to solve or even to schematize the problem. But with no time limit (for instance by giving the problem as homework) students produce fine results. This suggests that such process-oriented activities may not be well suited for testing by means of time-restricted written tests. Also, in contrast to the preceding schematic solution, teachers were more likely than students to feel the need to produce formulas such as,

6	-1	0	1	2	3	4	5	6	7	8	9
N	2	6	6	6	24	44	60	132	258	438	834
T	2	8	14	20	44	86	146	278	536	974	1808

Figure 4. One teacher's schematic solution to the "rat problem."

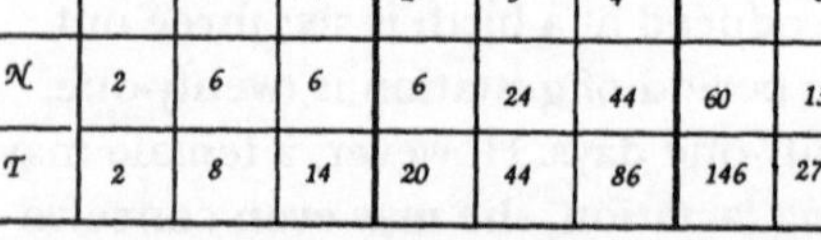

$$A_{n+3} = A_{n+2} + 3;\ A_n = 2;\ A_0 = 8;\ A_1 = 14.$$

An approach that is different from the preceding ones uses graphs and matrices (which are part of the curriculum). For example, Figure 5 shows a graph that represents the growth of the rat population and a matrix that can represent the graph. Another possibility is to look at the nature of the growth process. For example, comparing the number of rats period by period, the growth factor on the long run is equal to 1.86. This leads to the formula:

$$An = 44*1.86^{(n-3)}$$

(We leave it to the reader to integrate and generalize the different solutions, an activity that can be seen as the top level of mathematization.)

Figure 5. A matrix-based solution to the "rat problem."

Goals in Mathematics Education

De Lange (1989) has articulated goals for upper secondary mathematics education. As a guide for test developers, the following partitioning scheme is used to list key mathematical ideas to be operationalized: (i) general goals, (ii) global description, (iii) concrete goals, and (iv) specific abilities.

The *specific abilities* are easiest both to describe and to assess. Students must have the command of certain basic techniques and tools which are not aims in themselves but which are essential for the attainment of concrete goals. A description of these specific abilities is required in order to give the boundary conditions for the attainment of the general and global goals. The description of these specific abilities and tools has a purely mathematical character and does not allude to the goals to be pursued and the students' activities. The following are examples of specific goals:

- The student is able to add matrices (if possible).

- The student is able to multiply matrices (if possible).

- The student is able to draw a histogram, a box plot, stem and leaf plot.

The *concrete goals* relate to specific areas of instruction. They interrelate and attach meaning to the subject areas from the lower level, the specific abilities. Instructional activities are most often directed towards concrete goals. Some concrete goals might be

- to describe the meaning of sum, product, and powers of matrices in their problem context;

- to read, interpret, process, and analyze information and to describe it in a graph or matrix, if suited;

- to analyze which visual representation is most suited for a given situation, taking into account the class width, the same class width or different ones, the necessity of frequency density graphs, and so on.

The *global characterization* of the subject areas provides a sketch of the domain, prepares the formulation of concrete goals, connects them with the general goals, and gives some indication of the interrelatedness of the various areas. For example, a goal description in statistics might be the following:

> Data visualization plays an important role in preparing students for future studies in several fields. This knowledge is also necessary for the students to become intelligent citizens. Many decisions, both by individuals and by groups, are based on graphical representations of statistical data. Interpretation of data and their visual display are especially needed for information found in books, magazines, and on television. Critical judgment should be a key factor. For this reason students should be able to construct visual representations of data by themselves, and read and interpret graphics that are taken from other publications.

General mathematical goals pertain to permanent qualities, skills, abilities, modes of thinking, and so on, that are not restricted to one specific area of mathematics. For that reason one can formulate those goals in a more general way. Globally, one can state that the student should demonstrate the ability to solve problems with mathematical tools, or to describe problems in a mathematical way and communicate them to others. This means that a student should be able to

- identify the relevant information, relations, and structures;

- formulate and visualize a problem in different ways;

- use the basic skills for the different areas in mathematics;
- use standard methods for solving problems from different fields;
- value and judge the use of mathematics in applications;
- make connections between problems and mathematical concepts, relations, and structures;
- interpret the results within the context after solving a problem and analyze the result in a critical way;
- use research and reasoning strategies;
- recognize isomorphic ideas in different situations; and
- adjust and refine a model after careful analysis.

The highest level of goal description, and at the same time the most neglected one in relation to assessment, is the *general goal description*. Also, to help students acquire knowledge, skills, and insight in the above-mentioned goals and to help them to develop a good attitude toward mathematical work, they should have the opportunity

- to be creative in suggesting solutions;
- to use knowledge and skills in a flexible way;
- to work in a systematic and organized manner;
- to generalize results;
- to judge critically both the input, the solution process, and the output of a problem;
- to estimate outcomes;
- to develop appreciation for mathematics;
- to develop self-confidence by building confidence in mathematical abilities; and
- to work in an interactive way, using different media.

This model of a goal description can be represented as shown in Figure 6. Also, in order to complete the schema, we need some additional components—instructional activities, tests, and didactics—to serve as vertical connectors between the different levels of our schema (Figure 7).

Figure 6. A goal description.

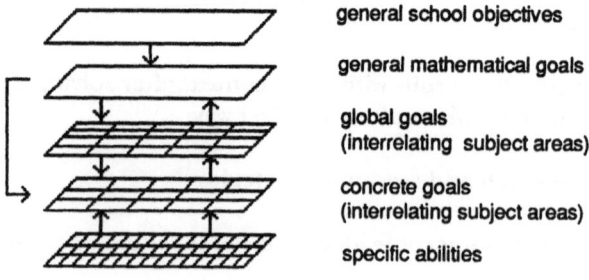

Figure 7. Additional components of goal descriptions.

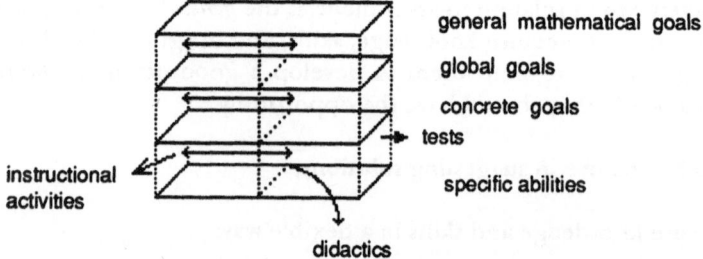

Often the clearest way to describe the orientation of this scheme is to describe instructional activities with each category. These activities are carried out at all levels and are closely related to the chosen didactics. For example, currently popular instruction and didactics in the Netherlands are characterized by a tendency toward realistic mathematics education curricula, as described by Freudenthal (1983), Treffers and Goffree (1985), and Streefland (1987). In the United States, similar methods have been called reconstructive mathematics education by Cobb (1991) and others. One distinction between instructional activities and didactics is that the latter is an ideal formula, whereas the former are carried out in classroom practice. This goal description was developed and written after years of experimentation and after implementation of the new mathematics curricula in the Netherlands. It is unwise to develop goals before the desired curriculum has proved its qualities in wide-ranging field experiments. Only then can one write a realistic goal description.

ASSESSMENT TO ACHIEVE HIGHER-LEVEL GOALS

Testing forms the third vertical connection between the goal planes. Commonly, testing is restricted to the lower level only (basic techniques and concepts), with a few excursions to the plane of the concrete goals. This is an undesirable situation, since testing is most important with regard to the general and concrete goals and their mutual relationships.

During experiments that eventually led to new curricula in the Netherlands for upper secondary students, we were confronted with two serious problems. Time-restricted written tests (not multiple choice) were considered to be improper for testing, especially for testing the higher-level goals. (One only has to think back to the rat problem.) Secondly, under any conditions, designing proper tests is very difficult. When designing tests was left to the teachers, only 15 percent of the exercises really tested at other than the lowest level. Consequently, we started our developmental research to test new formats, adopting the following guiding principles:

- Tests should be an integrated part of the learning process, so tests should improve learning.

- Tests should enable students to show what they know, rather than what they do not know. We call this positive testing.

- Tests should operationalize all goals.

- The quality of the test should not be dictated by its possibilities for objective scoring.

- Tests should be practical enough to fit into current school practice.

Later these criteria were also used in the test development for elementary school level (see Streefland, this volume). The results of our efforts are best shown in examples of actual exercises and exams from experiments in the Netherlands and the United States, and from national examinations in the Netherlands.

National Examinations in the Netherlands

In the Netherlands, there are final nationwide examinations at the end of four, five, or six years of secondary education. Roughly, the six-year curriculum prepares the student for the university, the five-year curriculum for higher vocational training, and the four-year course for the lower vocational level. Otherwise, students tend to start working right after their secondary education. Since 1985, we have had two new curricula for the four-year course, and the experiments (carried out by OW&OC) that led to the new curricula showed with great clarity the problems of achievement testing. The first nationwide examination on the new curricula took place in May 1987. Before 1987, examinations were of the time-restricted, written-test kind (TRWT) with "open" questions and no multiple choice. The open questions were, in fact, very closed because both the answer (a number, a graph) and the solution left no degrees of freedom whatever. Since 1987, modest progress has resulted in a somewhat more open final examination, still of the TRWT kind but more problem-number and text-oriented.

In the summer of 1990, two new curricula for the five-year stream were also introduced: one (B) for the more mathematically gifted students, and one (A) for those who are going to use their mathematical skills in their nonmathematical profession or schooling. The first experimental examination on these curricula (in May 1991) showed that some progress is being made. That is, the exercises are

- more open to multiple solution strategies and results;

- assessing higher-order thinking skills; and

- giving the students the freedom to solve problems at their own ability levels.

A typical final examination in mathematics (A or B) takes three hours. The exam consists of about some five big problems with approximately twenty questions and perhaps six pages of text. (Because of space limitations, we show here only some of the shorter problems.)

An example problem from the A-level examination

One should know, when looking at the exercise shown in Figure 8, that the curriculum does not cover differentiation of functions, even though students study the changes of real phenomena in a discrete way (which prepares in an excellent way for calculus). Instead of the graph of the derivative of a function, the students are accustomed to the discrete apparatus called an "increase diagram." So, the first question is very straightforward and operationalizes only the lowest level. The other question was a new, desired, and long-awaited addition because it involved communicating mathematics, drawing conclusions, and finding convincing arguments—all activities that are too often not very visible in mathematics tests and examinations. Nonetheless, when such new questions were introduced, many teachers were surprised and didn't know what to think of this new development—even though the experiments had given a fair indication of the new approach. Some student responses to the questions in this exercise include the following:

> I would wait for four years and then catch 20,000 kg per year. You can't lose that way, man.

> If you wait till the end of the fifth year, then you have a big harvest every year: 20,000 kg of fish; that's certainly no peanuts. If you can't wait that long, and start to catch one year earlier you can catch only 17,000 kg, and if you wait too long (one year) you can only catch 18,000 kg of fish. So, you have the best results after waiting for five years. Be patient, wait those years. You won't regret it.

Figure 8. A problem from the A-level examination.

If no fish are caught, the number of fish will increase in the coming years. The graph shows a model of the growth of the number of fish.

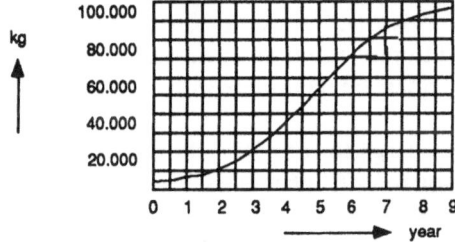

Draw an increase diagram with intervals of a year, to start with the interval 1-2.

The fish farmer will wait some years before he's going to catch or harvest the fish. After the first catch, he wants to harvest every year the same amount of fish as in the first year, and as much as possible. After every catch the number of fish increases again according to the graph.

What would you advise the fish farmer about the number of years he has to wait after planting the fish? The amount of fish that he will catch every year?

Give convincing arguments.

An example from the B-level examination

The students who sit for the B-level exam are preparing for a higher technical vocational school. In general, they need quite a bit more formal and abstract math for their job or schooling. Some three-dimensional insight would also be useful. The exercise shown in Figure 9 gives some insight into what we expect from these students. This exercise is less open than the previous one, but it invokes an understanding of three-dimensional geometry that is somewhat different from traditional approaches. The main difference is that the students need to apply their knowledge to a real-world problem—and there are few bridges to cross the gap between mathematics and its applications. In this exercise, the photo is real, the tower is real, the dimensions are real, the problems are realistic, and the mathematics has some substance. Figure 10 shows the

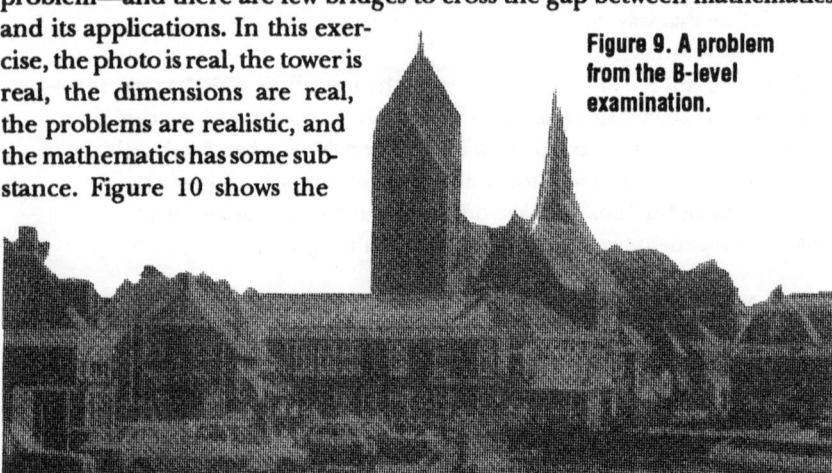

Figure 9. A problem from the B-level examination.

worksheet (left) and two different solutions from students, and Figure 11 shows a solution to the last question, complete with a sketch.

Look closely at the large tower of the church in the picture. The floor plan of this tower is a square of 6 by 6 meters. The roof is formed by four equally sized rhombs. The lowest vertices of these roof parts are at a height of 18 meters above the ground. The top is at a height of 26 meters. The other four vertices of the roof are at a height of 22 meters, each on the axis of symmetry of the sidewalls.

On the worksheet you see the beginnings of a drawing of the tower, in the so-called engineers' projection. Finish the drawing of the tower.

The upper gaps in the walls of the tower are the reverberation holes. Behind those gaps hangs the bell that is rung every half hour. The quality of the sound depends on the shape and volume of the bell room. The floor of this room is at a height of 12 meters above the ground. The ceiling can be constructed at a height of 20, 22, or 24 meters.

Draw to scale 1:100 in one figure the shape of the three possible ceilings. The ceiling is placed at a height of 22 meters.

Compute the volume of the bell room.

Figure 10. Solutions given by two students.

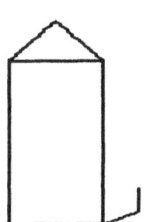 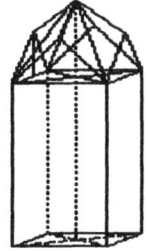 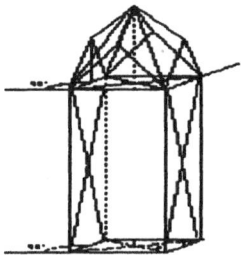

Figure 11. A solution to the question, "Compute the volume of the bell room."

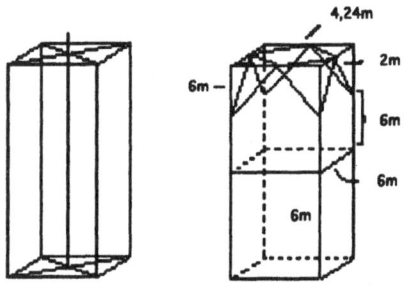

Overall results from students suggest that, if the trend continues toward more open examinations, there will definitely be an effect on the teaching and learning of mathematics. As in many other countries, the teacher (or school) is judged by how well the students perform on their final exam. This leads to test-oriented learning and teaching, but if the test is made according to our principles, this disadvantage (test-oriented teaching) will become an advantage. It will be very difficult and time-consuming to produce proper tests, but in the Netherlands, tests are produced by a government-funded independent institute for achievement testing (CITO), which, in turn, tries to cooperate as much as possible with others involved in mathematics education, especially with OW&OC.

The teacher remains a key factor in the reform of testing. He or she has to accept wholeheartedly the changing emphasis on more open-ended, complex problems. Teaching will become more difficult and complex as well. The teacher will lose some authority because of smart solutions by students. He or she will do less telling and will interact more with the students in the discussion of solutions. So, even if the test producers succeed in making better achievement tests, the teacher remains the most critical factor. The teacher deserves a lot of attention and help in designing school tests. Fewer restrictions will give the teacher a wide variety of possibilities in test design and administration. Thus test problems can become more exciting and rewarding for the student, but they will also become more difficult for teachers to invent and grade.

School tests

For testing the lowest-level goals, the TRWT (excluding multiple choice) remains a very good tool. Even under time restrictions students can perform well on higher levels—if other proper questions are satisfied. The next example (Figure 12) is taken from a 50-minute school test that consisted of three problems (de Lange and Kindt, 1986). The students were taking the A-curriculum and were 16 years of age. Of course, this exercise bears all the marks of a TRWT. It is relatively closed and has a series of short questions to guide the students. It would be interesting to find out what had happened if we had posed only the last question. A more open question is shown in Figure 13.

Figure 12. Example from a school test, A-level curriculum.

Here you see a crossroads in Geldrop, The Netherlands, nearby the Great Church.

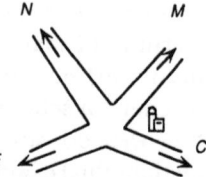

In order to let the traffic flow as smooth as possible, the traffic lights have been regulated so as to avoid rush hour traffic jams. A count showed the following number of vehicles had to pass the crossroads during rush hour (per hour):

$$A: \text{from} \begin{array}{c} M \\ N \\ E \\ C \end{array} \left(\begin{array}{cccc} 0 & 40 & 200 & 30 \\ 30 & 0 & 80 & 50 \\ 210 & 60 & 0 & 60 \\ 30 & 40 & 80 & 0 \\ M & N & E & C \end{array} \right)$$

The matrices G1, G2, G3, and G4 show which directions have a green light and for how long. 2/3 means that traffic can ride through a green light for a period of 2/3 minute.

$$G_1 \; \begin{array}{c} M \\ N \\ E \\ C \end{array} \left(\begin{array}{cccc} M & N & E & C \\ 0 & \tfrac{2}{3} & \tfrac{2}{3} & 0 \\ 0 & 0 & 0 & 0 \\ \tfrac{2}{3} & 0 & 0 & \tfrac{2}{3} \\ 0 & 0 & 0 & 0 \end{array} \right) \quad G_3 \; \begin{array}{c} M \\ N \\ E \\ C \end{array} \left(\begin{array}{cccc} M & N & E & C \\ 0 & 0 & 0 & 0 \\ 0 & 0 & \tfrac{1}{2} & \tfrac{1}{2} \\ 0 & 0 & 0 & 0 \\ \tfrac{1}{2} & \tfrac{1}{2} & 0 & 0 \end{array} \right)$$

$$G_2 \; \begin{array}{c} M \\ N \\ E \\ C \end{array} \left(\begin{array}{cccc} M & N & E & C \\ 0 & 0 & 0 & \tfrac{1}{3} \\ 0 & 0 & 0 & 0 \\ 0 & \tfrac{1}{3} & 0 & 0 \\ 0 & 0 & 0 & 0 \end{array} \right) \quad G_4 \; \begin{array}{c} M \\ N \\ E \\ C \end{array} \left(\begin{array}{cccc} M & N & E & C \\ 0 & 0 & 0 & 0 \\ \tfrac{1}{2} & 0 & 0 & 0 \\ 0 & 0 & 0 & 0 \\ 0 & 0 & \tfrac{1}{2} & 0 \end{array} \right)$$

How many cars come from the direction of Eindhoven during that one hour? And how many travel towards the city center?

How much time is needed to have all lights turn green exactly once?

Determine G = G1+G2+G3+G4, and thereafter T = 30G. What do the elements of T signify?

Ten cars per minute can pass through the green light. Show in a matrix the maximum number of cars that can pass in each direction in one hour.

Compare this matrix to matrix A. Are the traffic lights regulated accurately? If not, can you make another matrix G in which traffic can pass more smoothly?

The example shown in Figure 13 is typical in the sense that it shows that a very simple problem may not have a very simple solution. The teacher has to carefully consider the students' different arguments and has to accept a variety of solutions. At first glance, many students—and teachers as well—tended to find class A better than class B because of the many scores in the 80s. However, some students argued that class B is definitely better because only two students did poorly. On the other hand, some argued that you cannot say anything because the average of both classes seems to be the same. But the median is definitely better for class, and how about the mode? So it boils down to the question what do you mean by better?

Figure 13. A simple problem with a variety of solutions.

The test results of two math classes are presented in a stem-leaf-display:

CLASS A		CLASS B
7	1	
7	2	34
4	3	
55	4	
4	5	
1	6	5
1	7	12344668
9966555	8	114
97	9	1

Does this table suffice to judge which class performed best?

School Tests in the Netherlands and the United States

School tests offer other exciting possibilities. Some of them were explored in the Netherlands and the United States in recent experiments.

Two-stage tasks

In two-stage tasks, the first stage is carried out like a traditional timed written test. The students are expected to answer as many questions as possible, and within a fixed time limit. After having being graded by the teacher, the tests are handed back to the students, while the scores are disclosed. The second stage follows. Provided with information about their mistakes, students repeat their work at home, without any restrictions. This task gives the students the opportunity to reflect on their activities in the first stage of the task.

The essay

In the American and Dutch experiments, a certain kind of essay question was very successful. The students were given an article from a magazine with lots of information in tables and in numbers. The students were asked to rewrite the article making use of graphical representations. The article discussed the problem of overpopulation in the Republic of Indonesia. One of the graphics designed by students (Figure 14) shows clearly the large differences in the spread of the population over the islands. The left bar on each island shows the area of the island as a percentage of the total area. The right bar shows the population of the island as a percentage of the total population. It is clear that Java is overpopulated and that there is more space on some of the other islands.

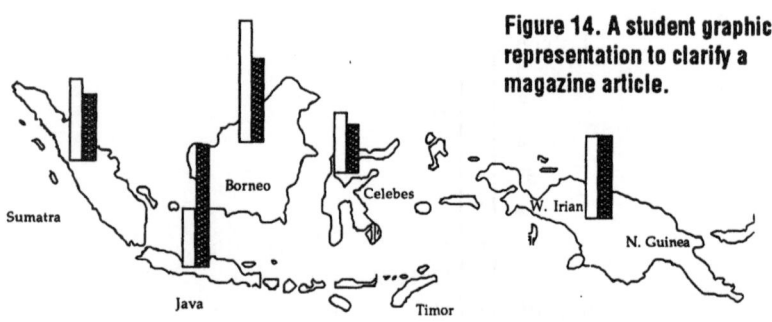

Figure 14. A student graphic representation to clarify a magazine article.

The test test

One of the more promising new ideas gives students the task of creating a test. They are given the directions shown in Figure 15. The results of such a test are surprising. Students are forced to reflect on their own learning process, and the teacher obtains an enormous amount of feedback on his or her teaching activities. It is too early to draw conclusions about this kind of testing, but in 1992 we will publish results from experiments at an American school. A ninth-grade girl studying data visualization produced the test question shown in Figure 16. Developing a critical attitude was one of the general goals.

Figure 15. "Design a test" test.

This task is a very simple one. By now you have worked through the first two chapters of your book and have taken a relatively ordinary test. This test is different. It has one question: Design a test for your fellow students that covers the whole booklet. You can start your preparations now. Look in magazines, papers, books, and so on, for data, charts, and graphs that you want to use. Keep in mind that students should be able to complete your test in one hour. You should know all the answers.

Figure 16. A student-produced test question on data visualization.

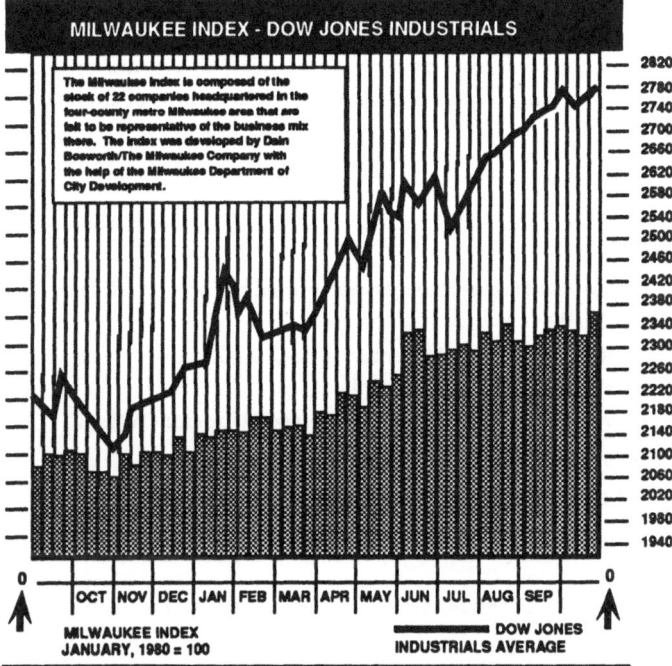

1. What have the makers of this graph done to save space?

2. What impression does this give the reader?

3. What could be done to make the graph less deceiving?

Initial results of this form of testing indicate that the students are hampered by their traditional teaching and learning background. They test lower skills, and they often conform to the chapter partition and do not integrate the different parts. But on the positive side we noted that almost all exercises were different in contexts and subjects and had a fair degree in common as far as mathematics without context is concerned.

FOR THE FUTURE

We are convinced that there are some interesting developments becoming visible. We see evidence that open-ended questions requiring higher-level thinking promote conceptual mathematization. However, the barriers to wide application are massive. We believe that multiple-choice questions should be abolished, but we must design clever open-question tests, and come up with alternative ideas in testing (keeping in mind the principles discussed in this chapter). We must investigate the actual effects on learning and teaching when using "proper" tests and evaluate their practicality. We must encourage test developers to develop such tests, and we must design an innovation strategy to convince teachers and parents and politicians of the merits of such testing.

REFERENCES

Ahlfors, L.V. (1962). On the mathematics curriculum of the high school. *Mathematics Teacher*, 55(3), 191-195.

Balacheff, N. (1985). *Social interactions for experimental studies of pupils conceptions.* IDM-TME, Bielefeld; 98-114.

Cobb, P. (1991). Reconstructing elementary school mathematics. *Focus on Learning Problems and Mathematics.* Vol. 13(2); pp. 3-22.

de Lange, J. (1987). *Mathematics: Insight and meaning.* Utrecht: OW&OC.

de Lange, J. (1989). *Eindtermen wiskunde.* Utrecht: OW&OC.

de Lange, J., and Kindt, M. (1986). *Matrices.* Culemborg: Educaboek.

Doise, W., and Mugny, E. (1984). *The social development of the intellect.* Oxford: Pergamon.

Freudenthal, H.F. (1973). *Mathematics as an educational task.* Dordrecht: Reidel.

Freudenthal, H.F. (1983). *Didactical phenomenology of mathematical structures.* Dordrecht: Reidel.

Goffree, F. (1985). *Ik was wiskundeleraar.* Enschede: Stichting voor de Leerplanontwikkeling (SLO).

Klamkin, M.S. (1968). *On the teaching of mathematics so as to be useful.* Educational Studies in Mathematics 1.

Kolb, D.A. (1984). *Experimental learning.* Englewood Cliffs, NJ: Prentice Hall.

Lewin, K. (1951). *Field theory in social sciences.* New York: Harper and Row.

Streefland, L. (1987). Free productions of fractions monographs. In J.C. Bergeron et al. (Eds.), *Psychology of Mathematics Education-1,* vol. I. Montreal: University of Montreal.

Thom, R. (1973). Modern mathematics: Does it exist? In A.G. Howson (Ed). *Developments in mathematics education.* London and New York: Cambridge University Press; pp. 194-209.

Treffers, A. (1986). *Three dimensions.* Dordrecht: Reidel.

Treffers, A., and Goffree, F. (1985). *Rational analysis of realistic mathematics education.* Psychology of Mathematics Education 9. OW&OC: Utrecht.

9 Thinking Strategies in Mathematics Instruction: How Is Testing Possible?

Leen Streefland

INTRODUCTION

The query in the title suggests other questions. While testing refers to instruction, thinking strategies seem to be the learner's privilege. How are instruction, testing, and thinking interrelated? Our questions raise the problematic issue of evaluating instruction (Mislevy, 1992) and, specifically, mathematics instruction. It is not enough to identify thinking strategies; if they are worth being taught, they have to be justified by general instructional goals, and tools must be developed to somehow diagnose and measure them. Which general goals and what thinking strategies? Globally formulated general goals may allow a bird's eye view of the intended mathematics instruction, but these are still unsatisfactory; more concreteness is needed. In this chapter, goals and thinking strategies will be discussed, while respecting the close connection between teaching and learning.

To give some idea of what we mean by thinking strategies, we first consider and analyze two examples. We then examine general goals suggested by developments in the Netherlands (Proeve van een National Programma) and compare them with those in the United States and the United Kingdom. In all three countries, one notices an unmistakable turn from reproductive to (re)constructive learning and an increased emphasis on such tools as thinking strategies. Thinking strategies are reconsidered, followed by a definition of heuristic mathematics education of the type expected to create a climate in which pupils will develop general strategies.

We will then examine a number of examples to show that strategies for problem solving lend themselves to the design of heuristic mathematics education, and describe five tenets of teaching and learning that are associated with a more realistic approach to heuristics mathematics education.

Next, we focus on testing in realistic mathematics instruction, which entails some serious restrictions. In particular, because thinking strategies are general goals that should pervade all mathematics instruction and are expected to become part of a student's permanent repertoire, they cannot be adequately tested by isolated test items. We then look at an instructional process that supports the general goals of mathematics instruction and discuss implications for future research.

WHICH THINKING STRATEGIES? WHAT MATHEMATICS INSTRUCTION?

An Experiment: Realistic Instruction in Fractions

For two and a half years, we experimented with realistic instruction in fractions. At the end of that time, we gave a test whose aim was twofold: (i) to compare the instructional effects of the new prototype with those of other courses, and (ii) to critically review, and if need be, to revise the new course. One test item asked the following question about sharing peppermint candy:

> Three peppermint sticks are fairly shared by a group of 4 children. Seven peppermint sticks are fairly shared by a group of 10 children. Compare the groups to find out in which group a child gets more, in the group of 4 or in that of 10? How much more? (Streefland, 1988a, p. 382.)

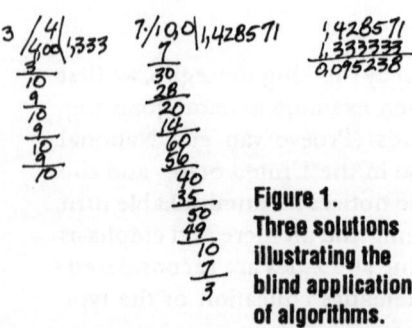

Figure 1. Three solutions illustrating the blind application of algorithms.

The results showed that 15 percent of the pupils in the control group (about 200) divided 4 by 3 and 10 by 7. This led to results in terms of thirds, sevenths, twenty-firsts and to lengthy computations such as those shown in Figure 1.

What was wrong here? The mathematical context of the problem was not correctly understood, if it had been identified at all. Our diagnosis was *algorithmic blindness,* which results from children's individual and isolated experiences with long sequences of sums, rote exercises, and algorithmic standard rules emphasized by mechanistic didactics (Lohman, 1987; Treffers, 1987a). Schoenfeld (1989) describes this phenomenon as "an endless stream of disconnected learning" (p. 100). The relations at the heart of the problem, though rather obvious, are wrongly mathematized, betraying a deficient view of what division means in mathematics (Bell, 1981; Hart, 1981). A particularly ominous symptom is the application of long division to this problem.

Does this mean that the item is not appropriate for testing thinking strategies? It is true that the children whose solutions we reported were satisfied by mock solutions, but this does not answer our question, because the quoted solutions simply reflect the nature of previous instruction. These solutions are witness to a rigid attitude that some action, steered by recipes and rules, must be performed on the numbers in the problem. Such solutions underline our insistence that identifying the mathematical context is an important characteristic of mathematical activity.

As shown in Figure 2, the solutions of children in the experimental group provided a striking contrast. These two children identified "ratio" in the mathematical context of the problem and applied the ratio table as an appropriate tool. One of them found the relative difference in the form $2/_{40}$. Such divergent reactions to the same problem cast doubt on whether tests are a means to uncover and judge thinking strategies. But what does this experience prove? Certainly, mathematics instruction reaps what it has sown (Lohman, 1987)—a simplistic statement if compared with the assumed complexity of the relation between teaching and learning. Or is the relationship itself rather obvious, while attempts to identify and uniquely define it are responsible for the complexity? In either case, on the strength of such examples, we may posit that if tests for thinking strategies must be developed, this must be accomplished within the framework of mathematics instruction. But let us first turn to another example.

Figure 2. Two ratio solutions from the experimental group.

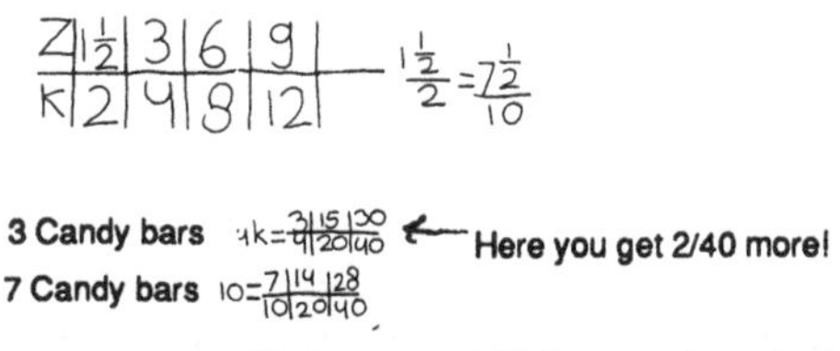

A Second Example: A Doubling Problem

Sixth graders were asked to respond to the "grains on the chessboard" problem (Treffers, 1987a):

The inventor of a chess game was awarded a royal recompensation. He asked for a quantity of grain, to be determined in the following way: one grain on the first field, two on the second, four on the third, eight on the fourth, and so on, going on by doubling up to the 64th. "A trifle," his king thought, "He asks for only a sackful of grain." Was the king right?

The sixth graders working on this problem quickly developed the following table as a result of collective efforts:

field number	1	2	3	4	5	6	7	8	9	10
number of grains	1	2	4	8	16	32	64	128	...	...

One of the pupils fired a warning shot. "One can stop at the 32nd field, and then double the whole." The teacher confronted the group with this proposal. A majority of "pros" rose against a minority of "cons." Yet, one objection rose above the flood. "I don't think it is correct; look, for instance, for the fourth field, there are 8 grains on it, half of them is 4, but there are less than 4 grains on the second field." Everybody was convinced by this reasoning (Streefland, 1980).

On closer inspection, this student's reasoning was remarkable. In fact, this pupil simplified the proposition, accepted the erroneous statement as though it was correct, and reasoned backward from the data in the table to a contradiction, in order to invalidate the proposal. His solution processes exemplify the types of thinking we would like students to develop. In addition to correct identification of the mathematical context, we shall consider these processes—or rather their expressions—in the context of general goals of mathematics instruction. To be sure, our examples also serve to characterize the kind of mathematics instruction we have in mind. We advocate realistic instruction, instruction that strongly relies on the pupils' contributions, their own constructions and productions. This type of instruction requires the availability of mathematical tools, interaction, and clear connections to existing knowledge for the purpose of structuring the mathematics that is to be produced (Treffers, 1987a; Streefland, 1989; Cobb, Wood, and Yackel, 1988, 1989).

THINKING STRATEGIES AND REALISTIC MATHEMATICS INSTRUCTION

General Goals

Mathematics instruction is changing in the western industrialized countries, and there is little need to repeat time and again that "mathematics counts," to admonish youth to "choose exact sciences," and to call for "agendas for action." Development is on-going; agendas are being translated into standards in the United States (Commission on Standards for School Mathematics, 1988), national curricula (National Curriculum Council, 1989) and our Dutch Proeve (Treffers, 1989). All of these call for instruction that goes beyond the traditional reproduction-directed teaching of arithmetic. General goals are reformulated in terms of desired abilities and attitudes. In the Proeve, for example, goals include developing students abilities to relate real problems to mathematical content; to search for connections, rules and patterns; and to use and to verbalize searching and thinking strategies, such as trying, experimenting, simplifying, observing, exaggerating, estimating, and generalizing (Treffers, 1989). Additional goals include developing a mathematical approach, applying a variety of strategies and methods, interpreting mathematical results within their original context, developing mathematical language, and reflecting on one's own mathematical activity and that of others. Similar goals have been formulated for the United Kingdom and the United States.

While the cited publications agree on basic skills and on the importance of estimates, they disagree concerning the learning of algorithms. While the United Kingdom would abolish algorithms, and the United States would do some pruning, the Netherlands favors maintaining those acquired by progressive schematization. There is general agreement on reproducible knowledge such as that pursued by traditional instruction as a long-term goal, although the emphasis has shifted to applicability. As far as skills are concerned, the reproduction-directed tests will remain important, although on behalf of goals such as those just mentioned, there is a demand for tests above the level of stimulus and response. However, higher levels of thinking are difficult to test when the goals change from reproductive to (re)constructive learning. Relating, searching, using and verbalizing, trying, experimenting, and so on, require more open problems, which are more difficult to score.

The shift from product to process goals in mathematics instruction entails moving between two poles:

acquiring	developing
skills	thinking strategies
routines	general abilities
ready knowledge	

This broad spectrum must be covered by mathematical instruction if it is our aim to develop tests for thinking strategies and for a mathematical attitude that expresses itself as an intensive and long-lasting characteristic of pupils' mathematical activity. This requires that a great many organizing and structuring activities, such as those mentioned among the general goals, be left to the learners.

Though the envisaged general goals are to be pursued throughout the curriculum, the manner in which this may be done can be shown by examples, provided they are offered in their didactical context. Some examples have already been given; other activities to promote thinking strategies will now be discussed.

Teaching Thinking Strategies

Exploration

Some thinking strategies can be and have been taught by the way of tricks: to solve the set problem algebraically, look for what is unknown, which is called x (and y), and for its interrelations! Once the trick has been understood it becomes an algorithm. But rather than telling the trick, one can heuristically guide the learner to (re)construct it. Such instruction is apt to promote an attitude of searching, developing, and constructing.

In the American literature the term "higher-order thinking skills" is taxonomical, as are Bloom's (1956) "analysis," "synthesis," and "evaluation." Sometimes one speaks about acquiring "mathematical power," or "acquiring a mathematical point of view" (Kaplan, Yamamoto, and Ginsberg, 1989; Schoenfeld, 1989). For our purposes, classifications of strategies are premature. First we want to identify process characteristics of mathematical activities, which may then be classified.

Thinking strategies evolve from mathematical activities. Acting is being continued in thinking; it becomes mentalized. Although the word "strategy" may suggest otherwise, it is not meant to imply consciousness; as described by Hadamard (1945), someone's mathematical abilities can manifest themselves without the person being aware of them, and before being consciously applied they may have changed into mere tactics (Freudenthal, 1983).

Thinking strategies are acquired by learning, which is not restricted to the acquisition and the improvement of insight but includes change of basic attitudes toward new problems. How can this learning be promoted by teaching? Instructing students the way computers are programmed is not promising, as it ignores the part pupils should have in the production of their own strategies. Reinvention under the teacher's guidance proves more useful, for example,

- learning column multiplication and division according to strategies of estimating and clever reckoning, which in the long run, consolidate into the algorithms (Treffers, 1987b); and

- learning to handily compare, order, and compose distribution problems and their solutions, which gradually develops into the relative comparison of ratios, and the addition and subtraction of fractions with common denominators (Streefland, 1988a).

In these cases, local discoveries evolve into standard rules and tricks; unconscious thinking strategies develop into conscious tactics. With regard to the progress within mathematics, the student is working on a higher level. As soon as acting on a new level has become familiar, ever higher levels can be achieved. For the learner this means a reappraisal of the level reached before. Let us illustrate this using an example from a course on fractions:

If 4 children share 3 bars of chocolate, how much does each of them get?

There are many ways to proceed: first, two bars of which each gets one half, and then the third with one quarter for each; or for each of them in turn one quarter of every bar; and so on. The matter operated on is the (imagined) bars and their pieces, and the operations are those of dividing and attributing. In spite of the terminology used the parts involved are pieces rather than fractions (first level). This changes as soon as the mutual relations between the parts are being considered (second level). At an even higher level, the pupils are operating on relations originating from the distribution. These relations can be elicited by free production tasks such as the monographs shown in Figure 3 (third level).

The methods used to produce such monographs (equivalence, applying commutativity, and so on) lead to the conversion to algebra of the system of fractions (Streefland, 1987), which is carried by the pupils themselves: "... the operational matter of the lower level becomes a subject matter on the next level" (Freudenthal, 1973). Not until the child is able to reflect on its own activity is the higher level accessible and can a reappraisal of level take place, as happens in the case of the fraction monograph; it

Figure 3. A fraction monograph is a symbolic representation of relations discovered on a lower level in the mathematization process.

$$\frac{3}{4} = \frac{1}{2} \quad ; \quad = \frac{1}{4} + \frac{1}{2} \quad ; \quad = \frac{2}{4} + \frac{1}{4} \quad ;$$

$$= \frac{1}{4} + \frac{1}{4} + \frac{1}{4} \quad ; \quad ; \quad = \frac{2}{8} + \frac{2}{8} + \frac{2}{8}$$

$$= 3 \times \frac{1}{4} \quad ; \quad = \frac{1}{4} \times 3 \quad ; \quad = 3 \times \frac{2}{8}$$

$$= 1 - \frac{1}{4} \quad ; \quad ; \quad = 1 - \frac{2}{8} \quad ; \text{ etc.}$$

becomes a concrete source for algebra (which is at the first level again).

Earlier we highlighted four remarkable features of solving procedures: inverting, simplifying, accepting erroneous statements, and identifying mathematical context. Our provisional term for such features will be strategies. But more important than the term are the conditions and circumstances that favor their development.

Polya's systematic heuristics

Problem solving is one of the broad fields dominated by thinking strategies, also known as heuristics. George Polya was the first to systematically study thinking strategies on a large scale, and to support their usefulness by innumerable examples. In a number of books, Polya attempted to teach problem solving, leading his readers by the hand in order to help them understand the ways of thinking that serve as roads to discovery and invention. He gives specific suggestions such as (i) consider a special case of the problem, (ii) look for an appropriate similar problem, (iii) invert the order, (iv) assume that the problem has been solved, (v) make a drawing, and so on (Polya, 1945). Both form and content of the approach (for instance, inverting versus drawing) provide a rich variety of ways to solve the problems: strategies, rules of thumb, know-how, and whatever may be understood as heuristics. For sources Polya looked to himself as a problem solver as well as to the history of mathematics.

Euclid's deductively systematized geometry is a one-way road of argumentation: given to be constructed construction, hypothesis to be proved proof. It is an outstanding example of system constraint, against which Polya raises objections. This order, he argues, is just opposite to the natural order of invention. His story of the origins of deduction differs from Euclid's. Polya mentions Pappus, who centuries after Euclid pointed to a kind of heuristics, called analysis, where things to be constructed or theorems to be proved are supposed as though they were already found or proven. Besides this backward solving or regressive arguing, Pappus considers the synthesis of forward solving or progressive arguing. For geometric constructions Pappus recommends the Greek surveying tradition, which is analysis: "Suppose what has to be done as achieved" and "draw a hypothetical figure which assumes all conditions of the problem as fulfilled"! (Polya, 1945, p. 131). As a matter of fact, long before Polya, traditional textbooks dealt with construction problems in a regimented heuristic way, starting with an analytic drawing like Pappus' hypothetical figure. However, this heuristic approach did not extend to theorems and proofs.

Perspective shifting

Kruteskii (1976) talks about rapidly switching from a direct to a reverse train of thought and the flexible performance of mental processes. Freudenthal (1978, 1979, 1990) speaks of change of sight or change of perspective, while distinguishing several kinds of "shifting":

> Change of sight, a complex field of strategies with the common feature that the positions of what is given and what is sought for (data and unknowns) in a problem or field of knowledge are—partially—interchanged; including the recognition of wrong changes of sight. (1978, p. 7).

In the context of geometry this "field of strategies" includes choosing a locally different view point, interchanging what is given and what is asked for, and inverting the order of steps in a construction. Moreover, Freudenthal points to four other big strategies:

- Identifying the mathematical context.

- Developing mathematical language (rather than getting it imposed) above the ostensive and the linguistically relative level, in particular at the level of conventional variables and functional description.

- Grasping the degree of precision that is adequate to a given problem.

- Dealing with one's own activity as a subject matter of reflection in order to reach a higher level.

The process versus the product

However marvelous Polya's trail-blazing work may be, it does not adequately answer our problem. Freudenthal (1983) suggests that the most important drawback concerns the interaction of author and readers. Rather than having readers thinking along with the author, an author thinking along with his readers would be most desirable didactically, though it is difficult to realize. Nevertheless, thinking along with the pupils is what matters, both in mathematics instruction and in testing for higher abilities.

It would be helpful to know more about the origin and the development of strategies. Yet what now looks like a written record is rather an abridged and smoothed revision of the original historical process, thanks to an activity of reorganizing, restructuring, redefining, abstracting, generalizing, unifying, and so on. To be sure, all these verbs refer to strategies that have driven the historical learning process. Polya (1945) was fully aware of the drawbacks of smoothed out mathematics, which disavowed that its historical development occurred

> ... heuristically, by searching, at random and intentionally, by finding by serendipity or systematically. This then is heuristics: the scribblings, as opposed to the clean copy as it is printed. (p. 40).

This, indeed, is what we are looking for. It is in the scribblings that the mathematical activity of the problem solver is adequately reflected, and where its characteristics surface as symptoms—formal symptoms of mathematical activity as I called them (Streefland, 1980), which have to be interpreted in their relation to the mathematical tools used, and within the individual learning process or personal learning history.

Lists of heuristics are problematic. One argument is that such a list asks to be memorized, like information being fed into the memory of a computer. Is this the way to educate problem solvers? Catalogues of strategies, as drawn up after Polya, have never been meant to be memorized (Davis and Hersh, 1980). Schoenfeld (1987), for instance, preferred

> ... to create a microcosm of mathematical culture. Mathematics was the medium of exchange. We talked about mathematics, explained it to each other, shared the false starts, enjoyed the interaction of personalities. In short, we became mathematical people. (p. 213).

He and his students were acquiring "mathematical power" and a "mathematical point of view." What really matters are the pieces of scrap paper as opposed to the fair copy, including the mental ones of the pupils. Teachers need to think along with the pupils, as well as to help them to become conscious about the strategies applied in their mathematical activities (Freudenthal, 1990).

Gradually the attention has shifted from learning to teaching, from thinking strategies to their symptoms as observed in learning processes, and consequently to heuristics—not in the form of a checklist for learning mathematics, but as guides for heuristic mathematics instruction.

Instructing and testing heuristically

A lesson involving the story "Grains on the Chess-board" (Treffers, 1987a) reflected this kind of instruction. A mathematical idea familiar to the teacher was to be grasped by the pupils, through a series of decisions and choices. The teacher would cautiously step in wherever the pupils' proposals and contributions asked for it, and would facilitate discussion about compounding units with comments such as the following:

> *The counting time:* 1 grain per second, which is 64 seconds for the seventh, that is, about a minute, 64 minutes for the 13th, that is about one hour, and so on. The one who is charged with the 64th would need about 250 billions of years!
>
> *The needed volume:* the eleventh field gets 1024 grains, that is, about a thousand, weighing 30 grams, or filling a measuring glass of 50 cm^3. The total grain would fill a cube with an edge of 8 km.

The shift involves thought-experiments in the classroom and constantly grasping the degree of precision that is adequate to solving the problem:

> In his mind the teacher has prepared a clean copy, which he expects the learner to produce, and even a somewhat vague series of scribblings leading to the clean copy—a plan, which in the actual experiment must be modified according to the student's cooperation. This is what from olden times they called heuristic instruction, quite unlike the "modern problem solving," which can mean anything from letting the student muddle up to having him tied to leading strings. (Freudenthal, 1983, p. 4).

For the teacher it means thinking along with the pupils. So mathematics instruction becomes more than a mere objective body of knowledge and techniques.

PROBLEMS THAT TEST FOR GENERAL GOALS

In this section, we will discuss how to test for general goals, by what kind of problems, and where to find such problems, beginning with a series of examples.

Some Examples With Variations

First example: Buses. Passengers stepping on and off buses at stops has proved an appropriate informal context for adding and subtracting in elementary Dutch textbooks. Figure 4 shows two different versions of the bus stop problem. Note that the change at the stop can be described by pairs of numbers (a suitable access to negative numbers?). A more open variant is obtained if the number of passengers before the stop is also omitted in the data.

Figure 4. Buses as a context for adding and subtracting.

Second example: Changing money. Among other instruments, tests are being used in the MORE (Methoden Onderzoek Rekenonderwijs) project, a comparative study of instructional methods. One of the items shown in Figure 5 gives 15 florins in a purse, three objects costing 7, 6, and 2 florins, respectively, and various choices for the amount of change a buyer might receive. The instructions are to buy one item and cross out the correct change. Interesting variants of the problem include (i) omitting the correct amount of change or (ii) omitting the prices on the merchandise.

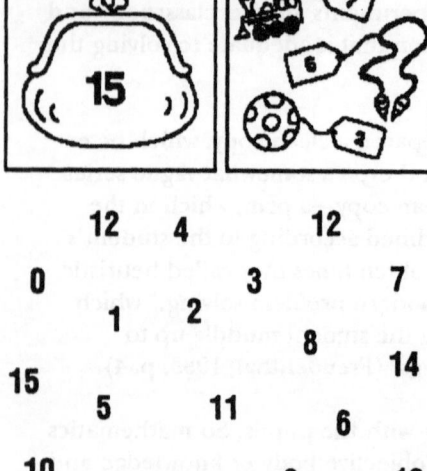

Figure 5. Another context for addition and subtraction: making change (Van den Heuvel-Panhuizen).

Third example: Sharing fairly. Distributing has proved highly productive as an access to fractions, in history as well in teaching (Streefland, 1988b). For instance: If four children share a bar of chocolate, each gets one quarter of a bar. After a while such distributing activities can be inverted: How many children shared how many bars if John got a quarter of a bar? After three bars for four children one may ask for the origin of $1/_2 + 1/_4$ bar.

Fourth example: Division with remainder. What is the result of the division 6394 + 12 = ? Various results are proposed,

a. 532

b. 533

c. 532 rem. 10

d. 532.84 rem. 4

e. 532.833333

f. about 530

and students are asked to invent appropriate stories for each.

Fifth example: Scale. Once scale has been introduced, it applies to a variety of situations. Figures 6 and 7 ask students to think about scales in direct and reverse situations.

Sixth example: An experiment on contents. Cans, barrels, and bottles of different shapes and capacities are filled, unit by unit, steadily with constant velocity (see IOWO, 1976); as a thought-experiment this situation involves serious mathematics, starting from predictions, which are supported by estimates and qualitative graphs such as those in Figure 8. Relating the changing cross-section to speed with which the fluid rises provides a remarkable change

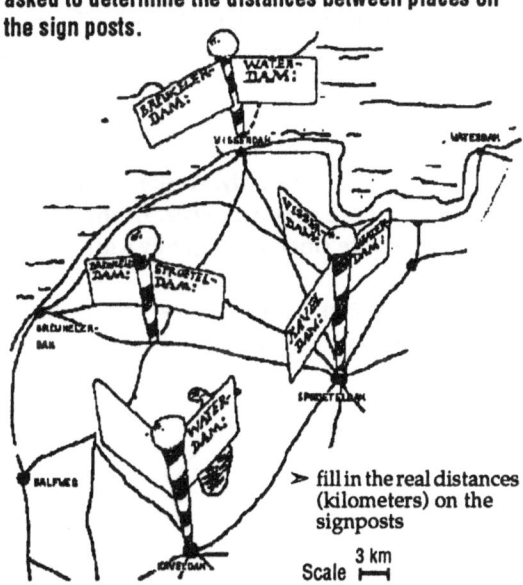

Figure 6. A scaling exercise in which students are asked to determine the distances between places on the sign posts.

➤ fill in the real distances (kilometers) on the signposts

Scale 3 km

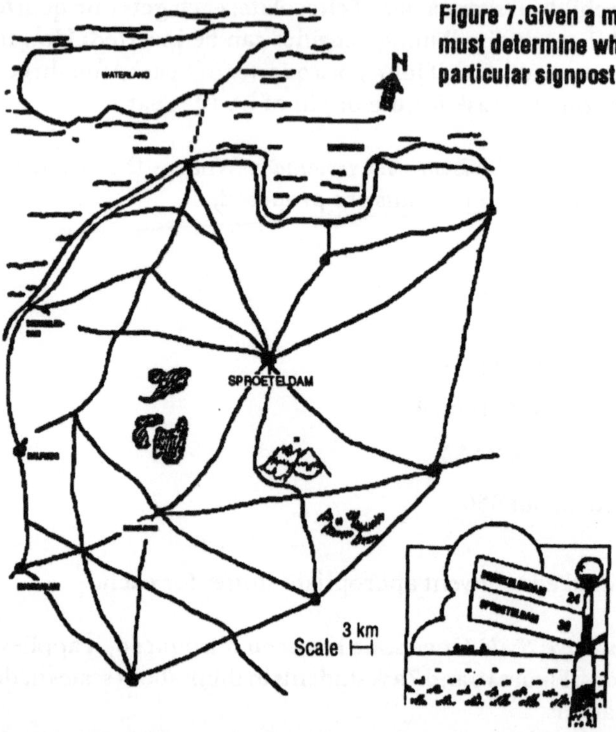

Figure 7. Given a map, students must determine where to place a particular signpost.

of perspective, which gives rise to functional ideas. (We may mention here the Italian mathematician Cavalieri (1598-1647), who compared volumes by the principle of the "mounting" cross-sections.) Pupils are asked to consider the sphere-like bottle. Qualitative reasoning on speeds may run as follows:

> At first the level will rise, and as the sphere gets wider the level will rise more slowly. At the middle, the opposite happens as the sphere gets narrower. In the neck, the level rises at the same rate since it is equally wide.

The same story can be read in an unsophisticated graph. After a few (thought or actual) experiments like this, the problem can be inverted:

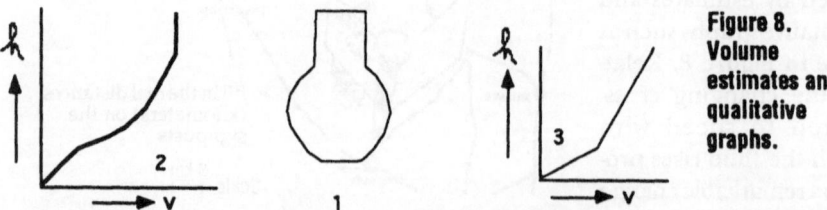

Figure 8. Volume estimates and qualitative graphs.

given a graph, the question is asked, What does it tell about the kind of container? There may also be a choice of bottles and graphs to select from.

The Examples More Closely Considered

The examples we have considered all have a common element: a direct problem in the first instance is afterwards inverted. Of course, this is no new idea. Traditional arithmetic applied it in splitting quantities as a counterpart to adding, indirect sums like 7 + ? = 12 were derived from direct ones, and similar ones were derived for the other operations. It was, however, a restricted and hardly paradigmatic repertoire, (see Freudenthal, 1978). But let us review the examples more closely.

Buses. Of course the variants of the direct sum can be solved by trial and error, but they can as well challenge pupils to change their perspective and solve the problems more systematically, that is, by expressing the difference by means of number pairs. Unlike the purely arithmetical indirect sums, these problems ask for comparing states or situations, which stimulates fresh thinking. Indeed, the required mental actions are meaningful on account of the context, which strongly suggests comparisons and even allows for a variety of solutions.

Changing money. According to some authors, heuristic means not yielding to obstacles (Polya, 1945; Streefland, 1988a; Lesh, 1990). As shown in Figure 9, a student must sometimes improvise. If prices are omitted, new comparing situations arise, which ask for more imaginative reflection than traditional indirect sums require.

Figure 9. A student adds the correct amount of change when it is missing from the given options.

Sharing fairly. Given the share per person, the quest of the original distribution problem refers the solver to the context source. At the same time the results, 1 for 4, 2 for 8, and so on, anticipate the future subject of equivalent ratios and fractions. The compound share anticipates the addition of fractions with different denominators (Streefland, 1988b). All of these examples illustrate the iterative function of reflection in long-term learning processes: prospect and retrospect included in one activity (Kilpatrick, 1985).

Division with remainder. This operation does not acquire its full meaning until diverging results can be fitted to appropriate contexts. This task deserves attention. Research has revealed that pupils often don't know what to do with remainders. (Carpenter, Lindquist, Matthews, and Silver, 1983.)

Scale. Starting one's solution with an attempt to place the signposts may be considered as a symptom of testing hypotheses, provided the distance to indicated cross-roads is methodically being investigated. Starts at the cross-roads show strategical changes of perspective. In both cases, the actual strategy can only be diagnosed on the strength of paper-and-pencil work or by close observation.

Contents. The direct (thought or actual) experiments involve all three dimensions of the cans and so on, on which content depends. Because of the special attention put to the variable height, the changing cross-section also draws attention and comes to the fore. As a consequence, attention will shift from the changing height to the cross-section, and their interdependence (or connection) is reflected by the graphic image. Conversely, if a graph within this context is given, it can be translated into a story of change that is related to the shape of a bottle.

Design of Heuristic Problem

Change of perspective

The common features in the preceding series of problems can hardly be accidental, but why were all these problems given the same form? A posteriori this fact seems to betray a common strategy of design, although this does not imply that the designers were a priori conscious about it (Treffers, 1987a). So what does this strategy mean?

Against the background of the big strategy of change of perspective, which is applied to create mathematics and solve mathematical problems, another change of perspective took place: a strategy within mathematics was lifted to the didact level, where it has resulted in the design of problems, assignments, and queries that may incite strategical behavior. One may even posit that as a didactical strategy, change of viewpoint was used as a building block of a theory of design for heuristic mathematics instruction.

This means that activities like designing mathematics instruction (lessons, courses, programs), developing tests for general goals, and (developmental) research can be interpreted as heuristic processes, where changes of perspective are applied as strategies. Their successful application reflects a didactical attitude that does not emerge spontaneously but develops in the course of the activities. Observing long-term individual learning processes occupies a central place in developmental research (Streefland, 1988b). It is useful for gaining a full view of the phenomenon of change of perspective, both by the teacher and the learner:

> Learning processes are marked by a succession of changes of perspective, which should be provoked and reinforced by those who are expected to guide them. (Freudenthal, 1990.)

Change of linguistic level

Are there more thinking strategies that can be raised to design strategies? I think linguistic level has this characteristic. Freudenthal, talked about developing mathematical language above the ostensive and the linguistically relative level, in particular at the level of conventional variables and functional description (1973, p. 2). This strategy involves making students conscious of mathematical relationships and facilitating their verbal communication, as can be seen in the following two examples.

Sharing fairly. "In a restaurant 24 children order 18 pancakes. Distribute the pancakes among those children." As shown in Figure 10, one student, Ann, drew the children sitting at one table. She soon changed her drawing and put the children at two tables. She could have gone farther, but she got fed up with drawing and asked for help, which was provided. The linguistic level could be changed by telling the children, "Make a sketch (a symbolic picture) showing that 24 children sit around a table where 18 things are to be distributed. Make a suitable sketch for two or more tables!" Making symbolic pictures is a way to get pupils heavily involved in the production of symbols and other tools to do mathematics, which includes making them conscious about mathematical relationships.

Content. Among the barrels and bottles of the (thought or actual) experiments there may be cylinders. Linguistic style can develop from

Figure 10. A student's symbolic picture for the fair sharing problem.

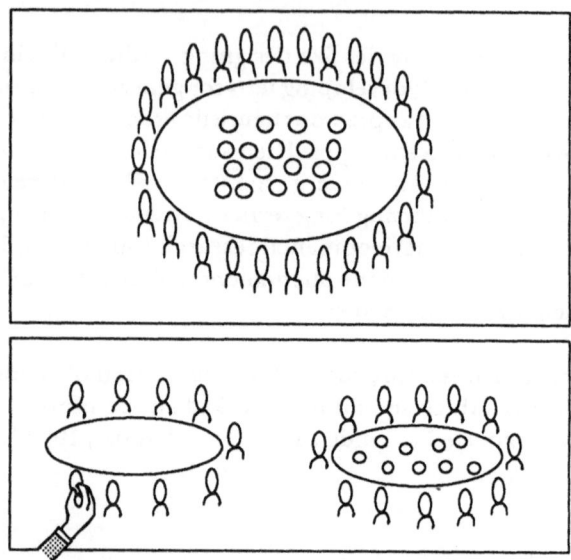

descriptions such as, "If *this* measure is added, the water rises *that* high," up to introducing decimal fractions in order to draw up and read tables such as this one:

number of deciliters	1	2	3
water level (in cm)	1.7	3.4	5.1

By experimenting, the pupils establish a basis for operating insightfully with decimal numbers. They literally experience that $1.7 + 1.7 = 2 \times 1.7$ is not 2.14, which is a common mistake among pupils who underwent a rule-directed training. In the above approach, notions on decimal numbers and operating behavior develop simultaneously. In the wake of such multifarious experiences, pupils themselves can develop algorithms for the operations by progressive schematizing (see Treffers, 1987b).

Figure 11. Graphical solutions.

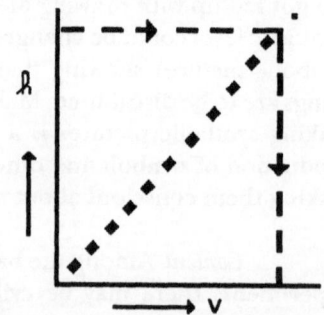

As shown in Figure 11, a graph may facilitate another change of perspective, that is, from the repetitive addition of the measuring unit (or multiplication) to the division of the height of the cylinder by it. If measured in such units—1.7cm in the present case—the height of the cylinder represents its content in deciliters. This insight can subsequently be verbalized and finally be put into an algebraic formula.

REALISTIC INSTRUCTION RECONSIDERED

In the examples above, the design of problems (and of larger parts of courses) is the source of thinking strategies. Change of perspective is only one such thinking and design strategy characteristic of realistic (heuristic) instruction. Before discussing strategies for realistic instruction, let us consider some of its goals.

The Goals of Realistic Instruction

The indicators of a mathematical attitude are many. A few of its characteristics are:

- creativity;

- organizing and structuring the data and phenomena in the problem;

- discovery and use of connections, analogies and isomorphisms, that is, similarities in context or in structure;

- modeling, abstraction, generalization, and transfer;

- change of approach, of viewpoint, of thinking level (Skemp, 1979);

- forming and changing hypotheses;

- circumventing obstacles (for instance by adding lacking information);

- breaking away from data in order to transgress fixed limits;

- being prepared for new problems and situations; and

- being able to reflect on one's own and other's thinking and acting.

These are also the goals of realistic mathematics instruction. To develop tests for these traits requires knowledge of children's thinking, as we discuss next.

Some Strategies for Realistic Instruction

The grand strategy by which mathematics teaching and learning are related to each other and understood in their mutual relation is couched in the principle of permanent change of viewpoint in both instruction and in instructional research. To take the pupils' viewpoint means, as we have shown, trying to have them disclose their thoughts. This allows us to think along with them and to stimulate them to express their strategies of acting and thinking and to become conscious of them.

How do we do this, and what does it mean for instruction? First of all we repeat what has been stated earlier, namely, that evaluations, tests, and any other tools for determining the effects of instruction, must belong to the instructional process itself (Cobb et al., 1989, Carpenter and Fennema, 1989). In this context, let us look at the tenets of realistic mathematics instruction embodied in the Wiskobas program, as formulated by Treffers (1987a; Gravemeijer, van den Heuvel, and Streefland, 1990), that is, constructing by concretizing.

Let us consider division, for example. If division is to be more than a formal operation, the meaning of the remainder must be included in the learning matter. An example, borrowed from a lesson from a refresher course for primary school teachers (Dolk and Uittenbogaard, 1989) will illustrate how to do this. It is a model lesson in which 8- to 9-year- olds, who were never taught division, are asked to construct solutions for two division problems:

> Tonight, 81 parents will visit the school. The tables seat six persons each. How many tables are needed?
> A coffeepot holds 7 cups. If each parent is to get a cup, how many pots have to be prepared?

By schematizing at various distances from reality, the children construct their solutions, keeping the building bricks for the long-term learning of division. They are learning to provide the remainder with a meaning that fits the context, such as, an extra table has to be reserved for the "remaining" parents. Figure 12 shows two student solutions to the first problem. The second question is answered by similar considerations.

Realistic instruction offers the pupils the opportunity to produce their mathematics by means of their own constructions. This is accomplished by explaining events at bus stops, determining changes and prices, acting in distribution situations, describing the results, comparing different situations, devising situations for given distribution results, developing methods for the arithmetical operations (such as the division with remainder), thinking up situations for prescribed operational results, and finding distances at cross-roads and cross-roads if the distances are given.

All activities of the pupils are embedded in the context of concrete situations (Treffers, 1987a; Streefland, 1988a). These are chosen not arbitrarily but with a constant eye on what is to be learned: adding and subtracting natural numbers, equivalence and comparison of fractions and ratios and related operations, and so on. Connections are made from real situations to the resulting mathematics and, wherever possible, the other way round, as in the division with remainder and the content problems. Indeed, deriving the mathematics of real situations asks for its counterpart, the realization of mathematical matter, which means interpreting and explaining it within a fitting context. Placing the signpost in the partially mathematized reality of the map mentally in order to determine the crossings at distances equal to those indicated on the signpost and so on, is another example where consciousness about the change of perspective plays an important role.

Figure 12. Two student solutions to the table problem.

Let us emphasize it once more: what matters is not merely constructing, but constructing by concretizing. That is, realities are chosen as a source of mathematical ideas, operations, and structures which are to be learned, and once this level has been reached, they are not discarded but rather are cultivated as fields of application.

Developing tools for the transition from concrete to abstract

To bridge the distance between real problems and their mathematical counterpart the pupils need adequate tools. These are forged by the pupils themselves in a process of developing mathematical ideas that turn the concrete into the abstract. For example, the pupils themselves develop methods for the operations on natural numbers, and later on decimal fractions, by progressive schematizing. They develop schemes for seating arrangements in order to reduce large distribution situations to more manageable proportions, which are again progressively reduced. They develop tools to compare and order ratio situations, and to determine their relative differences, as in the proportion table. By further schematizing, their efforts will eventually lead to more standard methods.

Attention should not be restricted to the means of schematization. There is a wide variety of tools and instruments, as well as symbols, patterns, (visual) models, notations, and so on. In the case of the seating arrangements, the construction of a symbol that visually refers to the sharers *around* and the pancakes *on* the table is critical to the development of ratio and fraction ideas. Schematizing (Figure 13) takes place against the background of the story on seating arrangements. This context situation so forcefully influences the schematization that it serves as a model situation. To say it even more emphatically: the context itself functions as a *situation model* (see Streefland, 1988b).

Even from this brief exposition it must be clear that raising one's level of thinking from concrete to abstract, as understood here, strongly differs from what is known as Bruner's modes of representation. Rather than the concrete material (whether palpable or mentally represented), it is the distance from (or nearness to) the real situation that counts. Guided by the teacher, the pupils move from the concrete level to that of the formal procedures and knowledge of the subject matter when it is systematized, through informal approaches and strategies of their own. The aids and tools they develop are to detach them from the concreteness and to garnish the intermediate level, where material, borrowed from the context, undergoes all kind of transformations (shortening, streamlining, and symbolizing), until rules and formulas are produced. It is the way to prepare for and eventually to realize the intended abstractions. As the learning process goes on, levels are being reappraised: for instance, what once has been abstract

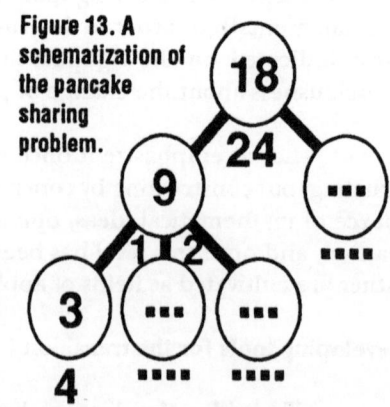

Figure 13. A schematization of the pancake sharing problem.

knowledge of number will become a concrete source of algebra (Vanden Brink, 1985).

Free productions to promote reflecting

Reflection on one's learning process can be incited by suitable assignments for free production, such as making a booklet for next year's class that contains easy, average, and difficult sums (see Streefland, 1988b), or thinking up a test on a certain subject or theme. Besides revealing states of affairs in individual learning processes, such assignments may reflect the received instruction. Developmental research has proved their influence on helping students to realize higher-order goals.

Interactive instruction and learning as a social activity

The open, realistic approach is apt to loosen the pupils' behavior. Reality as a source and a domain of application helps problems become concrete and imaginable. As everybody becomes involved, reactions of different qualities and levels are elicited. By developing mathematical tools together pupils can progress together, even though there is much personal freedom in the construction and production process and in the different levels of schematizing. There is enough room during the lesson for every pupil's personal contributions: solutions can be paralleled; ideas exchanged; approaches on different levels of schematizing considered and compared; arguments criticized, refuted, amended, supported; and ways to continue the learning process negotiated.

By promoting reflection, group activity contributes to the individual learning processes; in fact, it is the only way to arrive at genuinely individualized instruction. Interaction creates a social environment of learning, where beyond being accepted, the children's ideas and informal strategies and procedures are encouraged and exploited on behalf of the individual and collective activities of progressive mathematizing.

Intertwining learning strands to structure the mathematics

Fair sharing is at least as old as fractions. Traditional instruction, however, derives fractions from an impoverished form of fair sharing, that is, within the unit. With this material a straight course is sailed to the arithmetical rules for fractions. Distribution situations with more objects and more sharers are richer by far. By structuring the company of sharers, the distribution itself is structured and simplified. By seating the sharers at different tables, the children construct relations between portions, partitions, and sharers, which eventually produce the mental objects of fraction and ratio.

By this switch from objects to sharers, ratio and fractions are intertwined. Ratios put order on distribution situations using comparison: six bars for eight children is the same as three bars for four—everybody gets the same. The pattern of table arrangements works as an instrument to create equivalent tables, which concretizes the equivalence class for fractions.

Figure 14 shows another example in which learning strands are mixed (Streefland, 1985). Students are shown two cakes and asked: Which cake has more pieces of ginger? From which cake will a bite contain more pieces of ginger? By intertwining area with the counting of large quantities, learners become early familiar with density. The pupil faces two competing ideas: absolute numbers and density. To settle the dispute the teacher tries to make the learners conscious of these competing ideas. Natural connections foreshadow mathematical ones, which is the proper aim. The structure of the envisaged mathematical knowledge is from the start predesigned in the context and the mathematical activities. Intertwined realities and mathematics are of great help in applications.

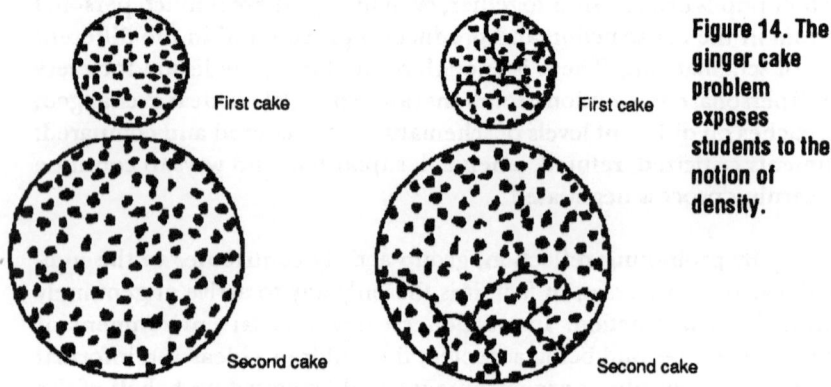

Figure 14. The ginger cake problem exposes students to the notion of density.

Let us repeat that realistic mathematics instruction has been interpreted here in the spirit of the Wiskobas project of the former Institute for the Development of Mathematics Education (IOWO). However, mathematizing in a process of guided rediscovery can be strongly emphasized, both with regard to horizontal-vertical mathematization and to the freedom of the learner. This holds for teachers and learners equally. Disregard for reality connections and being too focused on reality are opposite attitudes, which may distinguish algorithmically gifted from algorithmically less gifted pupils, respectively. Teachers should steer a clear course between the two.

sharing problems before instruction in fractions; counting large unstructured quantities; structuring quantities; and repeated addition, multiplication, and multiplication tables.

- Assignments for free productions such as "Think up as many sums with the result 24 as you can!"

- Inventing a problem if a story is given, such as, "A tourist drove from Amsterdam to Paris. After two thirds of the distance he had one quarter of gas left in the tank...."

- Looking for missing information, such as, "The broadcast says there is a queue of 5 km at Bottleneck Bridge. How many cars may be involved?"

- Cognitive conflicts such as that shown in the time-distance graph on a running competition pictured in Figure 16. The students are asked, (i) "Did the runners collide?" (a question suggested by the well-known inclination to interpret the graph as a path) and (ii) "After how many seconds/meters did they catch up?" (a question that may help in answering the first one). These questions may be repeated with the time-speed graphs shown in Figure 17.

- Inviting changes of geometrical perspective; for example, students are asked if the pictures in Figure 18 could all represent the same situation. They are also asked change-of-perspective questions such as, "Jan thinks that $1/2 + 1/3$ equals $2/5$. Is he right? Why or why not? What might he have thought?" and multiple-choice questions with no acceptable answer, or to think up their own multiple-choice questions with plausible looking answers.

Estimating combined with clever calculating.

Assignments that ask for combining former activities, such as the problem posed in connection with Figure 19: "In this ground plan of paths around a cross-like lake, a route from A to B along the paths must be as short as possible. How many such routes are there?" This problem appeals to knowledge about lattices, the Pascal triangle, and the paths model for multiplication, and thus can be relatively new to the pupil (Van de Kooij, 1989).

Testing in Realistic Mathematics Instruction

Indications for improved testing are found in the gene[ral] and basic principles of realistic mathematics instruction, and in th[e ...]tics. The MORE project (Gravemeijer et al., 1990) focuses on tests t[hat test] more than mere reproducible knowledge: Tests are intended con[tribute to] learning; to give pupils the opportunity to show their knowl[edge and] abilities, including solving procedures and strategies; to cover th[e ...] which does not include a demand for scoring objectivity; and t[o be] administered in the classroom. Written tests that satisfy these [criteria] have been developed, even for young children who have not yet r[eceived] arithmetical/mathematical instruction. As expected they also s[how] higher-order abilities. Figure 15 shows three items borrowe[d from the] MORE project.

Figure 15. Nonstandard test questions: problems that ask children to kee[p score,] candy bars, and create sums of 24.

Testing in realistic mathematics instruction i[ncludes:]

- Problems with a piece of scrap paper like th[e ... "Tim] and Mieke play together a game. They note [... .] What are the final scores? Figure it out! Wri[te ... in] the empty boxes! You may use the piece of [scrap paper for] the work."

- Problems that ask for anticipating activitie[s ...]

Figure 16. Students must resolve cognitive conflicts arising from this graph. Do these runners collide?

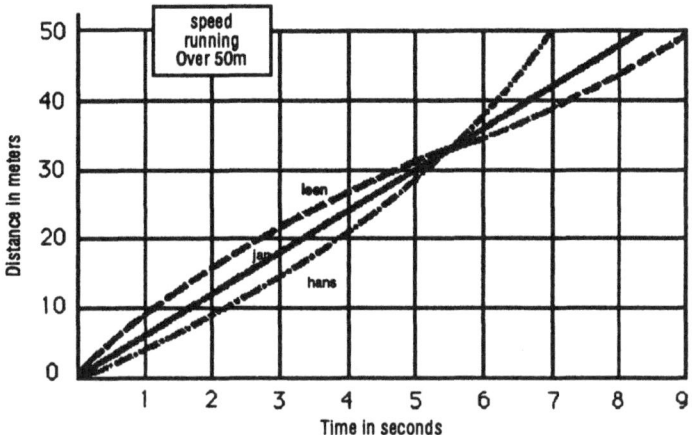

Figure 17. This time-speed graph prompts the question, Do the runners collide again?

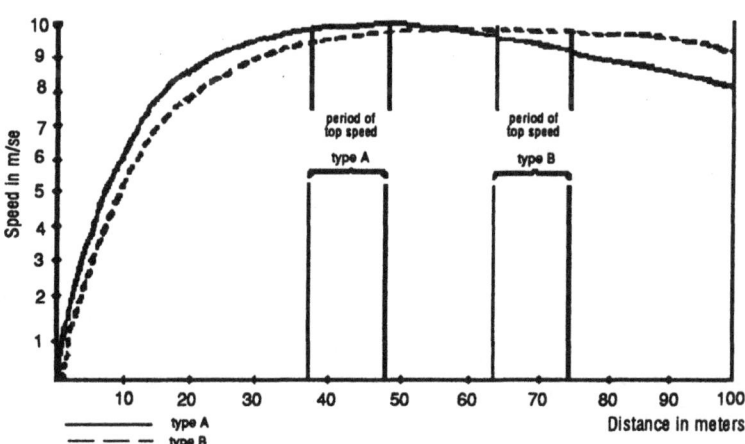

Figure 18. A question of perspective.

 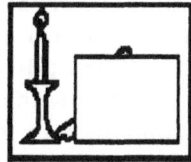  

In general, assignments in which two or more situations must be compared are very suitable for realistic mathematics instruction. This also holds for the comparison of one's own knowledge or point of view with a given situation or story, for instance in the form of

Figure 19. How many paths from A to B?

- A little thinking or interesting story, such as, "A rope ladder hangs on a ship overboard and just touches the water. During high tide the level of water raises 20 cm per hour. The distance between the rungs of the ladder is 30 cm. How many rungs will be under water after 5 hours?"

- Problems designed for testing on various levels, such as, "The train at the amusement park, shown in Figure 20, takes 10 minutes for the short ride. How long does it take for the long one?"

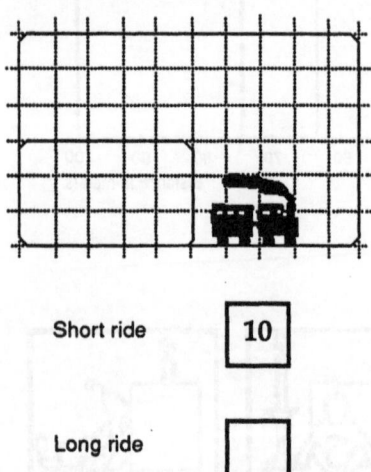

Figure 20. A problem inviting responses of various levels of sophistication.

Two-stage tests are alternative kind of evaluation (de Lange, 1987). The first stage is a traditional timed test, which is returned to the student with corrections and annotations added; the second stage (home-work to be delivered three weeks later) is an essay-like elaboration of the first-stage subject matter. Students used, critically analyzed, and refined their first-stage mathematical models, which raised their level of thinking through reflection. Other possibilities are log books kept by the pupils, as proposed by the NCTM Standards, and essay assignments on a given theme.

IMPLICATIONS FOR FUTURE RESEARCH AND DEVELOPMENT

Rather than artificially separating development and research from each other, attention should be paid to developmental research, which includes the development of prototypes of courses, and the shaping of theory on learning and teaching corresponding to the prototypes. A provisional course is composed of thought experiments that are to be checked and revised by means of the long-term class instruction experiment and the researchers' observations, including various forms of clinical interviews. Clinical interviews reveal the learning processes of pupils, teachers, and researchers, and are sources for adaptations of the course and the shaping of theory. New instruction, derived from the prototype, by other teachers and their classes, constitutes a reconstruction of the original experiment (see Streefland, 1988b).

Research in the United States, such as that of Cobb, Wood, and Yackel (1988, 1989) and Carpenter and Fennema (1989), is a first step on a road to a development that has taken place in the Netherlands in the past two decades (Treffers, 1987a). Although there is still a long way to go, this research adds new evidence on the fruitfulness of our approach. The Milwaukee project conducted by the National Center for Research in Mathematical Sciences Education (of Madison, Wisconsin), undertaken in cooperation with OW&OC of Utrecht State University, provides further support for the same realistic theory on which it is based.

These research programs are switching the emphasis in mathematics education: (i) from teaching to learning, (ii) from reproducing to reconstructing, (iii) from imposing mathematical structure to creating it through the analysis of realities (Freudenthal, 1983), (iv) from evaluational to developmental research, (v) from "objective" to intrinsic scoring, and (vi) from Bruner to Van Hiele levels.

We believe that, for higher-order abilities to result from mathematics instruction, one must first sow and nourish them. To accomplish this, mathematics instruction must satisfy specific conditions. It should be realistic, corresponding to the historical learning process of mankind. To enable pupils to perform such reconstructions, teachers should plan instruction that facilitates and exploits pupils' contributions—their own constructions and productions. So teaching itself becomes heuristic. Because it must take into account the pupils' efforts, it cannot be based solely on the fixed ideas of the teacher, and therefore it contributes to the idea of mathematics education as a complementary learning and teaching process (Freudenthal, 1983).

REFERENCES

Bell, A., et al. (1981). Choice of operation in verbal problems with decimal numbers. *Educational Studies in Mathematics, 12,* 399-421.

Bloom, B.S. (1956). *Taxonomy of educational objectives: The classification of educational goals; Handbook I: Cognitive domain.* New York: David McKay.

Carpenter T.P., Lindquist, M.M., Matthews, W., and Silver, E.A. (1983). Results of the Third NAEP Mathematics Assessment: Secondary School. *Mathematics Teacher, 76,* 652-659.

Carpenter, T.P., and Fennema, E. (1989). *Building on the knowledge of students and teachers.* University of Madison, WI: National Center for Research in Mathematical Sciences Education.

Cobb, P., Wood, T., and Yackel, E. (1989). Assessment of a problem-centered second grade mathematics project. In D. B. McLeod and V. M. Adams (Eds.), *Affect and mathematical problem solving: A new perspective.* West Lafayette, IN: Purdue University; pp. 6-7.

Cobb, P., Wood, T., and Yackel, E. (1988). Coping with the complexity of classroom life. Paper. West Lafayette, IN: Purdue University.

Commission on Standards for School Mathematics of the NCTM. (1988). *Curriculum and evaluation standards for school mathematics.* Reston, VA: National Council of Teachers of Mathematics.

Davis, P.J., and Hersh, R. (1983). *The mathematical experience.* New York: Penguin.

de Lange, Jan J. (1987). *Mathematics insight and meaning.* Utrecht: Research Group on Mathematics Education and Educational Computer Centre (OW&OC).

Dolk, M., and Uittenbogaard, W. (1989). De ouderavond. *Willem Bartjens, 9*(1), 14-20.

Freudenthal, H. (1973). *Mathematics as an educational task.* Reidel: Dordrecht.

Freudenthal, H. (1978). *Weeding and sowing.* Reidel: Dordrecht.

Freudenthal, H. (1979). How does reflective thinking develop? *Proceedings of the 3rd International Conference for the Psychology of Mathematics Education.* Warwick, UK: The University of Warwick.

Freudenthal, H. (1983). Is heuristics a singular or a plural? In R. Hershkowitz (Ed.), *Proceedings of the 7th International Conference for the Psychology of Mathematics Education.* Rehovot: Israel; pp. 38-50.

Freudenthal, H. (1983). *Didactical phenomonology of mathematical structures.* Reidel: Dordrecht.

Freudenthal, H. (1990). *Revisiting mathematics education.* Dordrecht: Kluwer.

Gravemeijer, K., van den Heuvel, M., and Streefland, L. (1990). *Contexts free productions tests and geometry in realistic mathematics education.* Utrecht: Research Group on Mathematics Education and Educational Computer Centre (OW&OC).

Hadamard, J. (1945). *The psychology of invention in the mathematical field.* London: Dover.

Hart, K.M. (Ed.). (1981). *Children's understanding of mathematics*. London: John Murray; pp. 11-16.

IOWO-Team. (1976). Five years IOWO. *Educational Studies in Mathematics, 7*(3), 285-289.

Kaplan, R.G., Yamamoto, T., and Ginsberg, H.G. (1989). Teaching mathematics concepts. In L. Resnick (Ed.), *Toward the thinking curriculum: Current in cognitive research*. ASCW Yearbook, pp. 59-81.

Kilpatrick, J. (1985). Reflection and recursion. *Educational Studies in Mathematics, 16,* 1-27.

Kooij, H., van der, Het eersts HAWEX-amen. (1989). *Nieuwe Wiskrant, 9*(1), 5-14.

Krutetskiiv V. A. (1976). *Psychology of mathematical abilities in schoolchildren*. Chicago: University of Chicago Press.

Lesh, R. (1990). Computer-based assessment of higher-order understandings and processes in elementary mathematics. In J. Kulm (Ed.), *Assessing higher-order thinking in mathematics*. Washington, DC: American Association for the Advancement of Science.

Lohman, D.L. (1987). Two implications of cognitive psychology for educational measurement. Address to the Division of Learning and Instruction Institute for Educational Research in The Netherlands. Utrecht.

Mislevy, R.J. (1992). Foundations of a new test theory. In N. Frederiksen, R. Mislevy and I.I. Bejar (Eds.), *Test theory for a new generation of tests*. Hillsdale, NJ: Lawrence Erlbaum Associates.

National Curriculum Council. (1989). *Mathematics in the national curriculum*. London: Author.

Polya, G. (1945). *How to solve it*. Princeton, NJ: Princeton University Press.

Schoenfeld, A.H. (1989). Teaching mathematical thinking and problem solving. In L. Resnick (Ed.), *Toward the thinking curriculum: Current cognitive research*. ASCD Yearbook; pp. 83-103.

Schoenfeld, A.H. (1987). *Cognitive science and mathematics education*. London: Erlbaum.

Skemp, R. R. (1979). Goals of learning and qualities of understanding. *Mathematics Teaching*, pp. 44-49.

Streefland, L. (1980). *Makro-strukturele verkenningen voor het wiskundeonderwijs*. Utrecht: Institute for the Development of Mathematics Education (IOWO).

Streefland, L. (1987). Free production of fraction monographs. In J. C. Bergeron et al. (Eds.), *Psychology of Mathematics Education-XI*, vol. I. Montreal: University of Montreal; pp. 405-410.

Streefland, L. (1988a). *Realistisch breukenonderwijs*. Utrecht: Dutch Research Group on Mathematics Education and Educational Computer Centre (OW&OC).

Streefland, L. (1988b). Reconstructive learning. In A. Borbas (Ed.), *Proceedings of the 12th Annual Meeting of the International Group for the Psychology of Mathematics Education*. Veszprém, Hongarije; pp. 75-92.

Streefland, L. (1989). Realistic mathematics education: What does it mean? Paper presented at the XIth Psychology of Mathematics Education-North American Chapter of the International Group, New Brunswick, NJ: Center for Mathematics, Science, and Computer Education, Rutgers.

Streefland, L. (1985). Search for the roots of ratio: Some thoughts on the long-term learning process (Towards ....a theory), Part II: The outline of the long-term learning process. *Educational Studies in Mathematics, 16,* 75-94.

Treffers, A. (1987a). *Three dimensions: A model of goal and theory description in mathematics instruction—The Wiskobas Project.* <Dordrecht: Reidel.

Treffers, A. (1987b). Integrated column arithmetic according to progressive schematisation. *Educational Studies in Mathematics, 18,* 125-145.

Treffers, A. (1989). *Proeve van een nationaal programma voor het reken-wiskundeonderwijs op de basis school.* Deel I Overzicht einddoelen, Zwijsen, Tilburg.

Vanden Brink, F.J. (1985). Class arithmetic books. *Proceedings of the 7th Psychology of Mathematics Education, North American Chapter of the International Group.* Columbus, OH: Ohio State University; pp. 282-286.

PART III: New Perspectives on Classroom-based Assessment

10 A Teacher's Struggle to Assess Student Cognitive Growth

Carolyn A. Maher, Robert B. Davis, and Alice Alston

INTRODUCTION

In the course of teaching a typical mathematics lesson, a teacher must make a myriad of decisions as he or she is attempting to meet the needs of the various individual students in the class. Little real-time information is available to the teacher to guide these decisions, and there is little time for deciding. What is involved here is a micro level of assessment, done in seconds or fractions of a second. Despite the necessarily hasty context in which this assessment occurs, it is precisely this level of evaluation of a student's work and needs that ultimately has the greatest impact on the student's progress.

In this chapter we report a study made by the authors, in cooperation with Linda, a sixth grade teacher, as she worked to help her students learn about fractions. During mathematics classes, Linda had the students work in small groups. A video camera recorded the discussion in one of these groups. In subsequent analysis of these videotapes, we focused on three main questions: (i) What representations did the students make for each mathematical situation, and how did these representations help them (or hinder them) in dealing with the situation? (ii) How did these representations change over time, as a result of conversations among students, experience with concrete materials (or other forms of experience), and teacher interventions of various sorts? (iii) How successful was the teacher in making correct identifications of student representations and in helping students to improve these representations whenever that was appropriate?

The three of us (and several colleagues, including Judith Landis, Amy Martino, Arlene Marasco, and Tom Purdy) study student ideas, because it would be difficult to assess how well the teachers are able to correctly recognize a student's ideas if we did not determine (using data that are not available to the teacher during the usual classroom lesson) what the student's ideas actually are. Portions of the student data in this chapter are taken from Landis's thesis (1990), from Davis and Maher (1991), and from Maher and Davis (1991). The main results of the study, which has been underway for three years (and still continues) are, first, the understanding of how student representations are used and how they grow, and, second, a large amount of evidence of how extremely difficult is the task of the teacher when he or she tries to identify a student's representations in order to help improve them. Here we deal mainly with the complexity of the assessment task that a teacher faces, moment to moment, in trying to make contact with the way that a student is thinking about a problem or situation.

ADDING FRACTIONS: CLASSROOM INTERACTIONS

The Two-Pizza Problem

During a previous lesson, Linda had moved around the classroom, interacting with students as they worked in small groups on problems involving fractions. She had judged that many students, including a boy named Brian, were having difficulty with improper fractions (fractions in which the numerator is not smaller than the denominator). So, she had made up a new work sheet, one that included the following problem:

> At Pizza Hut each large pizza is cut into 12 slices. Mrs. Wilson ordered two large pizzas. Seven students from Mrs. Wilson's class ate one piece from each of the pizzas. What fraction of the two pizzas was eaten?

In the episode that follows, Brian is working with one other boy, Scott, on the problem. The teacher comes by to see how the two boys are progressing and to help them if that turns out to be necessary. The episode begins with Brian explaining his solution to the teacher:

> Brian: Here ...
>
> Teacher: Do you, uh ...

Scott: I think we got it right.

Brian: I think I know I'm right.

Scott: Yeah, we think we know we're right.

Brian: So there's 24 slices in both pizzas, so Mrs. E. wants 7 students ... she took 7 students to Pizza Hut, so ... she's gonna give 'em one slice from each pizza so we would have, uh, 14 out of the 24, right, slices.

Teacher: All right, now let me ask you this. How do you get 24 slices in the one pizza, and 12 slices in the other? [Note that this is not what Brian had actually said. As we shall see presently —calling upon information that was not then available to the teacher—it is also not at all what Brian had actually done.]

Linda's tone here might be described as slightly disapproving and surprised. We return to the matter of her disapproving tone later; let us first consider the matter of her surprise.

Linda, who at this point has had no access to the videotapes, has misunderstood what Brian was actually thinking—a situation that our study is showing to be a very frequent occurrence in many classrooms. Linda's interpretation was influenced by two factors. In the first place, she herself had solved the problem incorrectly, using a single pizza as the unit, despite the fact that the problem statement explicitly called for using both pizzas together as the unit ("What fraction of the two pizzas was eaten?"). As a result, Linda's answer to the question is 14/12. When she hears Brian's answer of 14/24, she believes that he is mistaken. But there is another reason for Linda's interpretation of Brian's answer. To see what it is, we need to look at what had happened in a previous lesson on the addition of fractions.

A prior fraction-addition problem

Teacher: You had 3/8 of this pizza and 3/8 of that pizza. How much is that?

Brian: 6/16.

Teacher: But the pizza has only 8 pieces.

Scott: Is this right? 6/8?

Teacher: Why isn't it 16ths?

Scott: Because the pizza has 8 pieces and you can't change it.

In fact, as we will see when we investigate further, Brian had been solving fraction problems correctly and had been thinking seriously about each problem. Scott, on the other hand, mainly relied upon repeating the words of his teacher, and had been less likely than Brian to get correct answers. However, this information was not available to the teacher at the moment when she had to make some key decisions.

Brian consistently used concrete materials to make his initial representations of problems that involve fractions. This technique (which was part of the teacher's goal for the class) usually served Brian well. As he modeled this problem with wooden blocks, he could see before him three selected pieces out of a total of eight, and another three selected pieces out of a total of another eight. If one counted all selected pieces, one had 3 plus 3, for a total of 6 selected pieces. If one counted all available pieces, one had 8 plus 8, for a total of 16. The teacher had said nothing about the notion of a unit, so that Brian had to work out for himself which of the possible numbers should be taken into account, and in what ways. His first approach—which was probably basically correct—was to count the selected pieces as a fraction of the total available pieces, getting the answer six sixteenths. Of course, the question of "how much pizza do you have?" calls for a basic unit, and one obvious choice would be to use one whole pizza as this "unit."

Our teachers keep a log book on all lessons. Concerning this lesson in which Brian had given his solution (which appeared to the teacher to be adding). Linda had written:

> As I circulated the room, I saw students continued to have problems with mixed nos. & reducing. Some students continue to add denominators. Brian continues to add the denominators.

In view of this earlier experience, it was not at all unlikely that when, in the two-pizza lesson, expecting the answer 14/12, Linda heard Brian say 14/24, she quickly interpreted this as another instance of Brian "adding denominators." As we shall see, Linda was mistaken on two counts. First, Brian had not obtained the denominator 24 by some erroneous process; second, Brian had not worked with written symbols at all, but had dealt directly with an isomorphic representation of the problem in terms of pattern blocks.

A Student's Understanding of the Two-pizza Problem

What had Brian actually been doing? By studying the videotape, we can watch Brian and Scott at work, while the teacher is busy helping other groups of students and before she comes over to work with them. Brian and Scott had, available on their table, a variety of manipulatable materials that they could use if they wished. As they worked on the two-pizza problem, they used pattern blocks, wooden blocks of the colors and shapes as shown in Figure 1.

Figure 1. Pattern blocks.

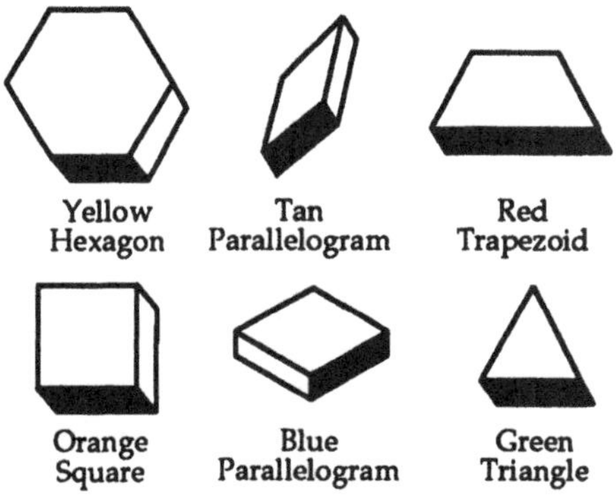

Yellow Hexagon   Tan Parallelogram   Red Trapezoid

Orange Square   Blue Parallelogram   Green Triangle

The dimensions and shapes of the blocks are so arranged that six of the green triangles fit on top of one of the yellow hexagons. Alternatively, two of the red trapezoids fit on the yellow hexagon. One blue parallelogram can combine with one green triangle to cover exactly the red trapezoid. (The reader can work out other arrangements that will fit exactly.) The tan parallelogram and the orange square are exceptions to this "commensurability" and do not fit onto anything else. There are many blocks of each kind available in the pile on the table.

> Brian: (reads the problem aloud, then picks up one of the blocks, a yellow hexagon) This is one pizza.
>
> Brian: (picks up several smaller pieces, apparently intending to indicate slices) No, this is one pizza (as he puts down two yellow hexagons). [Notice what he has done: His checking of his representation showed him that he would not be able to find twelfths of the hexagon—the smallest piece, the green triangle, is one sixth of the hexagon. So he has taken a step of surprising subtlety: He uses two pieces of wood (carefully chosen) to represent one pizza.]
>
> Brian: This is a pizza, here. (He is merely restating his new definition, presumably for the benefit of Scott, his partner.)
>
> Scott: Yeah. (He, too, picks up two hexagons.) These are the

two pizzas. [Notice that Scott has missed the subtlety of Brian's representation of one pizza by two hexagons. Scott is using one hexagon to represent one pizza. Matching your mental representation to someone else's is not merely difficult when the teacher is trying to do it, it is also extremely difficult when one student is trying to get a correct match to the representations of another student.]

Brian: OK [His posture and tone make it clear that he is not really responding to Scott; in fact he is really ignoring Scott.]

Scott: Yeah, this is one pizza. [It is possible that Scott has now adopted Brian's representation, and is using two hexagons to represent one pizza, but this is not entirely certain. The videotape does not provide conclusive evidence either way.]

Brian: (who has been fitting small green triangles, representing slices, on top of a hexagon.) This (picking up a red trapezoid) counts as three greens, OK? [Brian is still ignoring Scott and carrying through his own solution of the problem. Brian apparently wants to use the red trapezoid so as to have fewer pieces of wood to handle.]

Scott: Wait! I just figured it out! If you have twelve pieces and you have seven students getting a piece ... wait! ... chopped into twelve slices (virtually talking to himself at this point) ... each of the students getting one piece of these twelve ... There's seven students, right? So, for two pizzas that would be fourteen slices of this ... Brian, if you added it all together, and then you have eight slices left over ...

Brian: Just think about it. [Brian's tone seems to say either "Don't bother me, can't you see I'm busy?" or else, perhaps, Brian has realized that Scott's representation is wrong, and is asking him to reconsider it. In either case, Brian doesn't want to be interrupted in his own thought processes, as his manner makes very clear.]

Scott: You have eight left over ...

Brian: Keep thinking about it. [That is, "Don't bother me right now!"] So ... nineteen and nineteen is ...

Scott: Thirty six.

Brian: Thirty eight.

Brian: But there's only twenty four slices!

Scott: How do you figure "twenty four" slices? This is a pizza, Brian! This is twelve slices. [He shows two hexagons, so at this point he seems to have adopted Brian's representation.] This is one pizza.

Brian: Twelve (displaying two hexagons). Twenty four (puts down two more hexagons, for a combined array of four hexagons).

Brian: (changing the subject) OK ... how many boys are in the class? [Actually, Brian is now beginning to work on building another part of the data representation, the representation of the children who are to eat the pizza. He has not recalled correctly the statement of the problem.]

Scott: One, two, three, four, five, six, ... I think eight.

Brian: (repeating himself). How many boys are in the class?

Scott: What class?

Brian: Our class. [In fact, the problem had made no mention of the total number of boys in the class, but had said "seven students." Under the stress of constructing a representation for the entire situation, Brian has made an incorrect mapping of the problem data into the abstract representation that he is building. Indeed, the videotapes of various students and various classes show repeated instances of incorrect construction of representations, or incorrect mapping of data into a representation. Apparently the cognitive demands of building representations are considerable, and often test student ability to the limit. We return to this matter below.]

Scott: Why do you want to know about the number of boys?

Brian: Just count them!

Scott: Nineteen, all together. There's 6, ... thirteen boys.

Brian: Thirteen and thirteen, that's twenty six.

Scott: Briiiaaaan ...

Brian: Here's the pizzas. (He has in place four yellow hexagons, two for each "pizza," and he is beginning to cover them with small green triangles, representing slices.)

Scott: Brian .... Brian! Figure this, Brian!

Brian: I think I know it.

Scott: I already figured it out. You wouldn't want to do it, Brian. [While Brian has been trying to build a representation using the pattern blocks, Scott has been trying to work the problem out on paper. Scott's words, here, seem to mean: "You wouldn't want to solve this problem by working it out on paper, Brian!" It subsequently turns out that Scott's paper-and-pencil "solution" is in fact incorrect.]

Brian: (still working with the pattern blocks) Yes, I would! Get me two greens (from that pile) over there.

Scott: Sure, if you feel like doing the work, OK.

Brian: OK, what's the answer? [Having now almost completed his construction with the blocks, Brian is really mainly talking to himself, here. In effect he is saying: "OK. Now I see what the problem is. If I look at this right, can I see the answer?"]

Scott: You have to listennnn ...

Brian: So ... There's one pizza ... (Two hexagons, now covered with small green triangles to show twelve "slices.")

Scott: I'm gonna listen to your solution right now and then you're gonna listen to mine.

Brian: (who is still working on his concrete representation of the pizzas, using pattern blocks) Get me twelve more of those (he is referring to the small green triangles).

Scott: Here you go.

Brian: 1 ... 3, 4, 6, 8, 10, 12 ... Thank you. ... and ... here is another pizza!

Scott: Now keep in mind that you were wrong last time.

Brian: Keep in mind that I was right more times than you!

Scott: That's why you got the whole ditto wrong, and I got the whole ditto right! I had it right!

Brian: I wouldn't do it your way. [Presumably Brian is rejecting the paper-and-pencil calculation that Scott has completed.]

Scott: OK ... I want to watch your solution and see if it's the same as mine. [Brian has carefully assembled two shapes, each of which consists of two hexagons built from small green triangles. He has thus modeled the two pizzas, with twelve slices on each of them. It is important to notice that he has not yet started to model the children eating the two pizzas.]

Brian: I might be wrong.

Scott: No, I'm not saying that you're wrong. I want to see if it's the same as mine.

Brian: Here's the two pizzas (gestures toward the four hexagons). Now, everybody gets a slice out of this pizza (pointing to the first pair of hexagons). OK?

Scott: Not everybody! Only ... (Scott picks up the paper and starts to reread the statement of the problem.) "Seven students from Mrs. Wilson's class are to eat one slice from each of the two pizzas."

Brian: So ... seven ... [And here Brian does something truly stunning; as he works to model the next part of the problem—namely, the students eating the slices—he is just as concrete as he was in using the pattern blocks to model the pizzas themselves. He looks around the class, and points to individual students, naming the specific student who is to eat each slice!] So this (pushing one "slice" toward Scott, and one toward himself) is for you and for me; Ron (pointing to a student), Rav (pointing to another student), Jennifer (again pointing), Mary (pointing to Mary), and Melissa (pointing to Melissa). (As he names each child, he moves one "slice"—one green triangle—away from the first "pizza".) Here's the seven slices (gesturing to the triangles that he has moved away from the hexagons that represent the first pizza). Now take these from here (he goes through the same procedure, taking seven "slices" from the second "pizza"). (Rereading the problem.) "What fraction of the two pizzas was eaten?" Two, four, six, eight, ten, twelve, fourteen. So ... 24 out of 14. I mean, 14 out of 24. (He writes the fraction 14/24.)

Scott: No! You can't change that bottom number! You can't change the 12. It's 14/12. [By "bottom number" Scott of course means the denominator, which Brian has just written as "24". Note that Brian is making a concrete representation of the problem, and actually counting small green triangles (or their proxies); he is not working with written symbols. By contrast, Scott does work with written symbols, and appears to pay little attention to the pattern blocks.]

Interpretating the Teacher's Reaction

In the earlier lesson on adding fractions, the teacher was misreading Brian's actions when she wrote in her journal, "Brian continues to add the denominators." In fact, Brian was not thinking primarily in terms of written symbols, but was counting pieces of wood. In misinterpreting Brian's work, Linda was making an error that we encounter very frequently. Most teachers (and most people in general) have learned to think of mathematics primarily as the manipulation of symbols. Linda was (wisely) trying to get away from this tradition, and to think of mathematics as the description of reality. That is to say, one goal was to get the children to regard symbols as windows through which you could look and see some of the reality. Despite her intentions, Linda's own previous education betrayed her, and she did not describe Brian as "counting the wrong pieces of wood," but rather as "adding denominators," when, in fact, having written no fractions, he had no "denominators" at all. We would prefer to think of Brian as working to build up a mental representation for the problem, and to carry out a mapping of the present data into this representation (Davis, 1984). The physical pieces of wood provided him with a basis in perception that could become a foundation for his mental representation (Davis, 1986).

As we learned in Brian's discussion of the two-pizza problem with Scott, Brian corrected his process for making representations, and thereafter made correct representations. When the teacher came by (in the two-pizza lesson) and told him that his answer of 14/24 was incorrect, he reacted to her disapproval. Up to that moment the boys had been animated, interested, thinking hard about the problem, and working industriously to construct a solution. When Brian's work was rejected, the mood of both boys changed immediately. From that point on, for the rest of the lesson, they seemed disinterested. It looked as if they had given up serious thought, and were merely paying the minimum attention needed to be considered "good students." They no longer seemed deeply committed to finding a solution to the problem.

SO LITTLE INFORMATION, SO LITTLE TIME

Up to now our main point has been that a teacher, working in any typical classroom, needs to know what the various students are doing. This is an important kind of knowledge, and without it successful guiding of the students is nearly impossible. But this is a difficult kind of information for the teacher to collect. Students are not necessarily doing what we, their teachers, may assume that they are doing. To make matters worse, students are often imprecise in their efforts to tell us what they are doing. (For an extreme example, consider Schoenfeld et al., in press, where it took many hours of analysis of videotapes of one teacher tutoring one student, before the analysts were able to discover the major misunderstanding that was blocking the student's progress.)

We are well aware that some readers may feel that obtaining the kind of information that we seek on the subject of the basic ways that students are thinking is not merely difficult, but impossible. In a sense that is true—one is never going to have as much of this information, nor in as complete a form, as one would like. But in working with teachers who choose to videotape their classrooms, and to study the resulting tapes (and even to allow us to do so!), it seems clear that changes in teaching procedures do in fact result. Teachers also become more analytical in thinking over lessons that they have recently taught. The realization that the student may be thinking about a situation differently than the teacher expects is a powerful one. We lack space here to pursue in detail the changes that can be seen in teachers, but offer one example to illustrate a teacher's accommodation of new information.

A Lesson with Cuisenaire Rods

The teacher, Pat, was using Cuisenaire rods to introduce some second grade children to the idea of fractions. He held up a red rod (which is 2 cm. long) and said, "Suppose I call this rod one. Which rod should I call two?" About two thirds of the class responded by holding up a purple rod (4 centimeters long), which is the answer that the teacher expected. But a sizable group of students—about one-third of the class—instead held up the light green rod (which is 3 cm. long). The teacher had, of course, assumed that they were modeling the size property of numbers by using the lengths of the rods. Hence, since four is twice two, the teacher thought that the answer "purple" was the obvious and correct choice. Because we had been encouraging the teachers to analyze student thinking, Pat went home and thought about it. Could one-third of the class be wrong? Or was there another legitimate way of interpreting the question?

He quickly realized that there was. The positive integers do have the size property which usually forms the basis for their usefulness, but this is not the only property that they have. They also have order. "Four" comes immediately after "three" in our usual counting sequence. "Seven" comes after "six." And this order property is important—it, and not size, is the basis for using numbers as the street addresses of our homes, or for numbering the pages of books. Nor does either property imply the other. The letters of the alphabet have order but not size; so do the months of the year. On the other hand, vectors have size, but no order arrangement. So the students who hold up a light green rod are making use of a perfectly legitimate property of both the rods and the integers. In this case, however, it wasn't the aspect on which the teacher needed to build the lesson.

By asking every student to hold up the appropriate rod, the teacher got an instant reading on the answer of every child in the class. Since nearly everyone answered either light green or else purple, and since the line of reasoning leading to each answer is fairly straightforward, the teacher was able to get most of the information that one would wish to have. In other cases it is not always so simple.

THE DIFFICULTY OF BUILDING MENTAL REPRESENTATIONS

We saw that, when Brian was busily building up a mental representation (based on being able to look at an arrangement of wooden blocks), he made occasional errors on what were, really, quite simple matters. For example, he forgot that the two-pizza problem spoke of "Mrs. Wilson's class." Errors of this type are so frequent that they are really the rule, not the exception. To put the matter more generally, studying student (or adult) performance convinces us that building up a mental representation is an extremely arduous task. Because we are convinced of this, we are more eager than ever to recognize whatever representation a student may have built up. If it is correct, we want to build on it. If it is somewhat defective, we want the student to come to see the precise nature of the defects. If it is hopelessly wrong, we want the student to see for himself or for herself why it is wrong, and to be able to use that knowledge to build a better representation.

There are other teaching implications, too. Because we believe that a student who has been working to build a mental representation has created something valuable and fragile, we are very reluctant to interrupt a student's thinking. In this we differ sharply from the pedagogical practice of those who say that students should always be paying close attention to whatever the teacher may be saying. In order to build an elaborate representation, one must shut out most incoming signals, and engage in rather deep thought. We try to allow for this in our classes.

THE TEACHER AS COGNITIVE COACH

There is a definite theory of teaching that underlies the demand for teachers to know as much as possible about the thought processes of their students. Put briefly, this theory sees the teacher as a kind of "cognitive coach" who, like an athletic coach, studies the performance of the student and tries to work with the student in order to improve it. This usually means trying to know as much as possible about a student's mental representation of a mathematical situation and how the student is trying to make use of that representation. Where necessary, one then tries to help the student recognize weaknesses and make improvements. We would argue that this needs to be done, wherever possible, by letting the student see where his representation succeeds, and where it fails when it is brought into direct contact with reality. Unfortunately, teachers often fail to do this. Our videotapes also document many such failures, as we see in the following section.

Persuasion and Compliance versus Logic and Cognitive Growth

What happened when Linda, misunderstanding how Brian was thinking about the "two-pizza" problem, sought to get him to change his answer? We return to the transcript of the tape. Recall that the teacher has told Brian that his answer, 14/24, is incorrect, and that the answer should be 14/12. She does not confront him with a clear-cut reality, whose structure cannot be misunderstood. Instead, she tries to persuade him to think about the problem her way. The analyst's notes read as follows:

> Brian now seemed ready to abandon his solution. The teacher directed the students by correcting their work; she discarded Brian's solution [which had, in fact, been correct] by mentioning that there was no box big enough for such a "gi-huge-ic" pizza. [Note that the

availability, or unavailability, of boxes has nothing to do with the mathematical question that the children are supposed to be solving!]¹ Brian again tried to justify his work, but is interrupted by the teacher before he is heard:

Brian: Yeah ...

Teacher: OK, you, you're putting your 2 pizzas together and making one ... gi-huge-ic pizza.

Scott: (laughing) gi-huge-ic pizza

Teacher: OK, we can't have one gi-huge-ic pizza because there isn't a box that could carry it in to take it home. (See Note 1.)

Brian: No, just stick it in (mumbles) ... slices. [Brian is still trying to salvage something from his way of thinking about the problem.]

Teacher: We have to keep it separate. They have to go in two separate boxes. [Still irrelevant.]

No wonder so many people believe mathematics to be incomprehensible nonsense. They are not given appropriate opportunities to test their ideas against reasonable demands of reality—the reality of the problem, that is to say, and not the reality of the power relationships within a classroom. Linda, by using made-up words such as "gi-huge-ic" seeks to get the boys favorably disposed toward her (and does succeed in getting Scott to laugh, though Brian does not). In invoking box size, she is clearly trying in some way to persuade. Such suggestions have no logical relation to the task at hand, and do not help us to understand it.

There are other alternatives for getting a student to reconsider a proposed solution. One of the best is a direct test of a student's theory against reality. Another good alternative, suggested by David Page,[2] is for the teacher to ask the question that the student actually did answer: thus, if a student says that "eight times seven is 15," Page recommends that the teacher ask "How much is eight plus seven?" Students nearly always respond: "Oh! It should be fifty six!" (Davis, 1984).

DO TEACHERS NEED TO KNOW STUDENT'S REPRESENTATIONS?

Whether, in this brief space, we have convinced the reader, only the reader can say. But we have become convinced, from studying these videotapes, and from working with teachers in going over analyses of these tapes, that both teachers and students have much to gain from making efforts toward this very fine-grained analysis of how students are actually thinking about mathematical problems and mathematical situations. This is a valuable and important form of assessment, and in the right context it can be carried out successfully.

A Note on Heroes

Clearly, we argue for the importance of looking on a very minute level at how students think about mathematical situations, and at how teachers think that students think about mathematical situations. We see this as one of the most critical questions in mathematics education. It requires careful and insightful analysis—but even more, it requires teachers who will seek out the best possible data, even in instances where they may, given hindsight, wish that they had done something differently.

This is the spirit of the very best in science, seeking truth without placing blame. In our view, the real heroes of these studies are the teachers, people like Linda and Pat, who have worked so hard and so selflessly to get and to share the most complete possible data, even when they wish they had known earlier some of the things they only found out about later on.

NOTES AND REFERENCES

1. Much could be said in analysis of the teacher's sudden introduction of the idea of boxes to hold the pizza. Pedagogically, we do not like it, since we see it as an intrusion that is not central to this problem. But from a theoretical point of view, it is in fact important. For one thing, it is one instance of the role that previous personal experience and concrete materials can have on shaping one's ideas (see, for example, Dana's concern about making outfits that do not match, Maher and Martino, 1992). It also looks quite different when viewed from the perspective of Lesh's important theory of "model-eliciting problems" (Algebra Group e-mail Forum, JKAPUT@umassd.edu, 19 February 1992). Since these tasks start with real-world activities or situations, many different aspects of these situations can be considered to be relevant. Indeed, just this sort of "reality" consideration has led the teacher into one interpretation, and Brian into a different one. Unfortunately, in this instance they were not able to share their different perspectives effectively.

2. David Page, whose work should be far better known than it is, was one of the leading innovators in improving school mathematics in the United States in the 1950s and 1960s.

Davis, R.B. (1984). Learning mathematics: The cognitive science approach to mathematics education. Norwood, NJ: Ablex Publishing Corp.

Davis, R.B. (1986). The convergence of cognitive science and mathematics education. *Journal of Mathematical Behavior, 55*(3), 321-333.

Davis, R.B., and Maher, C.A. (1990). What do we do when we "do mathematics"? In R.B. Davis, C.A. Maher, and N. Noddings (Eds.), Constructivist views on the teaching and learning of mathematics. Monograph of the *Journal for Research in Mathematics Education.*

Landis, J.H. (1990). Teachers' prediction and identification of children's mathematical behaviors: Two case studies. Unpublished doctoral thesis, Rutgers University.

Maher, C.A., and Davis, R.B. (1990). Building representations of children's meanings. In R.B. Davis, C.A. Maher, and N. Noddings (Eds.), Constructivist views on the teaching and learning of mathematics. Monograph of the *Journal for Research in Mathematics Education.*

Matz, M. (1980). Towards a computational theory of algebraic competence. *Journal of Mathematical Behavior, 3*(1), 93-166.

Schoenfeld, A.H., Smith, J.P. III, and Arcavi, A. (in press). Learning: Z microgenetic analysis of one student's evolving understanding of a complex subject matter domain. To appear in R. Glazer (Ed.), *Advances in intructional psychology*, Vol. 4. Hillsdale, NJ: Lawrence Erlbaum Associates.

VanLehn, K. (1982). Bugs are not enough: Empirical studies of bugs, impasses, and repairs in procedural skills. *Journal of Mathematical Behavior, 3*(2), 3-71.

11 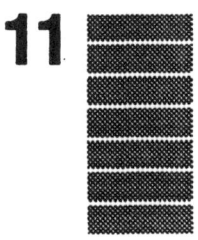 Assessing Understandings of Arithmetic

*Herbert P. Ginsburg, Luz S. Lopez,
Swapna Mukhopadhyay,
Takashi Yamamoto, Megan Willis,
and Mary S. Kelly*

INTRODUCTION

There is widespread agreement that mathematics should be taught as a thinking activity (see, for example, National Council of Teachers of Mathematics, 1989). Doing this requires that evaluators and teachers obtain information concerning students' thinking activities, their efforts at understanding, and their procedural and conceptual difficulties. Yet, too often, teachers appear to understand little of what mathematical thinking is all about; evaluators provide teachers with assessments that fail to illuminate thinking and understanding; and teachers themselves seem to possess few sound methods for obtaining information concerning thinking, particularly in the classroom, the setting where it is most important to do so.

Given this situation, it is essential to develop methods for assessing children's understandings of a variety of key mathematical topics, including whole number arithmetic. Note that we have referred to methods and understandings in the plural. This reflects our belief that, even for a subject as apparently simple as arithmetic, understanding is extraordinarily complex and many methods are necessary to assess it (or *them*). Some assessment methods may be useful for evaluators (school psychologists, math specialists, assessment specialists, and so on) while other methods entirely may be appropriate for teachers to use in the everyday classroom. Similarly, understanding is not a single thing but a multitude of processes and functions, the essence of which we are only now beginning to glimpse.

This chapter begins with a brief account of children's understandings of arithmetic. Then we describe several different types of assessment methods designed to achieve different purposes and to be used by persons with different roles in the educational system, including (i) screening, for teachers to obtain a preliminary portrait of children's levels of understanding; (ii) standard testing with probes, for evaluators to establish children's levels of difficulty in learning arithmetic and to gain preliminary insight into their understandings; (iii) clinical interview technique, for evaluators to assess the understandings of multiplication and fractions; and (iv) classroom assessments, for teachers to assess mathematical thinking and to promote it.

THE NATURE OF UNDERSTANDINGS

Although stressing diverse aspects of understandings, most accounts agree that the understanding of arithmetic involves far more than accurate computation and the "basics" of rote memory (see, for example, Pirie, 1988; Mack, 1990; Van den Brink, 1989). Our own view, drawing heavily on Vygotsky (1962), attempts to situate understanding in a larger psychological context. Our view pictures understandings as sense-making procedures involving at least the following:

- Informal knowledge, formal knowledge, and the development of connections within and between these domains.

- The role of the intermediary schema in promoting such links.

- Rules for transfer, generalization, and application of mathematical knowledge.

- Learning potential.

- Self-consciousness, verbal fluency, and metacognition.

- Higher-order beliefs, attitudes, and feelings.

Informal Knowledge, Formal Knowledge, and Connections

In the natural environment, the infant and then child is engaged in coping with the world and making sense of it. This process results in the child's construction of different forms of "spontaneous knowledge" (Vygotsky, 1962), including an "informal mathematics" having several features. Usually grounded in concrete reality and everyday motives, informal mathematics is personal, emotional, unsystematic, powerful, pragmatic, and often unreflective. The informal system does not usually involve written numbers or symbols, and it is typically not taught by adults. Thus, the three-year-old child determines that there are more candies here than there and enjoys the result.

Research has shown that the informal system is complex, containing such features as notions of more and less, principles of cardinal number, rules for generating counting numbers and for enumerating objects, and procedures and concepts for addition and subtraction. Moreover, the system is more powerful and widespread across cultures than was initially assumed; for example, primitive discriminations concerning numerosity begin soon after birth (Antell and Keating, 1983); virtually all four-year-old children are capable of simple and effective operations of addition and subtraction (Ginsburg, 1989); street children in Brazil engage in rather complex computations in order to sell candy (Carraher, Carraher, and Schliemann, 1985); and children in a variety of cultures, many of them unschooled, develop at least adequate systems of informal arithmetic (Saxe and Posner, 1983).

Typically, children already possess a rather complex and reasonably competent informal mathematical system when school introduces formal mathematics (or even when parents attempt to teach it at home during the preschool years). The formal mathematics of school is systematic, written, explicit, codified, and represents the accumulation of cultural wisdom. The child, already relatively skilled in informal mathematics, now attempts to make sense of this new body of material, which includes written symbols, standard procedures, explicit principles, and formal models.

Sometimes educators attempt to facilitate the process of sense making by introducing various "intermediary schemata" or " bridges." These are artificially constructed devices intended to promote connections or links between the child's informal mathematics and the formal mathematics taught in school. In the early grades, intermediary schemata usually involve "manipulatives" like Cuisenaire rods, Dienes blocks, or the Japanese method of tiles (Kaplan, Yamamoto, and Ginsburg, 1989). These manipulative devices are metaphors or models that provide a bridge between informal procedures like counting or ordering and formal concepts like written number or commutativity. Thus, the child sees that a 4 stick joined to a 3 stick produces a 7 stick, just as do a 3 stick joined to a 4 stick. The informal activities of perception and visual comparison of the sticks are then connected to the written statements $4 + 3 = 7$ and $3 + 4 = 7$. Eventually, children internalize the physical manipulatives in the form of visual and other kinds of imagery and no longer require the physical objects to serve as the intermediary schemata. What is important is not the manipulative per se, but the bridging function of the intermediary schema, be it physical or mental.

With or without the benefit of intermediary schemata, the child tries to make sense of the formal mathematics presented in school. "Understanding" of the formal mathematics seems to involve several features. Perhaps the chief of these is the formation of connections within and among informal mathematics, formal mathematics, and intermediary schemas. The child comes to interpret a given aspect of formal arithmetic in terms of various informal notions and procedures, intermediary schemata, and various other formal notions and procedures. Consider a very simple example. Suppose that the child encounters a school activity or lesson dealing with the idea that $2 + 3 = 5$. The child can deal with this situation in several different ways. On the one hand, the child can simply attempt to memorize the number combination as such and without any connection to anything else. Whether the child succeeds or fails in this attempt is of little interest; in either event, the performance involves simple rote memorization without understanding. On the other hand, the child may attempt to link the simple number combination with other aspects of mathematical knowledge. The child may link the combination with already available counting procedures, realizing that if a set of two elements is combined with a set of three, and the total is counted, five will be the result. Associated with this informal procedure may be informal knowledge to the effect that when two sets are combined, the result is larger than either.

The child may also connect the number combination $2 + 3$ with various formal ideas or procedures. Thus, the child may link the combination with operations on the number line: if you move forward three spaces from the number 2, you end up on the number 5. This, in turn, may be linked with formal principles such as commutativity, so that the child realizes that moving three spaces from 2 gives the same result as moving two spaces from 3.

The child may also link the number combination to an intermediary schema like unifix cubes. Thus, the child may realize that the numeral 2 corresponds to two cubes, that the numeral 3 corresponds to three cubes, that the + refers to combining the cubes, and that the numeral 5 corresponds to the result obtained. The child may also realize that the two cubes are just like two fingers, and that combining two and three cubes gives the same result as combining two and three fingers. In this respect, the cubes serve as a bridge between informal knowledge (counting, addition concepts) and the written symbols, concepts, and procedures of formal mathematics (the numerals 2, 3, the symbol +, the concept of commutativity).

Beyond Connections

Understanding involves more than links or connections among the various systems. Some other key features of understanding are the following:

- *Rules for transfer, generalization, and application of mathematical knowledge.* From the earliest days of psychology, "transfer" has been accepted as a test of understanding. If the child really understands something, he or she should be able to apply it to a somewhat new situation. If the child understands that $2 + 3 = 5$, she or he should be able to use that knowledge to determine the sum of two and three objects, to solve a "story problem" involving two and three objects, to realize that similar procedures can be used to solve a problem involving two hundred and three hundred objects, and to use similar procedures to solve problems involving $2 + 4$ objects. Conversely, according to Hatano (1988), the introduction of novel problems is a key factor in producing perplexity and conflict and thereby promoting the motivation for understanding.

- *Learning potential.* A related test of understanding is the ability to learn somewhat new material. Thus, if the child really understands that $2 + 3 = 5$, he or she should be able to learn, without a great deal of difficulty, that $2 + 4 = 6$, because similar principles and procedures are involved. In a sense, the learning of moderately new material is similar to generalizing existing procedures and concepts to moderately novel situations.

- *Metacognition.* This overused term refers, among other things, to various kinds of self-consciousness. The child who understands is aware of how he or she solves problems, can describe these procedures to others through the medium of language, can monitor thought processes and check them, and is generally aware of his or her mathematical thinking. Thus, the child who understands $2 + 3$ knows about the "counting on" procedure used to get the answer; can tell others how the process was executed; can monitor and check the process of counting on; and can describe how the process is related to the number-line model.

- *Higher-order beliefs, attitudes, and emotions.* It is important to situate understanding within the person. From a psychological point of view, understanding—and cognitive activity generally—

does not exist in a vacuum. Instead, understanding can be understood only within the larger psychological context of the person's functioning (Ginsburg, 1989)—the individual's beliefs, attitudes, shifting goals (Saxe, in press), and emotions. Thus, the belief that mathematics can make sense may be considered a prerequisite for understanding or perhaps an aspect of it.

- *The complexity of it all.* The various features described above—links among areas of knowledge, transfer and generalization, learning potential, and beliefs and attitudes—all contribute to understanding or may be considered aspects of understanding. And no doubt it may prove useful to consider other aspects of understanding not described here. There is no single or simple criterion of understanding, no discrete point when we can say that understanding is present or absent. Rather, the more the child moves away from rote memorization and mechanical applications of isolated procedures, and the more the child moves toward rich linkages, flexible generalizations, active learning, and productive beliefs, the more we can say that some degree of understanding is present.

ASSESSMENT FOR EVALUATORS

Given the preceding approach to the dynamic system of understanding, and focusing not only on higher-order processes (for example, metacognition) but also on "lower-order" processes (informal knowledge) and on the connections between them, it is appropriate that assessment techniques take different forms to assess different aspects of understandings. Our research group has developed several approaches to assessment that can be organized into broad categories.

First we will consider several approaches designed for evaluators—school psychologists, mathematics specialists, clinical psychologists—who are able to conduct assessments with individual children over a fairly lengthy period of time. The techniques we have designed for evaluators include standard tests, a system of probes, and clinical interview procedures. Then we will discuss several techniques that can be used by teachers in classrooms, including a screening procedure and various in-depth examinations of children's thinking.

Standard Test

There are relatively few standard tests of children's mathematical thinking, particularly tests that attempt to measure the complexities of understanding. In general, existing standard tests can provide reliable information concerning the ranking of students in various areas of mathematical performance. Thus, a standard test can reveal that a student is relatively weak in "computation" and relatively strong in "concepts." Standard tests of this type do not provide much useful information concerning the various strategies students employ to solve problems; they do not reveal whether or not an incorrect answer is simply the result of a minor misunderstanding of the question; and they provide little useful information concerning connections among various aspects of knowledge.

Recognizing these limitations of standard tests, Ginsburg and Baroody (1990) developed a standard test of mathematical thinking (Test of Early Mathematical Ability or TEMA) designed to accomplish several purposes. The main goal was to provide information concerning students' functioning in several key areas of mathematical work, both formal and informal. Most tests of mathematical thinking are not based on sound theories of mathematical knowledge. Indeed, most of the theories underlying these tests (for example, the KeyMath Test; Connelly, Nachtman, and Pritchett, 1976) involve little more than the notion that mathematical thinking involves procedures and concepts. Consequently, Ginsburg and Baroody developed a test that focuses on various aspects of informal and formal mathematical knowledge in children from preschool through approximately third grade. The test items were, in fact, selected from data-gathering procedures used in cognitive developmental research on mathematical thinking over the past twenty or thirty years. The test items deal with such matters as informal addition, the mental number line, the concept of cardinality, simple number combinations, alignment procedures for addition and subtraction, base ten concepts, and the like.

The TEMA is individually administered. The child is presented with problems involving concrete objects as well as written problems. The evaluator can use the results to obtain an overall ranking of the child relative to peers (based on national norms); can obtain separate scores describing the child's proficiency in informal mathematics compared to formal mathematics; and can examine the child's performance in such areas as informal concepts and procedures, and formal concepts and procedures.

In brief, the TEMA draws on current research to examine the child's relative performance, under standardized testing conditions, in key areas of mathematical thinking. We believe that this test provides more information than previously available tests. Its most important function is to encourage evaluators, and the teachers and others who are recipients of the evaluations, to think differently about children's mathematical thinking. The nature and content of the test force the evaluators and others to consider that mathematical thinking has informal and formal components, that strategies and concepts of particular types are involved, and that, in general, mathematical thinking involves much more than memorization of the "number facts." So long as standardized tests are employed, test developers should draw upon psychological research as the basis for their design and construction.

Probes

Clearly, ranking children in various aspects of mathematics performance—however interesting these areas may be—does not shed a great deal of light upon thinking processes generally and understanding in particular. Consequently, Ginsburg (1990) developed an organized system of probes to be used in parallel with the TEMA. The general idea was that after the TEMA had been given in the standard fashion, many examiners would find it useful to probe further into the thought processes that produced the observed performance, particularly in the case of errors. Most evaluators, however, have not had training or experience in assessing children's thinking. Consequently, Ginsburg attempted to provide examiners with a structured and comfortable procedure for probing the strategies and concepts underlying children's responses to the TEMA. We recognize that clinical interviewing is a more effective and difficult means for achieving the same purpose; yet because most evaluators are not prepared to engage in extensive clinical interviewing, organized probes are a useful first step.

The probes for each of the sixty-five items of the TEMA involve three main features. The probes first attempt to establish whether the child has understood the basic question. Often children produce an incorrect response because they have misinterpreted a minor feature of the question asked. The probes attempt to distinguish this situation from that in which children do not understand the question because they fail to comprehend the relevant concept. Next, the probes attempt to determine the strategies and processes used by the child to solve the problem. For example, in the case of mental addition, the probes attempt to determine whether the child used such procedures as counting on the fingers, mental counting on, or memorized number facts. Third, the probes attempt to establish learning potential. The issue is whether the child can learn the relevant material with a minimum of hints or whether more substantial teaching is required. In the first case, it is clear that the child is close to "understanding"; in the second case, the child is not. (After describing how to establish the child's level of understanding, the manual then goes on to recommend appropriate educational activities relevant for the material tested by each item of the TEMA.)

We believe that probes of the type described can be a useful supplement to any standardized test in mathematics and in other areas as well. The information elicited by the probes can be the first step in obtaining a more detailed picture of children's understanding. Systematic probes, however, are limited in flexibility and power; the clinical interview is a more effective but difficult procedure.

Structured Clinical Interviews

Clinical interviewing involves flexible questioning designed to uncover basic features of the individual's thinking. The questioning depends upon the individual's responses and may vary from person to person. The clinical interview technique is thus deliberately nonstandardized. Many researchers now believe that for the purposes of measuring cognitive process, the clinical interview is the method of choice. It is far more sensitive (and more difficult to administer successfully) than the standard test. (For a review of the logic behind the standard test and clinical interview, see Ginsburg, 1986.) Clinical interviews vary on a continuum of structure, with some being more planned than others. Here we describe relatively structured clinical interview procedures, first in the area of multiplication, and then, in the next section, fractions.

Interviews about multiplication

At an introductory level, multiplication can be defined as a problem of finding the total quantity of objects contained in a given number of groups each with same number of elements. Thus, a prototypical multiplication problem would be: "Find the total number of apples when you have four apples on each of three dishes." For a systematic understanding of this situation a learner has to know and operate with two different grouping systems. In multiplication—unlike addition and subtraction—the two numbers in the problem refer to different types of quantities. In this example, 3 is the number of groups and 4 is the number of objects in each group. (A confusing complication is that another type of grouping appears in the answer. In the place value system, the answer 12 should be understood as "one group of 10 apples and 2 individual apples.") By contrast, in addition (4+3) or subtraction (4-3), the two numbers refer to like quantities in each of the two sets (for example, 4 apples and 3 apples).

To assess learners' understanding of these aspects of multiplication, Yamamoto developed a structured interview using picture cards. At the beginning of this procedure, subjects who had already been taught multiplication at school were asked to teach the meaning of this operation to an imaginary child who had no idea of what multiplication is. The subjects—each seen individually—were then shown two sets of pictures, one set multiplicative (regularly grouped) and one nonmultiplicative (irregularly grouped). Within each set were three types of pictures, as illustrated in Figure 1. One type of picture showed distinct sets of objects as in the four blocks in each of three distinct groups (a quantity per set model as described above); another type of picture showed three lines of four blocks each (something like an area concept of multiplication); and a third picture showed a rabbit jumping past three groups of four blocks arranged in a line (a number line model). Altogether six pictures (three showing regular groupings and three irregular) were randomly placed in front of each subject. In addition to the pictures, blocks with and without numerals, toy dishes, and other materials were available to be used in dealing with the problem.

Figure 1

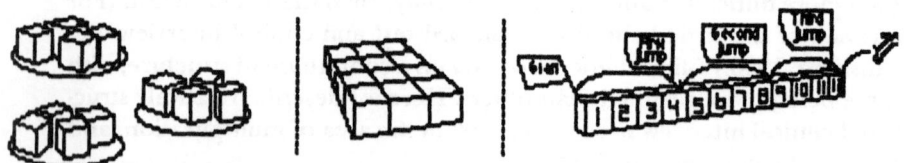

The subjects were asked to choose all pictures that could be used to teach the meaning of multiplication. A number of subjects chose some or all of the "regular" pictures and explained how they illustrated multiplication. In this case, after listening to their explanation, the examiner asked the following questions in reference to each picture chosen: (i) "Can you write down the multiplication equation which fits this picture?" (picture/notation relationship); (ii) "You wrote down three numbers here. What does this first number [usually 3 or 4] mean in the picture? How about the second number? What does this third number [usually 12] mean in the picture?" (referents of numbers); (iii) "You wrote number 12 for the answer. Do you think this 2 means something in the picture? How about this 1? If you think so, please color the part for this 2 in the picture with blue marker and color the part for this 1 with red marker" (base 10 grouping); (iv) "Can you show me 6 x 4, using these blocks and other materials here?" (application of model); (v) "You said all of these different pictures are 3 x 4 = 12. Can you tell me why?" (relationship among models). Although the interviews were done in a flexible manner, the procedure was structured in that these core questions were always asked of each subject.

When the "irregular" pictures were chosen, a different set of questions was used to determine the subject's reasons for selecting the pictures and to ascertain whether that subject had the flexibility to modify the irregularities so as to use the pictures in an appropriate way. In some cases, the subjects struggled to apply their forms of understanding to the irregularly grouped situations. Typically, they ended up with inconsistent or obviously absurd explanations and revealed their lack of conceptual understanding. In some cases, the subjects were able to modify the pictures and explain how they could then be used to deal with multiplication. This was typically done by regrouping the blocks into regular groups or by adding blocks to the groups that did not have enough of them.

Preliminary interviews included two groups of Japanese children living in the greater Tokyo area. One group had been taught the concept of multiplication (as described above) thoroughly and explicitly through the use of structured manipulatives, while the other group had received standard instruction, mainly based on typical Japanese textbooks, and focusing more on computation than concepts. Over half of the subjects in the manipulative group responded to all of the questions correctly with rich explanations referring to the pictured models. In a typical case the dialogue developed as follows:

I: You wrote down three numbers here [4 X 3 = 12]. What does this first number 4 mean in the picture [referring to the quantity per set model]?

S: Well, in this picture you always see four and only four blocks in each dish. That's what I mean by number 4.

I: O.K., What does this 3 mean then?

S: That's the number of these dishes.

I: Fine. What does this number 12 mean in the picture?

S: That's the answer.

I: Can you cover the part you mean by 12 with your hands?

S: Yes, like this [covering all of the blocks on the three dishes].

I: O.K. Well, you used two numerals in 12, 1 and 2 [circling each numeral with a pencil]. Do you think you can find a part for this number 2 in the picture? If you think so please mark the part in the picture with this blue marker.

S: Yes. [S marks two blocks in the first dish].

I: O.K. Now, do you think you can find a part for this number 1 in the picture? If so please mark the part with this red marker.

S: Yes. [S marks the remaining blocks in the three dishes].

I: Fine. Here we have some plastic blocks and toy dishes. Do you think you can make 6 x 4 using these materials? [The interviewer knows that the subject has never been taught 6 x N facts at school].

S: Yes, no problem. [S takes four dishes, counts blocks carefully, and constructs an appropriate configuration for the 6 x 4].

I: Do you know the answer for this problem?

S: Ah. . . Wait a second. [Pause/8 seconds]. 24.

I: That's right. How did you get the answer?

S: 6 + 6 is 12, right? Then I added 6 more and got 18. Then I added 6 again.

I: Very good.

The other students in the manipulative group had difficulty with only one question, the base 10 grouping.

The other group (mainly textbook instruction with no explicit concept teaching) responded roughly in two ways. Over half of the subjects showed a poor level of understanding, although they knew the number facts quite well. Here is a typical example:

I: You chose this picture to explain what multiplication is [referring to the quantity per set model]. Can you write down the multiplication which fits this picture?

S: [Pause/10 seconds]. That's 8 x 4 [writing it down].

I: O.K. Do you know the answer for this problem?

S: [Pause/3 seconds]. Yes, thirty . . . two.

I: That's right. What does this first number 8 mean in the picture?

S: [Pause/7 seconds].

I: Can you circle the part you mean by 8 with this pencil?

S: [S circles the contents of the two dishes].

I: O.K. Now, which is the part for this number 4 in the picture?

S: Here. [S circles the contents of the third dish].

I: O.K. Do you know how many blocks in this picture in total?

S: Yes, 12, of course. [S responded immediately with a relieved facial expression].

Other subjects (who seemed to be extremely intelligent) displayed high understanding levels, despite the fairly conventional instruction they received. This implies that average students may benefit from a manipulative approach stressing concepts while highly intelligent students can develop sound understanding even without such instruction.

This preliminary study showed that several aspects of understanding can be measured in a fairly structured way. Since the subjects were requested to respond to the specific questions on the pictured situations in their own words, mechanical applications of fixed types of "instructed" knowledge were almost impossible. Verbal questions were supplemented by visual and concrete materials. This helped less verbally proficient subjects both in comprehending the questions and in expressing their interpretations. In general, subjects with sound basic understanding were able to solve most of the interview questions, whereas those with little basic understanding tended to get stuck at the outset and did not get beyond the initial questions. Hence it was relatively easy to discriminate between levels of understanding. Further, understanding seemed to be independent of number-fact accuracy: some "high-understanding" subjects had learned less than half of the 100 number facts taught at school.

Interviews about fractions

Traditionally, assessment of fraction knowledge has focused on solving word problems and statements such as 1/2 + 2/3. The child can succeed on these tasks merely by applying appropriate calculational procedures and does not have to understand the underlying concepts. In recent work, Willis has used a structured clinical interview approach in an attempt to assess conceptual knowledge of fractions as well as calculational procedures. Her work drew on the previous investigations of Behr et al. (1983) focusing on children's underlying concepts and representations of fractions.

Willis' approach explores various aspects of the child's ability to make connections between informal and formal understanding. (Although the content is the addition of fractions, the same interview format may be applied to the assessment of fraction concepts ranging from partitioning, identification, comparison, and equivalence, to operations such as subtraction, multiplication and division.) Like the work of Behr et al. (1983), the fraction interview explores the child's use of various methods and models of representation, such as pencil and paper, Cuisenaire rods (continuous length model), pattern blocks (area model), chips (discrete model), and a geoboard (area model).

As the first step in the assessment process, the interviewer asks the individual child to write and solve a problem such as 1/3 + 1/6. To identify the method underlying the child's solution, the child is then asked, "How did you do it?" The response to this question may lead to additional hypotheses about the child's formal and procedural knowledge as well as the child's ability to demonstrate understanding through explanation. Questions are reworded or new questions asked, depending on previous answers. For example, if no response is given, the previous question may be reworded to say, "What were you saying to yourself as you did the problem?" Or a new question may be asked: "How would you explain to another child how to do this problem?" Responses to questions like these may include: "I added 1 + 1 and 3 + 6 and got 2/9"; "I added the 3 and the 6 and kept the one the same, and got 1/9"; or "I found that the common denominator was 6 because 2 goes into 6 three times. So, I multiplied the top and the bottom by 2, and then added 2/6 + 1/6 to get 3/6." (In this case, to further examine the child's understanding of equivalence, the examiner may then ask, "Can you show me 3/6 another way?")

The course of the interview often depends on the answer to the first question. When the answer is correct, a counter-suggestion technique may be used to determine the stability of understanding. This includes asking questions such as: "Why didn't you add the bottom numbers since you added the top numbers?" If the answer to the initial question is not correct, additional problems and probing questions are used to test hypotheses about the child's understanding.

The second step of the clinical interview examines the child's ability to demonstrate understanding of a problem using various representations of fractions. Because the child may be confused by the physical characteristics of one type of manipulative, he or she is allowed to represent a problem like 1/3 + 1/6 by choosing among the following: pencil and paper, Cuisenaire rods, pattern blocks, chips, or a geoboard. The child is asked to use each model to explain each step of the solution.

Third, the interviewer focuses on the child's ability to make connections between different models—for example, area, discrete sets, continuous lengths, and volume. The interviewer asks the child to describe similarities and differences between models.

The fourth part of the interview examines the comparison of fractions, and the understanding of differences between the whole number and rational number systems. The child may be presented with problems involving the value of increasing denominators. For example, after the child answers the written equation 1/3 + 1/6 correctly, the interviewer asks, "OK, now suppose I said the answer was 1/9, how could you prove to me that I was wrong?" The child is permitted to choose the type of representation used. The child's responses may include: "1/9 is smaller than 1/6, and when you add two fractions together you need to come up with a larger answer, so this is wrong." To test the child's confidence in the answer, the interviewer may then reply, "The 9 is bigger than the 6, so how can that be so?"

Next, the interviewer explores the child's understanding that commutativity applies to the addition of fractions as well as the addition of whole numbers. By noting whether 1/6 + 1/3 is solved immediately after the child has already solved 1/3 + 1/6, the interviewer determines whether the child used commutative principles or had to solve the problem step by step all over again.

In the sixth step, focus is placed on the child's ability to create an equation to represent a manipulative model. Deep understanding involves the ability to connect concrete manipulative models to abstract equations and vice versa. The interviewer demonstrates the problem 1/3 + 1/3 using a model such as pattern blocks and asks the child to write the equation. To determine if the child has difficulty with some manipulatives and not others, this procedure is repeated using different addition of fraction problems and different manipulatives.

The seventh part of the interview examines the child's ability to connect formal concepts and procedures to a story concerning an everyday situation. The child is asked, for example, to deal with a problem such as: "If Mary divided her garden into eight equal sections and planted peas in two sections and tomatoes in four sections, how much of her garden would be filled with vegetables?" Conversely, the child is also asked to create a real life-problem corresponding to a particular equation.

ASSESSMENT IN THE CLASSROOM

Classroom assessment procedures must involve methods that teachers can use under ordinary classroom conditions to obtain information concerning students' mathematical thinking. These techniques should allow teachers to determine the methods students use to solve problems, to identify the concepts and misunderstandings underlying students' work, and, in general, to gain insight into the different thinking processes students use in learning classroom mathematics.

Classroom assessment techniques are essential because good teaching requires sensitive assessment. Effective teaching requires knowledge of what students know and do not know, how students approach problems, and how they react to educational activities. Although assessment information can make a real difference in the classroom, teachers have not had available the tools to conduct assessment most relevant for their own needs. Our goal is to provide teachers with practical methods for assessing their students' understanding in the classroom.

We note from the outset that these methods cannot be employed in a mechanical fashion. Sensitive assessment requires rigorous theorizing and analysis on the part of teachers. This stringent requirement does not deter our efforts: good teachers perform sensitive and intellectually demanding analyses of their students literally every day of the school year, and their analyses are sometimes more complex than those of which some psychologists are capable.

Two serious problems face classroom teachers interested in assessing children's understandings of arithmetic. First, there are few, if any, available testing instruments that can be used with *groups* of students to assess understanding. Second, few, if any, available instruments provide information of direct relevance to classroom instruction. With the support of the Fisher-Landau Foundation, Lopez and Kelly are attempting to address these needs by developing group-administered screening measures designed to help teachers of grades one through three identify key aspects of understanding and the cognitive processes and strategies students use to solve arithmetic problems.

Although they cannot provide in-depth assessments, screening instruments can provide information that can help teachers identify and then correct problems in mathematical learning before they become serious. Information from screening measures can lead to specific educational recommendations; it can also help the teacher to adapt instruction to the student's needs instead of referring the student for further evaluation or special education placement. In more technical terms, teachers can utilize a "pre-referral preventive approach" to educational intervention rather than a remedial approach.

The screening procedure involves two steps. First, a whole class is given a series of mathematics tasks in a paper-and-pencil format. The tasks are designed to assess faulty understanding of procedures and to provide as much insight as possible into the processes and strategies youngsters use. Second, smaller groups of students are given follow-up tasks to probe further their understandings of these skills and concepts. The follow-up is guided completely by the results of the screening; only the areas in which the student

Figure 2. Sample report form for math screening tests.

FISHER-LANDAU FOUNDATION
Teachers College • Columbia University
MATH SCREENING
GRADE 3

Name _____

Circle the student's raw score for each screening subtest and note the category in which the student's score falls to determine if follow-up is needed.

	Administer Follow-Up	Passing Criterion
A. Time	0 1 2 3 4	5
B. Number facts	0 1 2 3 4 5 6	7 8
C. Money	0 1 2 3 4 5 6 7 8	9 10
D. Calculation		
• Addition	0 1 2 3 4 5 6 7 8	9 10
• Subtraction	0 1 2 3 4 5 6 7 8	9 10
E. Problem Solving	0 1 2 3	4
F. Arithmetic Problems	0 1 2	3
G. Concepts	0 1 2 3 4 5	6 7

appears to have difficulty are examined further. The decision to follow up in a particular area is based on whether an individual student reaches a set criterion of number of items correct (see Figure 2).

Throughout the process teachers are encouraged to observe students' problem-solving strategies and to ask questions that probe their understandings. In general, the items progress from evaluating factual knowledge to evaluating procedures, comprehension, and the ability to think about and apply skills and concepts. The emphasis is on evaluating understanding, thinking, and applying skills rather than on facts.

The first grade screening focuses on the evaluation of informal mathematics skills and concepts such as the understanding of the number line, informal addition, reading numerals, the concept of more, and the ability to count out objects from an array. The second and third grade screenings include the assessment of both formal mathematical knowledge and informal mathematics. They include items that assess number and numeration, time, number facts, money, calculation, problem solving, and concepts such as place value and base ten. The procedures for administering the preliminary classroom screening and the follow-up tasks are the same at all three grade levels.

The preliminary classroom screenings are designed to identify common errors that suggest a lack of understanding of the skill and concept being examined. For example, some calculation items from the third grade screening assess the use of "buggy" procedures: If a student's answer to the problem 92 minus 38 is 66, the teacher is guided to hypothesize that the student subtracted the smaller from the larger and did not borrow. The following example from the third grade screening illustrates how a classroom teacher proceeds from the preliminary classroom screening through the follow-up.

Sam took the preliminary screening along with all his classmates. His teacher, Ms. Smith, scored the test and found that Sam did not meet the passing criteria for the number facts section. Among these items, Sam failed those from the N minus 0 family (for example, to the item 7 minus 0, Sam answered 6.) Like Sam, three other students seemed to have difficulty in the area of number facts. Ms. Smith decided to explore further their number fact understandings. She brought the students together and explained that they would all do some math work. Ms. Smith then distributed the follow-up test item (Figure 3). She read out loud to the students, as they read along: "Mike and Carlos have 4 candies. Sally asked them to give her some. Mike and Carlos did not give her any candy. How many candies do Mike and

Carlos have now? Select the box with the number sentence that illustrates this story and write the answer to it." Once the students had marked the boxes the teacher asked them to share their answers with the group. Sam said he selected 4 - 0 and that his answer to this operation was 3. The following dialogue took place between Sam and Ms. Smith:

Ms. Smith: Sam, how did you get that answer?

Sam: Well, I just knew it.

Ms. Smith: OK, but how would you prove to your friends that you are correct?

Sam: Well, I could say that 4 minus 0 is 3 because when you take away, it is like you have to take away something and that is 3.

Ms. Smith: Please, teach Robbie how to do the problem.

Figure 3. A follow-up test item for the third grade screening test.

Mike and Carlos have 4 candies.
Sally asked them to give her some.
Mike and Carlos did not give her any candy.
How many pieces of candy do Mike and Carlos have now? Mark the box with the number sentence that goes with this story. Then write your answer to the number sentence.

$4 - 1 =$ ☐
$4 + 0 =$ ☐
$4 - 0 =$ ☐
$4 + 1 =$ ☐

Sam: Well, you see, Robbie, you have the 4 and take away 0. Zero is like having nothing, so you have to think of another number to do the take away. I always think of the number 1 because it is the easiest. Then, you go 4 take away 1 and that is 3.

Ms. Smith: Sam, what does this symbol [pointing to 0] mean?

Sam: It is zero.

Ms. Smith: Yes, but zero is like having how many? Show me with these blocks what zero means.

Sam: Oh, that is easy, I give you all the blocks back. Zero is like having nothing.

It is apparent at this point that Sam could choose the correct number sentence and that he had an informal understanding of zero. It is the written subtraction procedure when it involved zero that he did not understand. It is possible now for Ms. Smith to develop a brief intervention to help Sam overcome the difficulty he is experiencing.

To date, teachers at all grade levels have attested to the value of these screening measures in the identification of students who are having difficulties with mathematics. In addition, they have found the information obtained on students' thinking and understanding to be of diagnostic and instructional value.

Observations

Ginsburg and Mukhopadhyay have been exploring observational techniques for use in the classroom. Our starting assumption is that assessment of thinking can occur only in classrooms in which thinking is encouraged. The traditional teacher who instructs students in standard algorithms and attends mainly to their correct and incorrect answers is not in a position to learn a great deal about understanding. By contrast, teachers who encourage students to engage in mathematical activities, to develop their own methods of solution, to discuss mathematical ideas and procedures, and to believe that their own approaches to learning are valued, can relatively easily learn a great deal about students' understanding. Putting it more bluntly, the teacher who does not encourage understanding cannot measure it in the classroom; the teacher who encourages it can learn a great deal about it.

We have been observing and working with one second grade teacher who encourages active learning in her classroom. Consider the following sequence of activities involving both good teaching and assessment.

The teacher wrote a simple computational problem, $9 + 7 = $ ___, on a large piece of paper in front of the class. She asked the children to solve the problem in their own way and to write down the answer. After this has been done, the teacher spent a good part of the math lesson exploring the children's strategies, for example: "I took 2 away from the 9 and that was 7. $7 + 7 = 14$. I add 2 more and I got 16." Or, "First I took the 7 and then I put up 9 fingers and I counted up 7 8 9 10 11 12 13 14 15 16. " Or, "I knew $10 + 7 = 17$, but 9 is 1 less than 10, so 1 less than that $= 16$."

She asked all of the children in the room to explain their method of solution. Sometimes, she asked them to describe it in writing. Throughout this process, the teacher established an atmosphere in which children are encouraged to share their strategies, to value the range of strategies arising in the classroom, and to verbalize strategies as explicitly as possible. In effect, the lesson involved sharing, valuing, and training in introspection. The teacher showed the class that there can be a variety of ways to obtain the correct answer. After exposing the children to a number of strategies, she asked them if they would like to choose a different strategy and if so why they would choose it. Some children preferred to continue in the strategy already chosen, while other children chose a new approach, often "because it is faster."

In the course of this exercise, the children provide very clear information concerning different strategies employed. With our assistance, the teacher had developed a simple scheme for coding the observed strategies. This scheme includes simple descriptions of procedures commonly observed in the research literature—recall of number facts; concrete counting involving fingers or other easily available objects; mental counting procedures, such as counting on; and various regrouping strategies, such as: "6 and 4 is 10 because I know that 4 and 4 is 8 and the answer is only 2 more than that." It is relatively easy for the teacher to record children's use of these strategies on a simple checklist, which provides a convenient record of the strategies the children were using. This is a clear improvement over the usual procedure in which the teacher simply attempts to remember the methods used by various children.

After the observation and discussion of strategies, the teacher then asked the children to create a story corresponding to the written problem 15 + 3, and a variety of responses were obtained:

I had 15 crayons. My mother gave me 3 more and I got 18.

One day there was a kid and she was 15. She wanted to know how old she was going to be when she was 3 years older. She added 15 + 3 and counted up 3 numbers and got 18. Today was her birthday. She was going to be 18. She had a party and she was very happy. The end.

The math problem: once upon a time there was a girl. She was in the second grade. She was not very good at math. Even plusses. One day she went to school. Her teacher said it is time for math. Then the teacher said the problem is 15 + 3 = blank. The whole class shouted "easy." So the girl said "easy." But it really wasn't. Then she said to the teacher "I cannot do this math problem." "Yes you can," said the teacher. The girl thought that she was mean, but she wasn't. Then she figured it out. From that day on she always felt good when she did math. She knew 15 + 3 = 18.

Stories like these provide many different kinds of information. Some stories show how children relate numerical statements to real life events. Thus, the birthday story relates numbers to ages and even describes the method of solution. Other stories, like "the math problem" reveal more about students' feelings toward mathematics and self-concept as learner than they do about the methods of solution. In both cases, however, useful information is obtained from a pedagogically valuable situation which has the virtue of linking language arts with arithmetic.

The Classroom Uses of Dumb Tests

Recently Ginsburg made the following observation concerning six sixth graders. All sixth graders in a school were given a standardized timed test involving a series of multiple choice items, mainly involving calculation. The students scoring exceptionally well on this rather dumb test were then feted at a party attended by teachers, the school principal, and administrators in the district. As part of the celebration, the students were asked to explain how they solved items on the test. It was interesting to observe that on almost every item of the test at least two or three different methods of solution were exhibited by the six students in attendance. The solution methods included tedious calculations, brilliant insights into the structure of the problems, and clever exploitation of the multiple-choice format. In other words, the students' explanations revealed a wide range of individual differences and a variety of strategies, some of them insightful, in dealing with a rather dumb test. Of course, these thought processes were evident only because the students were asked to talk about them; they are not usually revealed by the test which is usually scored so as to yield only the number correct.

The point of the story is that even a dumb test can be exploited as a useful assessment technique, provided children are encouraged to reflect on and reveal the solution processes employed. A corollary of this proposition is that teachers who are forced "to teach to the test" can turn the exercise into an interesting activity by encouraging students to make explicit and to discuss the methods of solution employed.

The Thinking Curriculum

Effective methods of assessment in the classroom—discussion of strategies, the creation of mathematical stories, the analysis of approaches to dumb tests, and other procedures we have not discussed here—all share a common feature. They are the by-products of classroom activities in which thinking is the focus and content of the curriculum. When the teacher focuses on mathematics as thinking, the inevitable result is discussion of procedures, self-examination and reflection, and the making public of mathematical thought processes. When mathematical thinking is the essence of the curriculum, the teacher naturally has available a good deal of material providing insight into students' thinking. In the thinking curriculum, thinking pervades the classroom and is thus relatively easy to assess.

ISSUES FOR THE FUTURE

Much remains to be done to develop effective techniques for the assessment of understanding. First, sound research and theory on mathematical thinking must underlie effective assessment techniques. Methods do not arise from nothing and do not stand alone; they depend on theory. To the extent that we have sophisticated research and theory concerning mathematical understanding, we have the basis for developing effective assessment techniques. As Mislevy (in press) has pointed out, much testing involves the use of twentieth century statistics combined with a ninteenth century psychology. Sensitive assessment requires a twenty-first century psychology of mathematical thinking.

Second, we need to develop a wide variety of new assessment techniques. Only two will be mentioned here. One is the development of effective clinical interview techniques suitable for the classroom. The clinical interview has proved to be an effective technique for research. Teachers are impressed by its power and wish to use it themselves in the classroom. Can this be done? Peck, Jencks, and Connell (1989) have made a start in this direction; much more needs to be done.

We also require methods for assessing higher-order attitudes and feelings concerning mathematics. In our view, beliefs, attitudes, and feelings about mathematics are fundamental in determining students' approach to the subject (Kaplan, Burgess, Ginsburg, 1988; Ginsburg, 1989a). In some sense, factors like these are more important than strategies, understandings, and misunderstandings in determining students' approach to classroom work in mathematics. Thus, it may be more important to know whether a student sees mathematics as a subject which makes no sense than to determine the particular strategy he uses to solve a problem. At the same time, we have available few methods for assessing these affective factors. We need to develop them.

Third, it does no good to assess children unless teachers understand and utilize the results. Teachers should be the primary consumers of assessment information. Indeed, teachers should be the primary conductors of assessments. Most assessments conducted by school psychologists, evaluators, and administrators provide little information of any value to classroom teachers. Yet the classroom teacher can have the most immediate effect upon children's learning. Assessments therefore should be made relevant to teachers, and teachers should learn to conduct assessments. But to appreciate these assessments and to use them well, teachers need to understand a good deal about children's mathematical thinking. It does little good to provide a teacher with information concerning strategy if the teacher does not understand what strategy is. Consequently, a major effort needs to be made in the area of teacher education. Teachers need to learn not only "math methods" but "children's methods"—that is, the ways in which children go about making sense of the world of school mathematics.

REFERENCES

Antell, S., and Keating, D. (1983). Perception of numerical invariance in neonates. *Child Development, 54,* 695-701.

Behr, M.J., Lesh, R., Post, T.R., and Silver, E.A. (1983). Rational number systems. In R. Lesh and M. Landau (Eds.), *Acquisition of mathematics concepts and processes.* NY: Academic Press.

Carraher, T.N., Carraher, D.W., and Schliemann, A.S. (1985). Mathematics in streets and schools. *British Journal of Developmental Psychology, 3,* 21-29.

Connelly, A.J., Nachtman, W., and Pritchett, E.M. (1976). KeyMath Diagnostic Arithmetic Test. Circle Pines, MN: American Guidance Service.

Ginsburg, H.P. (1986). Academic assessment. In J. Valsiner (Ed.), *The role of the individual subject in scientific psychology.* New York: Plenum.

Ginsburg, H.P. (1989). *Children's arithmetic: How they learn it and how you teach it.* (2nd ed.). Austin, TX: Pro Ed.

Ginsburg, H.P. (1989a). The role of the personal in intellectual development. *Newsletter of the Institute for Comparative Human Development, 11,* 8-15.

Ginsburg, H.P. (1990). *Assessment probes and instructional activities. The test of early mathematics ability,* (2nd Ed.) Austin, TX: Pro Ed.

Ginsburg, H.P. and Baroody, A.J. (1990). *The test of early mathematics ability,* (2nd ed.) Austin, TX: Pro-Ed.

Hatano, B. (1988). Social and motivational bases for mathematical understanding. In G. Saxe and M. Gearhart (Eds.), *Children's mathematics.* San Francisco: Jossey-Bass.

Kaplan, R.G., Burgess, P., and Ginsburg, H.P. (1988). Children's mathematical representations are not (always) mathematical. *Genetic Epistemologist, 16,* 7-14.

Kaplan, R.G., Yamamoto, T.A., and Ginsburg, H.P., (1989). Teaching mathematics concepts. In L.B. Resnick and L.E. Klopfer (Eds.) *Toward the thinking curriculum: Current cognitive research.* 1989 Yearbook of the Association for Supervision and Curriculum Development.

Mack, N.K. (1990). Learning fractions with understanding. *Journal for Research in Mathematics Education, 21,* 16-32.

Mislevy, R. J. (in press). Foundations of a new test theory. Princeton, NJ: Educational Testing Service.

National Council of Teachers of Mathematics. (1989). *Curriculum and evaluation standards for school mathematics.* Reston, VA: NCTM.

Peck, D.M., Jencks, S.M., and Connell, M.L. (1989). Improving instruction through brief interviews. *Arithmetic Teacher, 37,* 15-17.

Pirie, S.E.B. (1988). Understanding: Instrumental, relational, intuitive, constructed, formalized...? How can we know? *For the Learning of Mathematics, 8*(3), 2-6.

Saxe, G.B. (in press). Culture and cognitive development: Studies in mathematical understanding. Hillsdale, NJ: Lawrence Erlbaum Associates.

Saxe, G., and Posner, J. (1983). The development of numerical cognition: Cross-cultural perspectives. In H. P. Ginsburg (Ed.), *The development of mathematical thinking.* New York: Academic Press.

Vygotsky, L.S. (1962). *Thought and language.* Cambridge, MA: Massachusetts Institute of Technology Press.

Van den Brink, J. (1989). Transference of objects. *For the Learning of Mathematics, 9*(3), 12-16.

PART IV: New Types of Scoring and Reporting

12  Toward a Test Theory for Assessing Student Understanding

Robert J. Mislevy,
Kentaro Yamamoto, and
Steven Anacker

INTRODUCTION

The view of learning that underlies standard test theory is inconsistent with the view rapidly emerging from cognitive and educational psychology. Learners become more competent not simply by learning more facts and skills, but by reconfiguring their knowledge; by "chunking" information to reduce memory loads; and by developing strategies and models that help them discern when and how facts and skills are important. Neither classical test theory (CTT) nor item response theory (IRT) is designed to inform educational decisions conceived from this perspective. This chapter sketches the outlines of a test theory built around models of student understanding, as inspired by the substance and the psychology of the domain of interest. The ideas are illustrated with a simple numerical example based on Siegler's balance beam tasks (Siegler, 1981). Directions in which the approach must be developed to be broadly useful in educational practice are discussed.

Background

When schooling became mandatory at the turn of the century, educators suddenly faced selection and placement decisions for unprecedented numbers of students, of diverse abilities and backgrounds (Glaser, 1981). Numbers of correct answers to multiple-choice test items were used to rank students according to their overall proficiencies in domains of tasks. These rankings were used in turn to predict students' success in fixed educational experiences.

Classical test theory (CTT) emerged when Spearman (1907) applied statistical methods to study how reliable estimates of this overall proficiency would be from different test forms that might be constructed for the purpose. Extensions of this work led over the years to a vast collection of techniques for building tests and making decisions with test scores (Gulliksen, 1950); to an axiomatic foundation for statistical inference about test scores (Lord, 1959; Lord and Novick, 1968; Novick, 1966); and to sophisticated techniques for partitioning test score variance according to facets of items, persons, and observational settings (Cronbach, Gleser, Nanda, and Rajaratnam, 1972). It is important to note that, in all this work, the object of inference is overall proficiency—the test score, observed or expected—in terms of numbers of correct responses in a domain of items.

Item response theory (see Hambleton, 1989, for an overview) represented a major practical advance over CTT by modeling probabilities of correct item response in terms of an unobservable proficiency variable. IRT solves many equating, test construction, and adaptive testing problems that were difficult under CTT. Advanced statistical methods have been brought to bear on inferential problems in IRT, including sophisticated estimation algorithms (for example, Bock and Aitkin, 1981), techniques from missing-data theory (Mislevy, 1991), and Bayesian treatments of uncertainty in models and parameters (Lewis, 1985; Mislevy and Sheehan, 1990; Tsutakawa and Johnson, 1988). The underlying psychological model remains quite simple however, and as in CTT, the focus remains on overall proficiency in a domain of items. From the perspective of IRT, two students with the same overall proficiency are indistinguishable.

As useful as standard tests and standard test theory have proven in large-scale evaluation, selection, and placement problems, their focus on who is competent and how many items they answer can fall short when the goal is to improve individuals' competencies. Glaser, Lesgold, and Lajoie (1987) point out that tests can predict failure without an understanding of what causes success, but intervening to prevent failure and enhance competence requires deeper understanding.

The past decade has witnessed considerable progress toward the requisite understanding. Psychological research has moved away from the traditional laboratory studies of simple (even random!) tasks, to tasks that better approximate the meaningful learning and problem-solving activities that engage people in real life. Studies comparing the ways experts differ from novices in applied problem-solving in domains such as physics and troubleshooting (see, for example, Chi, Feltovich, and Glaser, 1981) reveal the central importance of knowledge structure—networks of concepts and interconnections among them—that impart meaning to patterns in what one observes and how one chooses to act. The process of learning is to a large degree expanding these structures and, reconfiguring them to incorporate new and qualitatively different connections as the level of understanding deepens. Educational psychologists have begun to put these findings to work in designing both instruction and tests (for example, Glaser et al., 1987; Greeno, 1976; Marshall, 1985, 1992). Again, in the words of Glaser, Lesgold, and Lajoie (1987),

> Achievement testing as we have defined it is a method of indexing stages of competence through indicators of the level of development of knowledge, skill, and cognitive process. These indicators display stages of performance that have been attained and on which further learning can proceed. They also show forms of error and misconceptions in knowledge that result in inefficient and incomplete knowledge and skill, and that need instructional attention. (p.81).

Paraphrasing Ohlsson and Langley (1985), Clancey (1986) summarizes the shift in perspective

> [to] describing mental processes, rather than quantifying performance with respect to stimulus variables; describing individuals in detail, not just stating generalities; and giving psychological interpretation to qualitative data, rather than statistical treatment to numerical measurements. (p. 391).

AN APPROACH TO MODELING STUDENT UNDERSTANDING

The modeling approach we are beginning to pursue can be encapsulated as follows:

> Standard test theory evolved as the application of statistical theory with a simple model of ability that suited the decision-making environment of mass educational systems. Broader educational options, based on insights into the nature of learning and supported by more powerful technologies, demand a broader range of models of capabilities —still simple compared to the realities of cognition, but capturing patterns that inform a broader range of instructional alternatives. A new test theory can be brought about by applying to well-chosen cognitive models the same general principles of statistical inference that led to standard test theory when applied to the simple model. (Mislevy, 1992)

The approach begins in a specific application by defining a universe of student models. This "supermodel" is indexed by parameters that signify distinctions between states of understanding. Symbolically, we shall refer to the (typically vector-valued) parameter of the student model as η. A particular set of values of specifies a particular student model, or one particular state among the universe of possible states the supermodel can accommodate. These parameters can be qualitative or quantitative, and qualitative parameters can be unordered, partially ordered, or completely ordered. A supermodel can contain any mixture of these types. Their nature is derived from the structure and the psychology of the learning area, the idea being to capture the essential distinctions among students.

Modeling Item-construction and Inference Problems

The modeling problem is delineating the states or levels of understanding in a learning domain. In meaningful applications this might address several distinct strands of learning, as understanding develops in a number of key concepts, and it might address the connectivity among those concepts.[1] Symbolically, this substep defines the structure of $p(x)$, where x represents observations. Obviously any model will be a gross simplification of the reality of cognition. A first consideration in what to include in the supermodel is the substance and the psychology of the domain: Just what are the key concepts? What are important ways of understanding and misunderstanding them? What are typical paths to competence? A second consideration is the so-called grain-size problem, or the level of detail at which student-models should differ. A major factor in answering this question is the decision-making framework under which the modeling will take place. As Greeno (1976) points out, "It may not be critical to distinguish between models differing in processing details if the details lack important implications for quality of student performance in instructional situations, or the ability of students to progress to further stages of knowledge and understanding."

The item construction problem is devising situations for which students who differ in the parameter space are likely to behave in observably different ways. The conditional probabilities of behavior of different types given the unobservable state of the student are the values of $p(x:\eta)$, which may in turn be modeled in terms of another set of parameters, say, β. The $p(x:\eta)$ values provide the basis for inferring back about the student state. An element in x could contain a right or wrong answer to a multiple-choice test item, but it could instead be the problem-solving approach regardless of whether the answer is right or wrong, the quickness of responding, a characteristic of a think-aloud protocol, or an expert's evaluation of a particular aspect of the performance. The effectiveness of an item is reflected in differences in conditional probabilities associated with different parameter configurations, so an item may be very useful in distinguishing among some aspects of potential student models but useless for distinguishing among others. Tatsuoka (1989) demonstrates the relationship between item construction and inference about students' strategies for subtracting mixed numbers.

The inference problem is reasoning from observations to student models. The model-building and item construction steps provide η and $p(x:\eta)$. Let $p(\eta)$ represent expectations about η in a population of interest—possibly noninformative, possibly based on expert opinion or previous analyses. Bayes' theorem can be employed to draw inferences about η given x via $p(x:\eta)$ $p(x:\eta) \propto p(x:\eta)$. Thus $p(\eta:x)$ characterizes belief about a particular student's model after having observed a sample of the student's behavior. Practical problems include characterizing what is known about ß so as to determine $p(x:\eta)$, carrying out the computations involved in determining $p(x:\eta)$, and in some applications, developing strategies for efficient sequential gathering of observations. As we have noted, analogous problems have been studied in standard test theory, and the solutions there, because they are applications of general principles of statistical inference, generalize to models built around alternative psychological models. The models are more realistic and more ambitious, but the formalism is identical.[2]

Previous research

Research relevant to this approach has been carried out in a wide variety of fields, including cognitive psychology, the psychology of mathematics and science education, artificial intelligence (AI) work on student modeling, test theory, and statistical inference. Cognitive scientists have suggested general structures such as "frames" or "schemas" that can serve as a basis for modeling understanding (see, for example, Minsky, 1975; Rumelhart, 1980), and have begun to devise tasks that probe their features (for example, Marshall, 1989, 1992). Researchers interested in the psychology of learning in subject areas such as proportional reasoning have focused on identifying key concepts, studying how they are typically acquired (for example, in mechanics, Clement, 1982; in ratio and proportional reasoning, Karplus, Pulos, and Stage, 1983), and constructing observational settings that allow one to infer students' understanding (for example, van den Heuvel, 1990; McDermott, 1984). We make no effort here to review these literatures, but point out that our work can succeed only by building upon their foundations. Our potential contribution would be to the structures and mechanics of model-building and inference. The following sections briefly mention some important work along these lines from test theory and statistics.

Modeling patterns in student behavior

The standard models of educational measurement are concerned solely with examinees' tendencies to answer items correctly—that is, their overall proficiency. Recently, however, models that focus on patterns other than overall proficiency have begun to appear in the test theory literature. Some examples that are relevant to educational applications are listed below.

- Mislevy and Verhelst's (1990) "mixture models" for item responses when different examinees follow different solution strategies or use alternative mental models. When a single IRT model cannot capture key distinctions among examinees, it may suffice to posit qualitatively distinct classes of examinees and use IRT models to summarize distinctions among examinees within these classes.

- Wilson's (1989b) "saltus model" for characterizing stages of conceptual development. This model parameterizes the differential patterns of strength and weakness expected as learners progress through successive conceptualizations of a domain.

- Falmagne's (1989) and Haertel's (1984) "latent class models for binary skills." These models are intended for domains in which competence can be described by the presence or absence of several (possibly complex) elements of skill or knowledge, and observational situations can be devised that demand various combinations of these skills. Also see Paulson (1986) for an alternative use of latent class modeling in cognitive assessment.

- Embretson's (1985) "multicomponent models" for integrating item construction and inference within a unified cognitive model. The conditional probabilities of solution steps given a multifaceted student model are given by IRT-like statistical structures.

- Tatsuoka's (1989) "rule space analysis." Tatsuoka uses a generalization of IRT methodology to define a metric for classifying examinees based on likely patterns of item response given patterns of knowledge and strategies.

- Yamamoto's (1987) "hybrid model" for dichotomous responses. The hybrid model characterizes an examinee as either belonging to one of a number of classes associated with states of understanding, or in a catch-all IRT class. This approach might be useful when certain response patterns signal states of understanding for which particular educational experiences are known to be effective. Instructional decisions are triggered by these patterns if they are detected, but by overall proficiency when no more targeted action can be provided.

- Masters and Mislevy's (1992) and Wilson's (1989a) use of the "partial credit rating-scale model" to characterize levels of understanding, as evidenced by the nature or approach of a performance rather than its correctness. These applications incorporate into a probabilistic framework the cognitive perspective underlying Biggs and Collis's (1982) SOLO taxonomy for describing salient qualities of performances.

These are the rudiments of models upon which concept-referenced achievement measures can be based. Applications to date have been fairly limited, and most have addressed one-to-many relationships between an underlying knowledge state and observable behavior. That is, a single (possibly unordered or multifaceted) variable has been used to characterize examinees, and performance on all items is modeled in terms of this variable. What is lacking from the point of view of the educator is a connection with meaningful real-world tasks, which are rarely segregated into these neat little sets. Rather, they often involve multiple concepts, connections among larger concepts, and transformations among alternative representations of a domain. While the simple tasks that characterize one-to-many domains are essential at early stages of learning, more complex tasks that involve multiple concepts in many-to-many relationships are needed to promote the integration among concepts that form the core of what is often called "higher-level learning."

Inference Networks

Recent developments in the context of probability-based inference networks (Lauritzen and Spiegelhalter, 1988; Pearl, 1988) offer a capability for integrating conceptual models of the type described above. These probability-based structures are attractive for educational measurement because they permit a coherent extension of the modeling approach and inferential logic of the new cognitive-assessment models mentioned above. To show how the approach might be applied in the educational setting, we first discuss an application in the setting of medical diagnosis.

MUNIN is an inference network that organizes knowledge in the domain of electromyography—the relationships among nerves and muscles. Its function is to diagnose nerve/muscle disease states. The interested reader is referred to Andreassen, Woldbye, Falck, and Andersen (1987) for a fuller description. The prototype discussed in that presentation and used for our illustration concerns a single arm muscle, with concepts represented by twenty-five nodes and their interactions represented by causal links.[3] A graphic representation of the network appears in Figure 1.

Figure 1. The MUNIN network: initial status (from Andreassen et al., 1987).

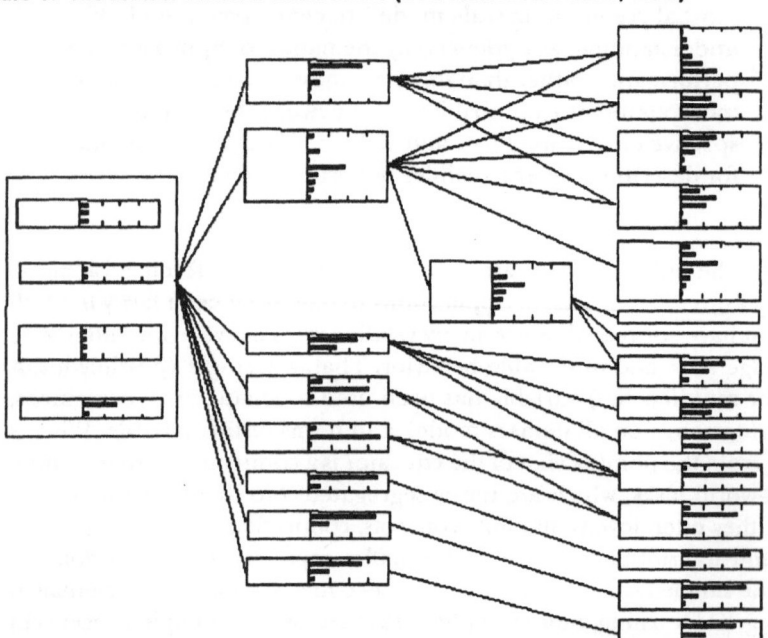

The rightmost column of nodes in Figure 1 concerns outcomes of potentially observable variables, such as symptoms or test results. These outcomes are the x vector in our earlier notation. The middle layers are "pathophysiological states," or syndromes. These drive the probabilities of observations. The leftmost layer is the underlying disease state, including three possible diseases in various stages, no disease, or "other" (a condition not built into in the system). These states drive the probabilities of syndromes. It is assumed that a patient's true state can be adequately characterized by values of these disease and syndrome states—our η parameter. Paths indicate conditional probability relationships, which are to be determined either logically, subjectively, purely empirically, or through model-based statistical estimation. In particular, the paths ending at observables represent $p(x|\eta)>$. Note that the probabilities of observables depend on some syndromes, but not others. The lack of a path signifies conditional independence. Note also that a given test result can be caused by different disease combinations.

As a patient enters the clinic, the diagnostician's state of knowledge about the patient is expressed by population base rates, or $p(\eta)$. This is depicted in Figure 1 by bars that represent the base probabilities of disease and syndrome states. Base rates of observable test results are similarly shown. Tests are carried out, one at a time or in clusters, and with each result the probabilities of disease states are updated. The expectations of tests not yet given are calculated, and it can be determined which test will be most informative in identifying the disease state. Knowledge is thus accumulated in stages, from $p(\eta)$ to $p(\eta: x_1)$ after observing the first subset of tests, to $p(\eta: x_1, x_2)$ after the second, and so on, with each successive test selected optimally in light of knowledge at that point in time. Figure 2 illustrates the state of knowledge after a number of electromyographic test results have been observed. Observable nodes with results now known are depicted with shaded bars representing observed values For them, knowledge is perfect. The implications of these results have been propagated leftward to syndromes and disease states, as shown by distributions that differ from the base rates in Figure 1. These values guide the decision to test further or initiate a treatment. Finally, updated beliefs about disease states have been propagated back toward the right to update expectations about the likely outcomes of test not yet administered. These expectations, and the potential they hold for further updating knowledge about the disease states, guide the selection of further tests.

Figure 2. The MUNIN network: after selected observations (from Andreassen et al., 1987).

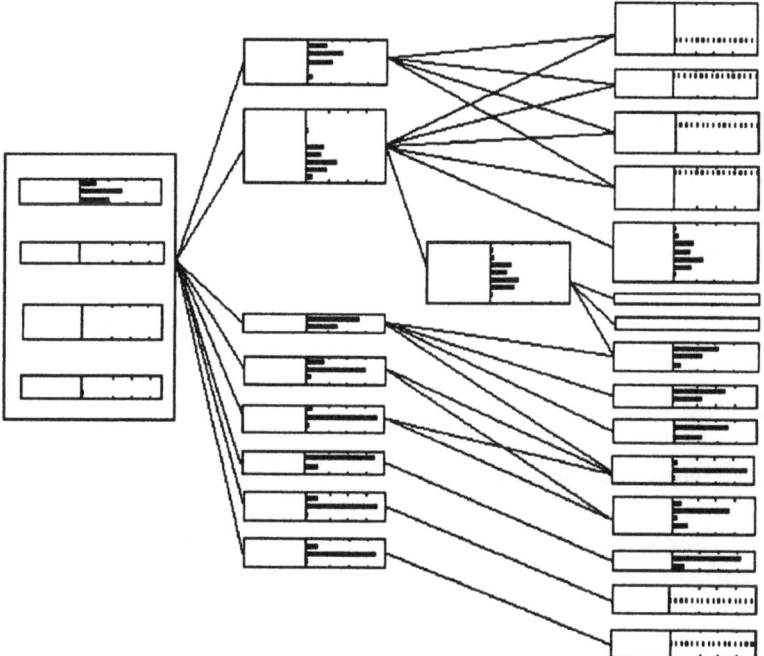

INFERENCE NETWORKS IN THE EDUCATIONAL SETTING

To see how the ideas underlying MUNIN apply to the educational setting, consider the following analogy:

Medical Application	Educational Application
Observable symptoms, medical tests	Test items, verbal protocols, teachers' ratings of levels of tests understanding, solution traces
Disease states, syndromes	States or levels of understanding of key concepts, available strategies
Architecture of interconnections based on medical theory	Architecture of interconnections based on cognitive and educational theory
Conditional probabilities given by physiological models, empirical data, expert opinion	Conditional probabilities given by psychological models, empirical data, expert opinion

The definitions of key concepts will be guided by theorized and observed stages of learning in the area, and the connections with observables will be expressed through measurement models such as those discussed above. The initialization of the probabilities in the network will be accomplished by one or more methods such as clinical analysis, with skilled interviewers assessing in detail the nature of students' understandings and relating these understandings to task performances; statistical analysis of data concerning selected models for portions of the larger network (Mislevy and Verhelst, 1990); or theoretical analysis, in which logic or theory provides expectations for outcomes under hypothesized cognitive states. After the initialization phase, connections can be updated periodically with the larger amounts of less precise data that will be accumulated as students provide information about the adequacy of the relationships embodied in the network and the accuracy of the baseline and conditional probabilities.

A Numerical Example

Siegler's balance-beam tasks

Kuhn (1970) emphasizes the central role that exemplars—or small, archetypical examples—play in science. Textbook examples are the vehicle through which students are acculturated to the concepts and relationships of a particular way of viewing a class of phenomena — a paradigm, in Kuhn's words. They function almost like parables or morality tales. New paradigms are introduced with new exemplars that introduce new concepts, highlight differences between the new paradigm and the old, and demonstrate how the new way of thinking solves problems the old way could not. Modeling the states of the electron in the hydrogen atom possesses this status in quantum mechanics.

Explaining children's understanding of balance-beam problems, an exemplar from developmental psychology originated by Piaget, is approaching the same status in test theory (for example, Kempf, 1983; Wilson, 1989b). Robert Siegler's balance-beam tasks yield data that are, on the surface, indistinguishable from standard test data, but there are two key distinctions:

- What is important about examinees is not their overall probability of answering items correctly, but their (unobservable) state of understanding of the domain.

- Children at less sophisticated levels of understanding initially get certain problems right for the wrong reasons. These items are more likely to be answered wrong at intermediate stages, as understanding deepens! They are bad items by the standards of classical test theory and IRT, because probabilities of correct response do not increase monotonically with increasing total test score. From the perspective of the developmental theory, however, not only is this reversal expected, but it plays an important role in distinguishing among children with different ways of thinking about the problems.

Attempting to study children's reasoning in a manner less subjective than Piaget's unstructured interviews, Siegler (1981) devised a series of balance beam tasks like the one illustrated in Figure 3. Varying numbers of weights are placed at varying locations on a balance beam. The child predicts whether the beam will tip to left, to the right, or remain in balance. Piaget's analysis of children's behavior on balancing tasks (Inhelder and Piaget, 1958), posits that a child will respond in accordance with his or her stage of understanding. The usual stages through which children progress can be described in terms of successive acquisition of the rules listed below.

Figure 3. A sample balance-beam task.

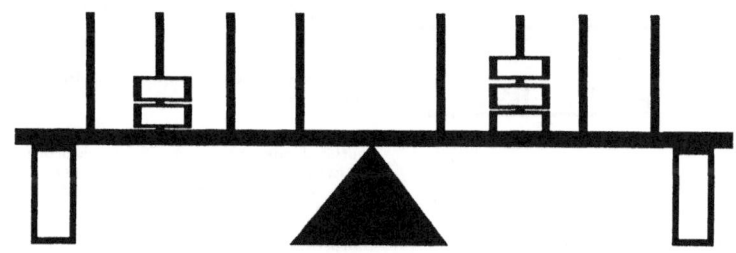

When the blocks are removed, will
the beam tip left, tip right, or stay flat?

Rule I: If the weights on both sides are equal, it will balance. If they are not equal, the side with the heavier weight will go down. (Weight is the "dominant dimension," because children are generally aware that weight is important in the problem earlier than they realize that distance from the fulcrum, the "subordinate dimension," also matters.)

Rule II: If the weights and distances on both sides are equal, then the beam will balance. If the weights are equal but the distances are not, the side with the longer distance will go down. Otherwise, the side with the heavier weight will go down. (A child using this rule uses the subordinate dimension only when information from the dominant dimension is equivocal.)

Rule III: Same as Rule II, except that if the values of both weight and length are unequal on both sides, the child will "muddle through" (Siegler, 1981, p. 6). (A child using this rule now knows that both dimensions matter, but doesn't know just how they combine. Responses will be based on a strategy such as guessing.)

Rule IV: Combine weights and lengths correctly (that is, compare torques, or products of weights and distances).

It was thus hypothesized that each child could be classified into one of five stages—the four characterized by the rules, or an earlier "preoperational" stage in which neither weight nor length are thought to bear any systematic relationship to the action of the beam.

Figure 4. Sample balance-beam items.

Item Type	Sample Item	Description
E		Equal problems (E), with matching weights and lengths on both sides.
D		Dominant problems (D), with unequal weights but equal lengths.
S		Subordinate problems (S), with unequal lengths but equal weights.
CD		Conflict-dominant problems (CD), in which one side has greater weight, the other has greater length, and the side with the heavier weight will go down.
CS		Conflict-subordinate problems (CS), in which one side has greater weight, the other has greater length, and the side with the greater length will go down.
CE		Conflict-equal problems (CE), in which one side has greater weight, the other has greater length, and the beam will balance.

Siegler developed the six types of problems shown in Figure 4 to distinguish among children at different stages of reasoning. Table 1 shows the probabilities of correct response that would be expected from groups of children in different stages, if their responses were in complete accordance the hypothesized rules. Scanning across the rows reveals how the probability of a correct response to a given type of item does not always increase as level of understanding increases. For example, Stage II children tend to answer CD items right for the wrong reason, while Stage III children, now aware of a conflict, flounder.

Table 1. Theoretical conditional probabilities: expected proportions of correct response.

Problem type	Stage 0	Stage I	Stage II	Stage III	Stage IV
E	.333	1.000	1.000	1.000	1.000
D	.333	1.000	1.000	1.000	1.000
S	.333	.000	1.000	1.000	1.000
CD	.333	1.000	1.000	.333	1.000
CS	.333	.000	.000	.333	.333
CE	.333	.000	.000	.333	1.000

A latent class model for balance beam tasks

If the theory were perfect, the columns in Table 1 would give probabilities of correct response to the various types of items from children at different stages of understanding. Observing a correct response to an S item, for example, would eliminate the possibility that the child was in Stage I. But because the model is not perfect[4], and because children make slips and lucky guesses, any response could be observed from a child in any stage. A latent class model (Lazarsfeld, 1950) can be used to express the structure posited in Table 1 while allowing for some "noise" in real data (see appendix for details). Instead of expecting incorrect responses with probability one to S items from Stage I children, we might posit some small fraction of correct answers p(S correct:Stage=I). Similar probabilities of "false positives" can be estimated for other cells in Table 1 containing 0's. In the same spirit, probabilities less than one, due to "false negatives," can be estimated for the cells with 1's. Note that inferences cannot be as strong when these uncertainties are present; a correct response to an S item still suggests that a child is probably not in Stage I, but is no longer proof positive.

Expressing this model in the notation introduced above, η represents stage membership, x represents item responses, and $p(x:\eta)$ are conditional probabilities of correct responses to items of the various types from children in different stages—a noisy version of Table 1. The proportions of children in a population of interest at the different stages are $p(\eta)$, and the probabilities that convey our knowledge about a child's stage after we have observed his or her responses are $p(\eta:x)$.

Siegler created a 24-task test comprised of four tasks of each type. He collected data from 60 children, from age three up through college age, at two points in time, for a total of 120 response vectors. We fit a latent class model to these data using the HYBRIL computer program (Yamamoto, 1987), obtaining the conditional probabilities—$p(x:\eta)$—shown in Table 2, and the following vector summarizing the (estimated) population distribution of stage membership:

$$p(\eta) = (\text{Prob}(\text{Stage}=0), \text{Prob}(\text{Stage}=I), ..., \text{Prob}(\text{Stage}=IV))$$

$$= (.257, .227, .163, .275, .078).$$

Table 2. Estimated conditional probabilities: expected proportions of correct response.

Problem type	Stage 0	Stage I	Stage II	Stage III	Stage IV
E	.333*	.973	.883	.981	.943
D	.333*	.973	.883	.981	.943
S	.333*	.026	.883	.981	.943
CD	.333*	.973	.883	.333*	.943
CS	.333*	.026	.116	.333*	.943
CE	.333*	.026	.116	.333*	.943

* Denotes fixed value

Note that different types of items are differentially useful to distinguish among children at different levels. E items, for example, are best for distinguishing Stage 0 children from everyone else. CD items, which would be dropped from standard tests because their probabilities of correct response do not have a strictly increasing relationship with total scores, help differentiate among children at Stages II, III, and IV.

Figure 5 depicts the state of knowledge about a child before observing any responses using the conventions of the MUNIN figures. For simplicity, just one item of each type is shown. The corresponding status of an observable node (that is, an item type) is the expectation of a correct response from a child selected at random from the population. The path from the stage–membership node to a particular observable node represents a row of Table 2.

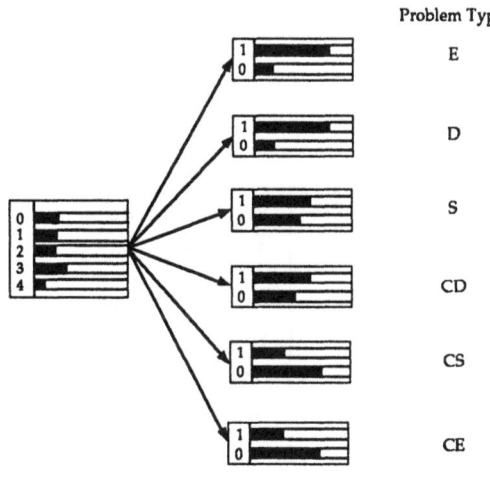

Figure 5. Initial state in an inference network for the balance-beam example.

Adaptive testing

Figure 5 represents the state of our knowledge about a child's reasoning stage and expected responses before any actual responses are observed. How does knowledge change when a response is observed? One of the children in the sample, Douglas, gave an incorrect response to his first S item. This could happen regardless of Douglas' true stage; the probabilities are obtained by subtracting the entries in the S row of Table 2 from 1.000, yielding, for Stages 0 through IV, .667, .973, .116, .019, and .057 respectively. This is the likelihood function for induced by the observation of the response. The bulk of the evidence is for Stages 0 and I. Combining these values with the initial stage probabilities $p(\eta)$ via Bayes theorem yields updated stage probabilities, p (η: incorrect response to an S item), for Stages 0 through IV, respectively, of .41, .52, .04, .01, and .01. Expectations for items not yet administered also change. They are averages of the probabilities of correct response expected from the various stages, now weighted by the new stage membership probabilities. The state of knowledge after observing Douglas' first response is depicted in Figure 6 (see appendix for details; also see Macready and Dayton, 1989.)

Figure 6. State of knowledge about cognitive level after an incorrect response to an S item.

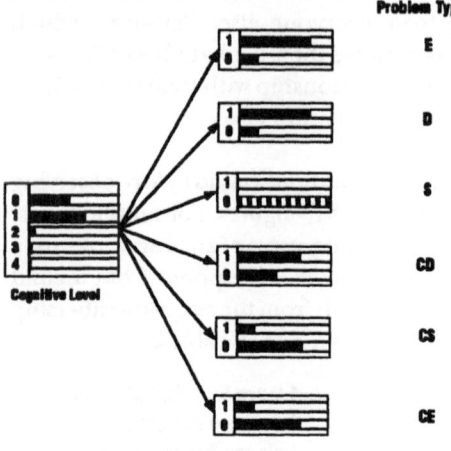

In a simulation of adaptive testing, we updated our knowledge about Douglas one response at a time, at each step looking at his actual response to an item expected to most substantially reduce our uncertainty about his stage membership. Figure 7 charts probabilities of stage membership for Douglas after each of the first 10 items, showing that we quickly converge to Stage 0.

Figure 7. Posterior probabilities of cognitive levels.

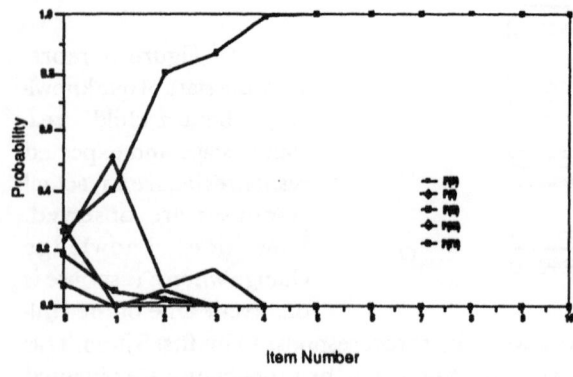

Extending the Paradigm

The balance-beam exemplar illustrates the challenge of inferring states of understanding, but it addresses development of only a single key concept. A major thrust of our proposal is to characterize interconnections among distinct lines of development. This section takes a small step in this direction by discussing a hypothetical extension to the exemplar, namely, the ability to carry out the arithmetic operations needed to calculate torques. For illustrative purposes, we simply posit a skill to carry these calculations out reliably, either possessed by a child or not. Obviously states of understanding could be developed in greater detail here.

Calculating and comparing torques to solve the "conflict" problems characterizes Stage IV. But if a child at Stage IV cannot carry out the calculations reliably, his pattern of correct and incorrect responses would be hard to distinguish from that of a child in Stage III. Although the two children might answer about the same number of items correctly, the instruction appropriate for them would differ dramatically. And children at any stage of understanding of the balance beam might be able to carry out the computational operations in isolation. The goal of the extended system is to infer both balance–beam understanding and computational skill. To make the distinctions among states of understanding in this extended domain, we introduce two new types of observations:

- Items isolating computation, such as, "Which is greater, 3 x 4 or 5 x 2?"

- Probes for introspection about solutions to conflict items: "How did you get your answer?"

Figure 8 offers one possible structure for this network. Others could be entertained, and in practice one would compare the degree to which they accord with observed data. To keep the diagram simple, only one balance–beam task each for an S and a CS task are illustrated. E and D items would have the same paths as the S task, and CD and CE tasks would have the same paths as the CS tasks. Also, the paths from Stage 0, I, and II indicators to balance beam tasks are not drawn in. The structure of paths, but not necessarily the values, would be the same as those connecting the Stage III indicator to those tasks.

Figure 8. Representation of an extended balance-beam network.

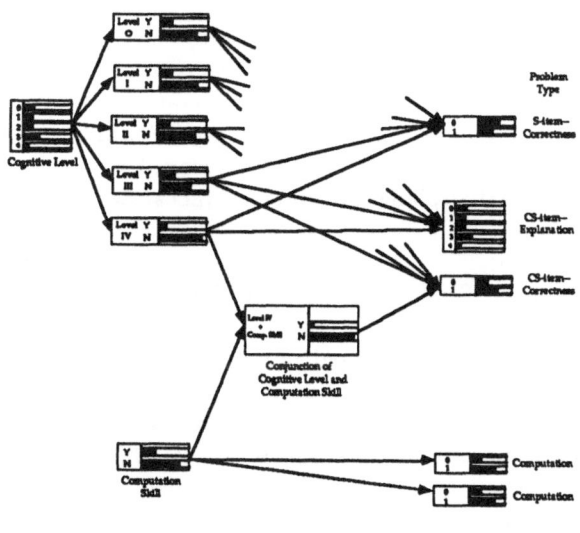

There are three kinds of unobservable variables in the system. The first group expresses level of understanding in the balance beam domain. It proves convenient to express stage membership in terms of dichotomous indicator variables for each stage, because of the special relationship of Stage IV to computational skill. Second is the ability to carry out the calculations involved in computing torques. The third concerns the integration of balance-beam understanding and calculating proficiency. Specifically, we posit an indicator for whether a child both is in Stage IV and possesses the requisite computational skills. Other features of the network worth mentioning are as follows:

1. The probabilities of the pure computation items depend on the unobservable computation variable only; they are conditionally independent of level of balance beam understanding.

2. The correctness aspect of an answer has only two possibilities, right or wrong, but an explanation can fall into five categories corresponding to levels of understanding. A Stage III child might give an explanation consistent with Stages 0, I, II, or III, but would not give a Stage IV explanation. Theory thus posits that the conditional probability of a Stage K response from a Stage J child is zero if $K>J$. Conditional probabilities for $K \leq J$ might be estimated from data or based on experts' experience. It may turn out, for example, that the most likely explanation for an E task from people at Stage IV would probably be a Stage II explanation: "It balances because both the weights and distances are equal."

3. For children in Stages 0 through III, both the right/wrong answers and the "How" answers to balance beam tasks depend only on level of understanding. Because they do not realize the connection between the problems and the torque calculations, their responses to the balance beam tasks are conditionally independent of their computational skill, even on items for which that skill is an integral component of an expert solution.

4. For children in Stage IV, right/wrong answers to conflict items depend on the understanding/computation integration variable, but "How" answers depend only on understanding. A child in Stage IV with low computational skill can thus be differentiated from a child in Stage III by his higher probabilities of giving Stage IV explanations and incorrect answers to pure computation problems.

EXTENDING AND CLARIFYING EDUCATIONAL MEASUREMENT PRACTICES

Connections with instruction can be forged more easily than with standard tests, because the focus is no longer on how many questions a student can answer, but how they are answered. In medical diagnosis, different diseases gave rise to similar results in certain tests; in education, so too can different approaches lead to similar test scores for students. But accounting for the patterns of performance, especially if probing adaptively, can pinpoint the areas that need attention to best improve performance.

Student reports can be provided at varying levels and highlighting different features of a student's status. Of particular importance to the student and the teacher are reports in terms of levels or stages of understanding of key concepts, since this is the level at which instruction is aimed. For the quality control purposes of administrators, however, one could predict a student's performance on a standard set of tasks in the domain, a "market basket" of tasks that, ideally, every student should eventually be able to handle.

Use of different strategies or mental models can be accommodated in an inference network. This can take the form of either a single strategy/mental model choice for all tasks in a class, as studied by Mislevy and Verhelst (1990), or strategy/model switching from one task to another (as in Snow and Lohman, 1984). The nature and the strength of inferences one can draw will depend on the potential observational settings. With rich information, such as verbal protocols or partial solutions, it may be possible to characterize the range of solution methods the student has available and the conditions under which he or she employs them.

Testing "higher-order thinking" can be accomplished by including unobservable nodes for connections among more basic facts or concepts, and observable nodes that correspond to tasks for which the relationships of interest are critical. Because such tasks might well be open-ended and approachable in a variety of ways, the possibility of alternative solution strategies would need to be built into the network.

Adaptive testing can be carried out among concepts, not just for a single concept. IRT applications of adaptive testing are based on the one-to-many relationships that are appropriate for determining overall levels of proficiency, but inadequate for understanding connections among concepts. The inference network facilitates stepping variously throughout a domain, gathering information about critical domains by presenting tasks that call for varying combinations of key skills.

Handling atypical knowledge configurations or observational patterns can be accomplished by incorporating nodes analogous to the "other" disease state in MUNIN or the catch-all IRT class in Yamamoto's (1987) Hybrid model. An "other" state of understanding is a mechanism for capturing observational patterns that do not accord with those specifically built into the network. A situation-sensitive student report might be generated in an instructional system when such a node becomes prominent, signalling that more intelligence than is embodied in the system is needed to figure out what this student is doing, and to decide what to do about it.

A Unified Conceptual Framework

Learning can be enhanced by a unified conceptual framework for instruction, testing, and reporting, because in such a framework coherent feedback loops can be constructed. This chapter has focused on the educational measurement aspect of a system built on this premise. The recent introduction of measurement models built around states of understanding, and of inferential techniques to connect such pieces into networks that describe domains of school learning, provide a foundation for improved educational practice.

NOTES AND REFERENCES

1.   A particularly interesting special case occurs when the universe of student models can be expressed as performance models (Clancey, 1986). A performance model consists of a knowledge base and manipulation rules that can be run on problems in a domain of interest. A particular model can contain both knowledge and production rules that are incorrect or incomplete; the solutions it produces will be correct or incorrect in identifiable ways. Here the parameter specifies features of performance models.

2.   Advocates of student modeling emphasize the qualitative aspects of student models. Our approach is compatible with this view as it is possible to build universes of qualitative models, indexed by parameters that distinguish their features. Our knowledge about a particular student's model is imperfect, however. It can be expressed in terms of probabilities expressing the plausibility of various models, given what has been observed. Probabilities are quantitative and admit to a calculus of manipulation. We might thus employ a quantitative model for our (imperfect) knowledge about qualitative student models.

3.   The ESPRIT team has generalized the application to address clusters of interrelated muscles in a network containing over a thousand nodes.

4.   This model assumes that the five states are exhaustive and mutually exclusive. Alternative models, such as those of Tatsuoka and Yamamoto mentioned earlier, could be used to relax these restrictions.

Andreassen, S., Woldbye, M., Falck, B., and Andersen, S.K. (1987). MUNIN: A causal probabilistic network for interpretation of electromyographic findings. Proceedings of the 10th International Joint Conference on Artificial Intelligence. Milan, Italy.

Biggs, J.B., and Collis, K.F. (1982). *Evaluating the quality of learning: The SOLO taxonomy*. New York: Academic Press.

Bock, R.D. and Aitkin, M. (1981). Marginal maximum likelihood estimation of item parameters: An application of an EM algorithm. *Psychometrika, 46,* 443-459.

Chi, M.T.H., Feltovich, P., and Glaser, R. (1981). Categorization and representation of physics problems by experts and novices. *Cognitive Science, 5,* 121-152.

Clancey, W.J. (1986). Qualitative student models. *Annual Review of Computer Science,* 1, 381-450.

Clement, J. (1982). Students' preconceptions of introductory mechanics. *American Journal of Physics, 50,* 66-71.

Cronbach, L.J., Gleser, G.C., Nanda, H., and Rajaratnam, N. (1972). *The dependability of behavioral measurements: Theory of generalizability for scores and profiles.* New York: Wiley.

Embretson, S.E. (1985). Multicomponent latent trait models for test design. In S.E. Embretson (Ed.), *Test design: Developments in psychology and psychometrics.* Orlando: Academic Press; pp. 195-218.

Falmagne, J-C. (1989). A latent trait model via a stochastic learning theory for a knowledge space. *Psychometrika, 54,* 283-303.

Glaser, R. (1981). The future of testing: A research agenda for cognitive psychology and psychometrics. *American Psychologist, 36,* 923-936.

Glaser, R., Lesgold, A., and Lajoie, S. (1987). Toward a cognitive theory for the measurement of achievement. In R. Ronning, J. Glover, J.C. Conoley, and J. Witt (Eds.), *The influence of cognitive psychology on testing and measurement: The Buros-Nebraska Symposium on measurement and testing,* (Vol. 3). Hillsdale, NJ: Lawrence Erlbaum Associates.

Greeno, J.G. (1976). Cognitive objectives of instruction: Theory of knowledge for solving problems and answering questions. In D. Klahr (Ed.), *Cognition and instruction.* Hillsdale, NJ: Lawrence Erlbaum Associates.

Gulliksen, H. (1950). *Theory of mental tests.* New York: Wiley.

Haertel, E.H. (1984). An application of latent class models to assessment data. *Applied Psychological Measurement, 8,* 333-346.

Hambleton, R.K. (1989). Principles and selected applications of item response theory. In R.L. Linn (Ed.), *Educational measurement* (3rd Ed.). New York: American Council on Education/Macmillan.

van den Heuvel, M. (1990). Realistic arithmetic/mathematics instruction and tests. In K. Gravemeijer, M. van den Heuvel, and L. Streefland (Eds.), *Context free productions tests and geometry in realistic mathematics education.* Utrecht, The Netherlands: Research Group for Mathematical Education and Educational Computer Center, State University of Utrecht.

Inhelder, B., and Piaget, J. (1958). *The growth of logical thinking from childhood to adolescence.* New York: Basic.

Karplus, R., Pulos, S., and Stage, E. (1983). Proportional reasoning of early adolescents. In R A. Lesh and M. Landau (Eds.), *Acquisition of mathematics concepts and processes.* Orlando, FL: Academic Press.

Kempf, W. (1983). Some theoretical concerns about applying latent trait models in educational testing. In S.B. Anderson and J.S. Helmick (Eds.), *On educational testing.* San Francisco: Josey-Bass.

Kuhn, T.S. (1970). *The structure of scientific revolutions* (2nd edition). Chicago: University of Chicago Press.

Lauritzen, S.L., and Spiegelhalter, D.J. (1988). Local computations with probabilities on graphical structures and their application to expert systems. *Journal of the Royal Statistical Society, Series B*, 50, 157-224.

Lazarsfeld, P.F. (1950). The logical and mathematical foundation of latent structure analysis. In S.A. Stouffer, L. Guttman, E.A. Suchman, P.F. Lazarsfeld, S.A. Star, and J.A. Clausen, *Studies in social psychology in World War II, Volume 4: Measurement and prediction*. Princeton, NJ: Princeton University Press.

Lewis, C. (1985, June). Estimating individual abilities with imperfectly known item response function. Paper presented at the Annual Meeting of the Psychometric Society, Nashville, TN.

Lord, F.M. (1959). Statistical inference about true scores. *Psychometrika*, 24, 1-18.

Lord, F.M., and Novick, M.R. (1968). *Statistical theories of mental test scores*. Reading, MA: Addison-Wesley.

Macready, G.B., and Dayton, C.M. (1989, March). The application of latent class models in adaptive testing. Paper presented at the annual meeting of the American Educational Research Association, San Francisco, CA.

Marshall, S.P. (1985, December). Using schema knowledge to solve story problems. Paper presented at the Office of Naval Research Contractors' Conference, San Diego, CA.

Marshall, S.P. (1989). Generating good items for diagnostic tests. In N. Frederiksen, R. Glaser, A. Lesgold, and M.G. Shafto (Eds.), *Diagnostic monitoring of skill and knowledge acquisition*. Hillsdale, NJ: Lawrence Erlbaum Associates.

Marshall, S.P. (1992). Assessing schema knowledge. In N. Frederiksen, R.J. Mislevy, and I.I. Bejar (Eds.), *Test theory for a new generation of tests*. Hillsdale, NJ: Lawrence Erlbaum Associates.

Masters, G., and Mislevy, R.J. (1992). New views of student learning: Implications for educational measurement. In N. Frederiksen, R.J. Mislevy, and I.I. Bejar (Eds.), *Test theory for a new generation of tests*. Hillsdale, NJ: Lawrence Erlbaum Associates.

McDermott, L.C. (1984). Research on conceptual understanding in mechanics. *Physics Today*, 37, 24-32.

Minsky, M. (1975). A framework for representing knowledge. In P.H. Winston (Ed.), *The psychology of computer vision*. New York: McGraw-Hill.

Mislevy, R.J. (1992) Foundations of a new test theory. In N. Frederiksen, R.J. Mislevy, and I.I. Bejar (Eds.), *Test theory for a new generation of tests*. Hillsdale, NJ: Lawrence Erlbaum Associates.

Mislevy, R.J., and Sheehan, K.M. (1990). Integrating cognitive and psychometric models in a measure of document literacy. *Journal of Educational Measurement*, 27, 255-272.

Mislevy, R.J. (1991). Randomization-based inference about latent variables from complex samples. *Psychometrika*, 56, 178-196.

Mislevy, R.J., and Verhelst, N. (1990). Modeling item responses when different subjects follow different solution strategies. *Psychometrika*, 55, 195-215.

Novick, M.R. (1966). The axioms and principle results of classical test theory. *Journal of Mathematical Psychology*, 3, 1–18.

Ohlsson, S., and Langley, P. (1985). Identifying solution paths in cognitive diagnosis. Technical Report RI–TR–85–2. Pittsburgh, PA: The Robotics Institute, Carnegie–Mellon University.

Paulson, J.A. (1986). Latent class representation of systematic patterns in test responses. Technical Report ONR–I. Portland, OR: Psychology Department, Portland State University.

Pearl, J. (1988). *Probabilistic reasoning in intelligent systems: Networks of plausible inference.* San Mateo, CA: Kaufmann.

Rumelhart, D.A. (1980). Schemata: The building blocks of cognition. In R. Spiro, B. Bruce, and W. Brewer (Eds.), *Theoretical issues in reading comprehension*. Hillsdale, NJ: Lawrence Erlbaum Associates.

Siegler, R.S. (1981). Developmental sequences within and between concepts. Monograph of the Society for Research in Child Development, 46.

Snow, R.E., and Lohman, D.F. (1984). Toward a theory of cognitive aptitude for learning from instruction. *Journal of Educational Psychology*, 76, 347–376.

Spearman, C. (1907). Demonstration of formulae for true measurement of correlation. *American Journal of Psychology*, 18, 161–169.

Tatsuoka, K.K. (1989). Toward an integration of item response theory and cognitive error diagnosis. In N. Frederiksen, R. Glaser, A. Lesgold, and M.G. Shafto, (Eds.), *Diagnostic monitoring of skill and knowledge acquisition*. Hillsdale, NJ: Lawrence Erlbaum Associates; pp. 453–488.

Tsutakawa, R.K., and Johnson, J. (1988). Bayesian ability estimation via the 3PL with partially known item parameters. Mathematical Sciences Technical Report No. 147. Columbia, MO: Department of Statistics, University of Missouri.

Wilson, M.R. (1989a). A comparison of deterministic and probabilistic approaches to measuring learning structures. *Australian Journal of Education*, 33, 125–138.

Wilson, M.R. (1989b). Saltus: A psychometric model of discontinuity in cognitive development. *Psychological Bulletin*, 105, 276–289.

Yamamoto, K. (1987). A model that combines IRT and latent class models. Unpublished doctoral dissertation, University of Illinois.

APPENDIX

EQUATIONS FOR THE LATENT CLASS MODEL

The Model

Let $\eta = (\eta_0,...\eta_4)$ denote the stage of understanding of a child, with $\eta_k = 1$ if the child is in Stage k and 0 if not. Let $\pi = (\pi_0,...,\pi_4)$ denote the population proportions of children in these classes; that is, $\pi_k \equiv p(\eta_k = 1)$. Let x_j represent a response to Task j, 1 if correct and 0 if not; j runs from 1 to 24.

The conditional probabilities of correct response are Prob($x_j = 1 \mid \eta_k = 1$), or P_{jk} for short. P denotes the matrix $((P_{jk}))$. A vector of item responses, $x = (x_1,...,x_{24})$ is assumed to have the following probability *conditional on stage membership*:

$$p(x \mid \eta_k = 1) = \prod_j P_{jk}^{x_j}(1 - P_{jk})^{1 - x_j}. \tag{1}$$

Similar expressions are assumed to hold for subsets of responses as well, regardless of the order in which they are observed.

The *marginal* probability of a response vector is an average of terms like (1), weighted by the population probabilities of stage membership:

$$p(x) = \sum_{k=0}^{4} p(x \mid \eta_k = 1)\pi_k. \tag{2}$$

Let X denote the matrix of response vectors of a sample of N respondents. For a generic pattern xl, let nl be the number of respondents producing this pattern. The probability of X as a function of P and π has the form

$$P(X \mid P, \pi) = C \prod_l p(xl)^{nl} \tag{3}$$

where C does not depend on P or π. Once X has been observed, (3) can be interpreted as a likelihood function, and maxima may be found with respect to P and π.

Because N is only 120 in the balance beam example, a number of constraints were introduced so that stable estimates would be obtained. Many could be relaxed or removed with larger samples. The results reported in Table 2 represent the best-fitting result among several models with similar numbers of constraints. The P_{jk}s that appear as .333 in Table 1 were fixed at that value. All four items of a given type were constrained to have the same P_{jk}s. For a given column, all P_{jk}s in cells that correspond to 1's in Table 1 were constrained to be equal to a single estimated value. Any cells in that column that correspond to 0's were constrained to its complement.

Adaptive Testing

The maximum likelihood estimates of P and π were treated as known true parameter values during simulated adaptive testing. The uncertainty in these values could be taken into account, but we have avoided the complication for this demonstration.

Before observing any responses from a given child, the expected value of the child's η is the population value π. The expected value of a response to a particular item j is obtained analogously to (2), simplified to a single, as yet unobserved, response:

$$p(x_j = 1) = \sum_k p(x_j = 1 \mid \eta_k = 1) p(\eta_k = 1)$$

$$= \sum_k P_{jk} \, p(\eta_k = 1).$$

(4)

Suppose that item g is administered to a particular examinee, and the value of x_g, either 0 or 1, becomes known. How is this information propagated through the network? First, using Bayes theorem, we update probabilities for the index parameter η. For $k = 0,...,4$.

$$p(\eta_k = 1 \mid x_g) = \frac{p(x_g \mid \eta_k = 1) \, p(\eta_k = 1)}{\sum_h p(x_g \mid \eta_h = 1) \, p(\eta_h = 1)}$$

(5)

This gives new probabilities that the examinee is in each of the possible stages. These are in turn reflected in new expectations for items not yet administered by replacing $p(\pi_k = 1)$ in (4) with $P(\pi_k = 1 \mid x_g)$ to obtain

$$p(x_j = 1 \mid x_g) = \sum_k p(x_j = 1 \mid \eta_k = 1) \, p(\eta_k = 1 \mid x_g).$$

(6)

This process can be repeated with additional items represented one at a time. Let x_s represent a partial response sequence; item $s+1$ is next administered to form x_s+1. Then

$$p(\eta_k = 1 \mid x_s+1) = \frac{p(x_s+1 = 1 \mid \eta_k = 1) p(\eta_k = 1 \mid x_s)}{\sum_h p(x_{s+1} = 1 \mid \eta_h = 1) \, p(\eta_h = 1 \mid x_s)}$$

(7)

and, for items not yet presented,

$$p(x_j = 1 \mid x_{s+1}) = \sum_k p(x_j = 1 \mid \eta_k = 1) \, p(\eta_k = 1 \mid x_{s+1}). \tag{8}$$

Selecting which item to present next and deciding when to stop depends on probabilities for π. In this paper we have addressed only the case in which no decision-making cost structure is available, and we address only the goal of minimizing uncertainty about π. This can be accomplished by minimum entropy adaptive testing. Entropy is a measure of randomness. For the five-class balance beam problem, the maximal value of entropy occurs when probabilities of all five classes are equal, and the minimal value occurs when the probability of one particular stage is one. The general formula for entropy after having observed x_s is

$$E(x_s) = -\sum_k p(\eta_k = 1 \mid x_s) \log [p(\eta_k = 1 \mid x_s)] \tag{9}$$

After having observed x_s, one can evaluate the expected entropy associated with the administration of any remaining item j as

$$E[x_s \cap (x_j = 0)] \, p(x_j = 0 \mid x_s) + E[x_s \cap (x_j = 1)] \, p(x_j = 1 \mid x_s). \tag{10}$$

The item that minimizes (10) is presented next.

It bears that these formulae assume both that the model is correct and the conditional probabilities are known with certainty. Violations of these assumptions generally degrade knowledge about an examinee's state, making (5) and (8) in particular overly optimistic. Work remains to be done, in studying the robustness of the approach to violations of the assumptions, learning how to minimize violations in practice, and modifying the model or the conditional probabilities to mitigate inferential errors in the presence of violations.

13 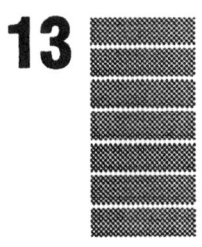 Interpreting Responses to Problems with Several Levels and Types of Correct Answers

Susan J. Lamon and
Richard Lesh

INTRODUCTION

The production and interpretation of model-eliciting activities (see Chapter 2) is a complex endeavor. Problem formulation and response interpretation within complex mathematical domains require a strong supporting framework. That framework should provide a blueprint for creating problems, a basis by which to interpret students' reasoning as they interact with the problems, and a guide to instructional decision making. In this chapter, we will use examples from the domain of ratio and proportion to explore some of the issues surrounding problem formulation and the interpretation of student thinking. A framework for problem formulation and scoring will be proposed, and levels of responses to a variety of problems will be discussed in light of that framework. We begin by discussing some issues that are arising as the mathematics education community considers alternatives to current assessment practices.

SOME CRITICAL UNDERSTANDINGS AND DISTINCTIONS

The campaign for a problem-solving curriculum during the 1980s taught us that early responses to pressures for reform sometimes create the illusion of novelty but stop short of producing substantial changes. Immature ideas and meager guidelines allow nearly any current practice to be rationalized to a good fit or to sanction cosmetic changes that fail to reform.

Interpreting Responses to Problems

Similar dangers are imminent in assessment reform.

The mathematics education world is alive with discussions of "authentic performance" and "alternate forms of assessment." Popular interpretations of assessment reform are captured in a few simple maxims, such as (i) use real, open-ended questions; (ii) don't use textbook problems; and (iii) don't use multiple-choice questions. Unfortunately, slogans or rally cries are frequently based upon a narrow interpretation of the needs motivating the reform. As we worked with teachers, testing and refining the guidelines for developing problem-solving situations that focus on deeper and higher-order understandings (presented in Chapter 2), we found several caveats necessary. Some of these critical understandings and distinctions follow.

Productive Open-ended Questions

Changing notions of learning and understanding have produced wide agreement that one of the characteristics of a good problem is that it captures as much as possible of the process that produced its solution. The desire to capture aspects of individual students' construction of knowledge has brought the open-ended question into vogue as an outlet for personal expression. However, merely providing the invitation for a student to expose a solution process is not a sufficient condition for creating good assessment items. The question in Figure 1, for example, was created by a fourth-grade teacher to elicit his students' thinking while reading graphs.

Figure 1. A graph-reading exercise and two typical fourth-grade student responses.

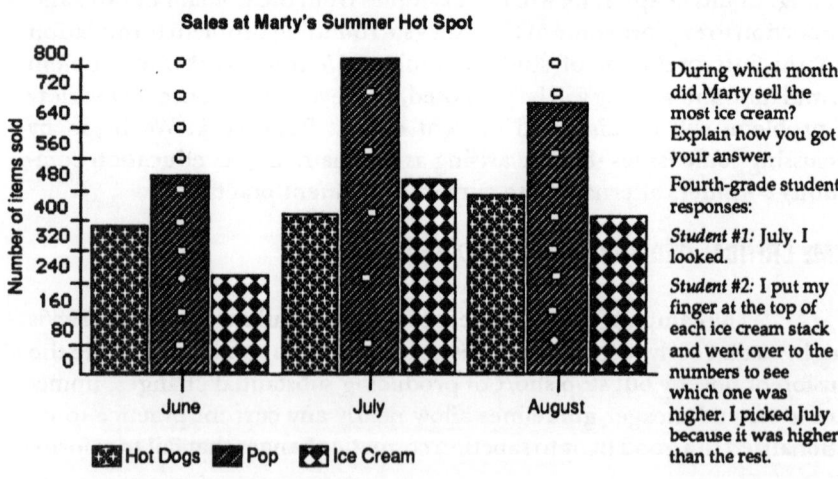

The typical response elicited by the question is a description of procedural events. Merely adopting an open-ended format was unproductive; attaching the command "Explain how you got your answer" failed to elicit students' thinking because the original question was procedural. The teacher gained no insight into the students' thinking because there was no need to reason, predict, describe, experiment, connect, or interpret.

Our guideline that problems should have more than a single level or type of correct answer refers to the substance of the elicited response and not merely to the form of the question. Essentially, it invites a broader conception of an open-ended question than most people currently hold. We have found several types of problems useful in eliciting deeper and higher-order thinking. These include (i) questions for which students must construct their own responses; (ii) projects that are of sufficient complexity that students must both ask and answer a series of questions in the process of carrying them out (see de Lange, this volume); and (iii) questions that may be solved at various levels of mathematical sophistication. The common property of all three types of problems is that they elicit and capture student thinking in a manner that allows it to be analyzed to reveal the underlying models. The third type, perhaps the least used of the three, is the focus of the latter part of this chapter.

Improving Textbook Problems

There are competing theories about the most productive approach to creating useful assessment items. Some people start with interesting situations and pose questions; others take existing problems and try to improve them. The "let's-make-it-better" approach frequently starts with textbook problems. In the absence of some intentional content objective, the meaning of "improvement" tends to be obscure. For example, some fourth and fifth grade teachers in an in-service course on questioning discussed the following typical textbook question on ratio and proportion:

> The ratio of boys to girls in a class is 3 to 8. How many girls are in the class if there are 9 boys?

After great deliberation, the teachers changed the problem to read:

> For every 3 boys in a class, there are 8 girls. How many students are in the class?

The consensus was that the second problem was better because it allowed for more than one correct answer and adjusted for the fact that fourth and fifth graders would not be familiar with the term "ratio." The teachers took the problem back to their classrooms with the goal of obtaining as much knowledge as possible about their students' thinking and deciding what might be appropriate for the next phase of instruction.

The results were disappointing. In a follow-up discussion, the teachers concluded that they had learned very little about their students' thinking because they had not given adequate attention to their purpose in asking the questions. They decided that they needed more specific information. For example, when students came up with the answer 11, did they realize that other class sizes were possible? Were they mindlessly performing operations on the given numbers, or did they realize why 11 was a real possibility? When students developed patterns by doubling and tripling each of the given numbers in turn, did they realize that the three-to-eight relationship was being preserved in the appropriate pairings?

The teachers later produced alternative wordings for the questions, depending on the conceptual underpinnings of ratio and proportion they wanted to highlight. In their efforts to understand their students' understanding, teachers first focused on the covariance of the numbers of boys and girls and the invariance of the boy-to-girl comparison. One teacher asked, "Could there be 25 people in the class? Why or why not?" She found out that half the students in her class considered 25 a possible class size because $3(3) + 2(8) = 25$. Another teacher observed that many of his students had set up a table and had added 6 boys and 16 girls across the second row of the table to arrive at 22 students as one of the possibilities. He probed their understanding with the remarks "I can think of other numbers that add to 22. How about $6 + 16 = 22$? Would 6 boys and 16 girls work just as well?" Too many students found nothing wrong with his suggestion!

Current trends in assessment tend to be clearer about desirable *forms* of classroom assessment than about the *substance* of assessment. As we begin to test open-ended questions with real students in real classrooms, we find that they are no panacea. Merely adopting new forms of assessment is no substitute for analyzing the goals we are assessing. Creating open-ended questions can be a futile effort in the absence of some intentional cognitive objective.

Improving Multiple-choice Questions

Multiple-choice questions were attacked as unsuitable assessment questions because of the likelihood of students obtaining the correct answer without thinking. An obvious solution was to create an open-ended version of the multiple choice question and the obvious replacement format was the fill-in-the-blank question. Unfortunately, this was another quick fix too hastily adopted. For example, consider the following multiple-choice question constructed by a middle school teacher:

> 5/15 = 3/9. Suppose I want these fractions to remain equal. If I change the number 15 to 24, does anything else have to change?
>
> (a) The 3 or the 9.
>
> (b) The 3 and the 5.
>
> (c) The 9 or the 5.
>
> (d) The 5 and the 9.
>
> (e) None of the other numbers.

The "improved" version of the question read:

> 5/15 = 3/9 and ?/24= 3/9

The entire sense of the question was changed! The teacher retracted the invitation to think and presented his students with a textbook exercise demanding a single correct answer. A question that asked students to investigate important numerical relationships was trivialized to one that merely required the application of a procedure. In the fill-in-the-blank version, students either (i) reduced 3/9 to 1/3 and then recognized that the missing number must be 8; or (ii) if they knew the cross multiple algorithm, solved the equation $9x = 72$.

If good questions are those that cause students to engage in higher-order thinking, all multiple-choice questions are not bad. Each needs to be carefully examined on its own merits. Some of the questions developed for the California Assessment Program (Stenmark, 1989), for example, provide outstanding examples of multiple choice questions that elicit good thinking. In judging the "goodness" of any problem, however, it is necessary to match question format with one's purpose in assessing. A critical distinction needs to be drawn between formats that elicit good thinking, and those that both elicit and capture higher-order thinking. Multiple-choice questions may elicit higher-order thinking, but because they do not require the thinker to record, videotape, or keep a written journal of the solution process, they have little value if one's goal is to gain insight into the kinds of models students are developing.

Process Goals or Heuristics

As problem solving became a major emphasis in mathematics education during the 1980s, it became fashionable to emphasize that the process one uses in solving a problem is at least as important as achieving the correct answer. For many mathematics educators, the four-step Polya process (Polya, 1945) became an important vehicle for encouraging students toward the investigative, creative, affective, and metacognitive goals of mathematics education. For others, the four steps became a routine—the phrases "understanding the problem," "devising a plan," "carrying out the plan," and "checking the solution" were typed into every solution space on every problem sheet their students received.

In working groups where we have asked teachers to write model-eliciting problems, teachers who have relied heavily on the four-step process experience a great sense of conflict with our rules of thumb for creating model-eliciting problems. As discussion turns to the nature of open-ended questions and the necessity of inviting a variety of response types, it becomes clear that prescribing the Polya routine is tantamount to consistently presenting students with problems that have a single type of correct response.

Process goals take students beyond facts and skills as products of instruction, but when they become the focus of instruction, they are really just rules—more nebulous than algorithms—but still rules. Problem solving is getting from givens to goals when the path is not obvious, and the method for doing that is by rule stringing. The means to an end becomes the end in itself, and the freedom to investigate, construct, manipulate, predict, test, clarify, and adapt is stifled.

Observations, Conclusions, and Clarifications

As we participate in the process of creating model-eliciting activities with other teachers and colleagues, we are deriving important lessons from the experience. Every failed attempt to make an algorithm of the problem formulation process, every dashed expectation, and every success in engaging students in high-level mathematical activity enhances our perspectives on assessment-related issues:

On problem formulation:

- The worth of questions, the manner in which responses are

interpreted, and the instructional value of the assessment depend on what one perceives as the intent of the assessment. Unless some content framework is operating, the focus shifts to form above substance, and assessment becomes a goal in itself.

- Good assessment items evolve; they are not written in a single session. An analysis of mathematical structure may provide an initial basis for devising problems, but the appropriateness of questions can only be judged when mathematical content interacts with students' interpretation of the problem situation and with their mathematical and experiential backgrounds. As students' unstable but developing models are revealed, problem revision is often necessary. A useful assessment item emerges after several rounds in a test-revise loop.

- There is no known method for making the problem-writing process shorter or easier; problem formulation is hard work. The payoff is that one good problem can assess and document learning better than many low-level questions can.

- There exists a dialectical relationship between the student's models and the model-eliciting activity. That is, the model a student uses in the course of interacting with the activity is a basis for inference concerning the appropriateness, depth, and level of refinement of that student's cognitive structures. At the same time, the quality of a student's model will constrain that student's command of the situation and limit his or her explanatory and predictive power. On the other hand, a student's exposure to various model-eliciting activities will enhance that student's models, while the activities to which the student's thinking has been confined constrain the development of models.

On cognitive goals:

- Cognitive goals cannot be identified with specific behaviors.

- Cognitive goals cannot be associated with the format of the assessment question.

- Cognitive goals cannot be associated with the correctness of the solution process. It is always possible to devise and teach noninsightful procedures for solving problems. Under these

- conditions, successful performance does not indicate that a model has been constructed (see Goldin, this volume).
- Because models can be adapted, refined, or integrated into larger models, cognitive objectives are like moving targets.

MODEL-ELICITING ACTIVITIES FOR RATIO AND PROPORTION

As careful analyses of content are completed in complex mathematical domains and children's models are investigated in relation to that content, frameworks will become available to guide problem formulation. Research that ties children's thinking to content and documents growth in competence from informal, intuitive concepts and strategies to more formal methods is still in its early stages, however (Hiebert and Behr, 1988). It is generally agreed that instruction is needed to support the learning of mathematically complex topics such as proportional reasoning because students do not develop the essential conceptual and procedural knowledge spontaneously (Hiebert and Behr, 1988), but the difficulty has been in finding insightful methods that encourage more than purely algorithmic solutions.

In the remainder of this chapter, we will use examples from some current instructionally based research whose goal is to find ways to build upon children's informal, additive, preproportional reasoning and facilitate its development into higher forms of reasoning (Lamon, 1992). In terms of our learning progress maps (see Chapter 14), the content of this research would be located around the outer rim of the base (or in the foothills) of the mountain called "ratio and proportion." Researchers have long known the kinds of model-eliciting activities appropriate for the central ring of the mountain—those involving orange juice mixtures, balance beams, scale drawings, and so on—but the smaller, more primitive sites, where children develop understandings about ideas central to proportional reasoning, have received little attention. We might think about these sites as points of entry to the world of ratio and proportion, places where children begin the model-building process.

Some General Structures

We contend that the work of designing instruction and assessment consists in creating a match among three critical elements: (i) cognitive objectives, (ii) model-eliciting activities, and (iii) children's existing models.

Cognitive objectives or models

Models are the ultimate goals in mathematics instruction, but they cannot be taught directly. For example, a proportion is a model for situations in which pairs of related quantities change without altering the relationships among those quantities. One might be able to teach students to mindlessly manipulate the algebraic symbols representing a proportion, but understanding of the relationships between the elements of a proportion cannot be taught directly. The coordination and understanding of such relationships occurs through the process of model building. In the domain of ratio and proportion, as in most complex mathematical domains, the outstanding research question is, What are the primitive mathematical ideas that will eventually mature into the models we would like children to have?

There is widespread agreement (Hiebert and Behr, 1988) that if research is to inform instruction, it is important to analyze mathematical structures and children's solution processes in light of the developmental precursors (or, sometimes, prerequisites) to the knowledge needed to function competently in a domain. These precursors or cognitive building blocks have been called by many names: key cognitive processes (Hiebert and Wearne, 1991), key informal strategies (Hiebert and Behr, 1988), theorems-in-action (Vergnaud, 1983), and central conceptual structures (Case and Sandieson, 1988). They are conceptions and procedures considered necessary, but not necessarily sufficient, for acquiring formal methods and meaningful knowledge, or at least they represent some immature form of the ultimately desired knowledge. It is very difficult to identify these cognitive building blocks in domains where mathematical ideas are complex and built upon a vast amount of prior knowledge and experience, but the operative theory is that if the ideas critical to understanding can be identified and made explicit in real-world phenomena, children's thinking can then be investigated and used to better inform instructional decisions.

Model-eliciting activities

When we are developing model-eliciting activities at the outer edges of a mountain, one concern should be that the activities allow students to experience the need for a new way of organizing knowledge, that is, to develop a rationale for climbing the mountain. Most often, this means that students need to be convinced that their current models are inadequate for the situation at hand. The middle-school mathematics program is especially demanding of new mathematical perspectives, and many students who are comfortable with whole number addition and subtraction models are reluctant to adopt new models to accommodate rational numbers and multiplicative situations.

At the same time, elementary model-eliciting activities should help students to develop the metacognitive capacity for detecting the conditions under which new models apply. Although students may begin to construct models in a very limited context (that is, their learning is situated), we would like students to "develop the kinds of sensitivities necessary to use relevant information in new situations" (Bransford, Franks, Vye, and Sherwood, 1986). Thus, these elementary model-eliciting activities are more than problems; they are prototypes. In particular, proportional reasoning is such a pivotal cognitive activity that model-eliciting activities in ratio and proportion facilitate access to most of the basic concepts of mathematics, science, and everyday problem solving.

Children's existing models

Technically speaking, model-eliciting activities are misnomers until students bring their existing knowledge to bear on those activities. (Even then, children have been known to construct correctly reasoned responses that have no connection whatsoever to the ideas someone else was trying to elicit. In such a case, the activity did not elicit the model at all!) For several reasons, the development of good model-eliciting activities is highly dependent upon the population for whom the activities are intended.

First, children do not always think in a manner consistent with adult thinking or rational task analyses, so the full range of processes they may use to respond to any given item cannot be predicted. It is always important to learn about the kinds of models with which children approach instruction because their existing models will influence the kinds of processes elicited by new activities and determine the appropriateness of instructional decisions. Even with a well-researched base of knowledge concerning prior models, teachers find unexpected results arising because learning often occurs while the student is interacting with a model-eliciting activity. It has also been observed that children use processes that are sufficient for solving a problem; these are not necessarily their most sophisticated strategies (Lamon, 1992). Thus, researchers and teachers need to go beyond reporting children's thinking to analyzing the kinds of models to which their thought processes are attached.

A FRAMEWORK FOR ASSESSING CHILDREN'S DEVELOPMENT OF RATIO AND PROPORTION

The following assessment framework for ratio and proportion is still in its formative stages. The cognitive objectives arose from an analysis of ratio and proportion tasks as well as extensive interviews with children. The goal of the analysis and interviews was to determine what sorts of experiences and understandings are critical for understanding ratios and proportions. The cognitive objectives that have resulted are probably necessary, but not necessarily sufficient, precursors to proportional reasoning.

Problems using a variety of contexts were formulated to assess children's preinstructional knowledge related to each cognitive objective. The resulting model-eliciting activities do not require that the student engage in proportional reasoning, yet they provide opportunities for the student to construct a primitive cognitive model that is likely to someday be integrated into a ratio and proportion model.

The following model-eliciting activities have been used in two ways. First, the activities have been used in clinical interviews with late fifth graders and early sixth graders to assess their preinstructional knowledge. In the course of an interview, it is frequently possible to document simultaneous learning. When a student is presented with activities concerning a single cognitive objective but in several different contexts, it sometimes happens that, if the student is able to respond, it is in only one context; then, the student may ask to return to a previous context because he or she has some new insights about it. Thus, learning is situated, it occurs in different contexts for different people (although some contexts elicit more reasoning than others), and its transfer to isomorphic situations in different contexts can be documented.

The second use of the model-eliciting problems generated under this framework is in instruction. The activities are currently being used in instructionally based research in a variety of formats (interviews, cooperative group work, individual concrete modeling situations, and whole-class activities) with middle school students and preservice and inservice elementary and middle school teachers in Milwaukee.

On the average, the following model-eliciting activities take students about fifteen minutes to complete. Questions have more than one correct answer, or else may be answered using any one of several levels of sophistication. When the activities are discussed in a small group for instructional purposes, learning occurs quite rapidly for both adults and children when group members present different perspectives on the activity and explain their thinking to the other group members.

Two of five cognitive objectives and their pertinence to ratio and proportion (outlined below) will be explained in detail, and then followed by examples of the model-eliciting activities built to elicit those cognitive objectives. Student protocols that best exemplify the preinstructional reasoning of beginning sixth grade students while engaged in the model-eliciting activities will then be interpreted.

Mathematical System: Ratio and Proportion

Elements: comparative indices of the form a:b conveying relative magnitude; pairs of related quantities

Relations: equivalence, equality, inequality

Operations: a:b ± c:d = (a ± b) : (c ± d)
c (a:b) = (ca : cb)
(a:b)/(c:d) = M(c:d) + ((a - Mc):(b - Md)),
where M = the minimum of the integral parts of the quotients {a/c, b/d}.

Cognitive Goals (6th grade): (i) accommodate covariance and invariance within a single activity; (ii) distinguish ratio-preserving and nonratio-preserving situations; (iii) distinguish relative and absolute change; (iv) construct ratio-preserving relations; (v) create a unit that consists of a pair of related quantities.

Distinguishing Relative and Absolute Change

Since Inhelder and Piaget began to study proportional reasoning in 1950s, research has continued to document children's difficulty in recognizing the need for multiplicative thinking. For example, children (and some adults) fail to see that additive strategies are insufficient for solving the problem shown in Figure 2. One typical explanation (this one from a 28-year-old) is, "Compare the length of the bases. Twelve inches is 4 inches longer than 8 inches. The height of the larger rectangle should be 4 inches longer than the height of the first. So 6 inches + 4 inches = 10 inches."

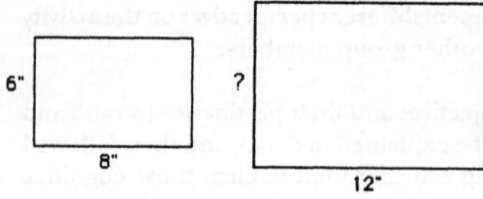

Figure 2. A problem requiring multiplicative thinking.

These two rectangles are the same shape, but one is larger than the other. How would you find the height of the larger rectangle?

The ability to reason proportionally depends on one's perception and accommodation of another interpretation of the word "compare." That is, another way to compare quantities is to relate them in such a way that interpreting the significance of one depends on the other. The problem in Figure 3 may be used to illustrate the distinctions between the additive or absolute notion of change and the multiplicative or relative interpretation. We can think about this situation in two ways. One answer is that both snakes will grow the same amount because the length of each will increase by 2 feet. This analysis represents an additive perspective. It compares the absolute change in length of the two snakes. A second perspective considers the relative growth of the snakes, or the amount of growth in relation to their present lengths. Over the next six months, Spot will grow two feet or half his present length, and Slim, two-fifths his present length. Using this analysis, Slim is closer to being fully grown.

Figure 3. A problem for which both additive and multiplicative reasoning are suitable.

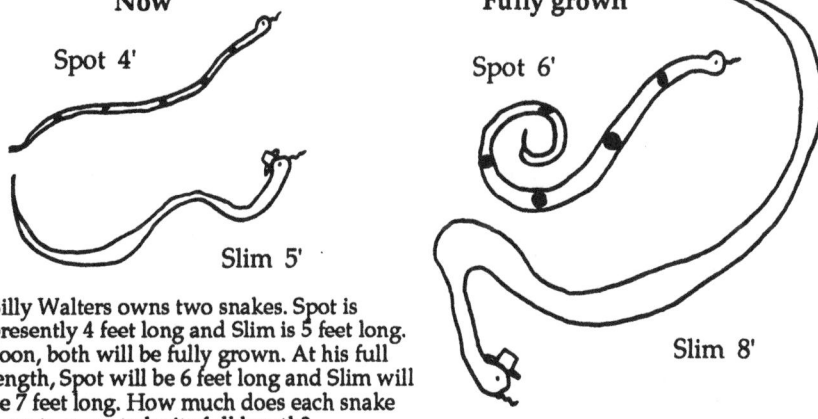

Billy Walters owns two snakes. Spot is presently 4 feet long and Slim is 5 feet long. Soon, both will be fully grown. At his full length, Spot will be 6 feet long and Slim will be 7 feet long. How much does each snake have to grow to be its full length?

Students need both these perspectives if their mathematical thinking is to advance beyond elementary arithmetic. In fact, a ratio is a comparative index that conveys the notion of relative magnitude. Ratios and proportional reasoning are critical to functioning in our scientific culture, and failure to develop multiplicative reasoning has serious ramifications in the secondary school curriculum as well as in everyday practical situations. Algebra, geometry, calculus, statistics, chemistry, physics, and biology all require proportional reasoning ability (Post, 1986), as does understanding unit pricing and inflation. Thus, an important cognitive objective is that students adopt a change model that accommodates both the additive and the multiplicative perspectives. The model-eliciting activities in Figures 4, 5, and 6 promote that cognitive objective.

Figure 4. Trees. A problem designed to elicit relative thinking.

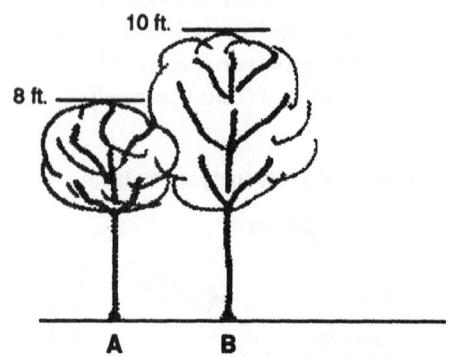

This picture was drawn five years ago when the heights of my two trees were 8 feet and 10 feet, respectively. Their present heights are 14 feet and 16 feet, respectively. Over the last five years, which tree grew most?

Follow-up question: Which tree grew most when you consider where it started five years ago?

The three activities merely use different contexts to assess the same ideas. The Families problem (Figure 5) is the most successful of the three for eliciting the desired model from beginning sixth graders; the Trees problem (Figure 4) is the most difficult context. The following student protocols from clinical interviews represent a typical range or responses to the Trees problem:

John: This one grew 6 feet and so did this one. They grew at the same speed.

Interviewer: Are you sure?

John: Yeah. Anyone can see it. They both grew 6.

Figure 5. Families: A problem designed to elicit relative thinking.

Here is a picture of the children in two families. The Jones family has three girls and three boys and the King family has three girls and one boy. Which family has more girls?

Follow-up question: Which family has more girls compared to boys?

Figure 6. Eggs: A problem designed to elicit relative thinking.

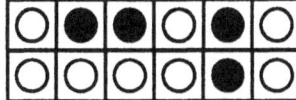

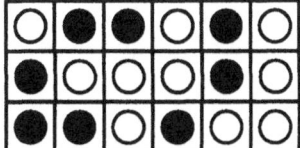

Here is a picture of two egg cartons. The first contains one dozen eggs (eight white and four brown), and the second contains one and one-half dozen eggs (ten white and eight brown). Which carton has more brown eggs?

Follow-up question: Which carton has more brown eggs compared to white eggs?

Interviewer: Do you think it matters that one tree started off taller than the other one?

John: No. Trees do that all the time. Some are always taller than others. But whether they're taller or shorter, they can both still grow 6 feet.

Dan: We need to find which one is a higher percentage of its height now. 8 feet ... 14 feet ... that's ... um ... six eighths ... that's ... that's 75 percent of its height there ... and this ... 10 to 6 ... that's less than that, right? ... that's a little less. The percentage here ... the percentage of the height it is ... O.K ... 60 percent.

Evan: Well, they both increased 6 feet, but this one climbed ... well ... It's this one (pointed to the 8-foot tree).

Interviewer: Well, tell me why you think that one.

Evan: 'Cause it was lower than this one, then it climbed higher ... well they both ... still ... this one's 2 feet higher. But ... well ... this one climbed 6 feet and this one climbed 6 feet ... It got higher. ... This one climbed higher but not *higher* ... I mean it climbed ... it's higher but it didn't climb more feet ... 'cause it was already higher. ... It didn't grow more, it's just higher. The other one grew more but it didn't grow higher than that one.

John is an absolute thinker. Even when it was suggested that he might consider the starting heights of the trees five years ago, John failed to think relatively. He has not met the cognitive goal of the problem and needs to spend some time discussing similar problems with his classmates so that alternative perspectives become available to him. Dan immediately interprets the trees' growth as a situation requiring relative thinking. He not only compares present height to starting height for each tree, but also computes the percentage each height is of its starting height. Though it is painfully difficult for Evan to express himself, the patient reader finds that Evan differentiates additive (absolute) change and relative change. He will grow more comfortable in talking about mathematical ideas as communication, both oral and written, is encouraged in his math class.

Several levels of sophistication are evident in the reasoning used by these children. If these children were asked to deal with relative thinking in several different contexts, the interviewer could be reasonably sure that either the student nearly always reasons multiplicatively, or is ready to reason multiplicatively in some situations, or is not ready to adopt the multiplicative perspective at all. The levels are most conveniently described by the following rubric which can be used for scoring all of the questions related to the same cognitive objective.

0: The student reasons additively.

1: The student reasons multiplicatively in some situations when prompted to consider a relative comparison.

2: The student reasons multiplicatively in some situations without prompting.

3: The student's initial response uses relative thinking.

4: The student thinks relatively and explains his or her thinking by making connections to other pertinent material or by translating to an alternate form of representation.

Accommodating Covariance and Invariance

Another notion central to advanced mathematical thinking is that, when all else is in flux, some dimension of a situation may remain constant. This notion lies at the heart of the scientific process. The notion of a controlled experiment, "keeping all other thinks constant" to isolate the effects of one variable (Karplus, Karplus, Formisano, and Paulsen, 1979, p. 47), is critical in this age of rapid scientific advancement. Thus, another important elementary model that students need to develop is a change model which recognizes that while quantities may be changing, relationships among them may remain invariant. To many children, this is not a foregone conclusion. Some children exhibit a strong tendency to dichotomize the notions of change and invariance; things either change or they don't.

A covariance/invariance model is necessary to understand both fractions and ratios and proportions. Some very basic fraction ideas can be troublesome to a student who has not constructed this elementary covariance/invariance model, for example, the notion that the fraction name "one fourth" can represent different amounts of pizza, depending on the given unit whole:

> Mrs. Thompson ordered two pizzas, a medium cheese pizza and a large pepperoni. Both pizzas were cut into four equal-sized pieces. Mary had a slice with cheese, and Mark had a slice of pepperoni. Did both children have the same amount of pizza?

The essential nature of a proportion also encompasses both covariance and invariance: both component measures of a ratio are varied but in such a way that their original relationship remains invariant. The following activities are designed to elicit a model that accommodates changing elements as well as invariant relationships within a single situation.

> Poker Chips: Here are 36 poker chips, 12 blue and 24 red.
>
> 4B   4B   4B
>
> 8R   8R   8R
>
> Here is another arrangement of the chips:
>
> 3B   3B   3B   3B
>
> 6R   6R   6R   6R

(i) Explain what changed between the first arrangement and the second. (ii) What didn't change? (iii) Show another arrangement of the chips that preserves the same relationship.

Candy Bars: In Mr. Trent's science lab, there are 3 people to each table. For mid-morning break, Mr. Trent put 2 candy bars on each table and told the students to split them fairly. "Before you start your snacks, though, I would like you to push all four tables together," he said. Presuming that you like candy, if you had your choice, would you rather get your share of the candy before the tables are pushed together, or after?

The following sixth-grade student protocols sample the kinds of reasoning elicited by the Candy Bars problem:

> Nancy: There is no way I would ever cheat my friends. I just

wouldn't do it. You couldn't get me to split up those candy bars. Most of them are to hard to split up. They never split just right. There's no way I would take a chance on cheating a friend of mine.

Missy: If there are two candy bars per table, splitting with 3 people you would get more than splitting with 12 people. You get a lot less if you have more people.

Ted: Well, they're really the same, 'cause if you reduce 12 and 8 down to its lowest thing, you get three two. It doesn't matter where the tables are.

Interviewer: Can you tell how much each person would get?

Ted: Yeah. Let's see. That would be three for two people. No, that would be too much. It is less than a candy bar for each person. It would be two thirds.

Kari: Either way, 4 people won't get a candy bar.

Interviewer: What about splitting the candy bars so that everyone gets some?

Kari: (Long pause. Kari draws the picture in Figure 7.) Each person would get a half and a sixth, however much that is. Before, each would get 2/3. I would say you would get more before because I don't think 1/2 and 1/6 will be that much.

Figure 7. Kari shows how to split two candy bars among three people.

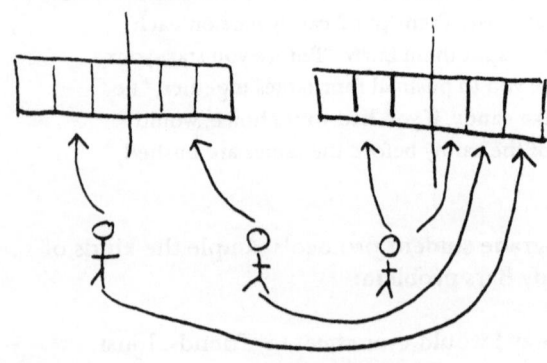

Interviewer: Can you figure out how much a half and a sixth is— exactly?

Kari: (Draws the picture in Figure 8.) Oh, I get it. You would get the same amount either way.

Interviewer: You're right. You would. Now suppose you had 20

Figure 8. Kari adds one-half and one-sixth.

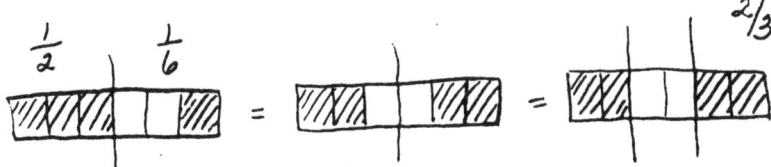

people and 30 candy bars. Would those 20 people get the same amount of candy?

Kari: You better be patient. It will take me a long time to draw that many!

Paul: It's better when the tables aren't pushed together because each person gets two pieces, two quarters. But wait, there's only three. Let's see. Do you have to have 4 quarters in a whole? O.K. Let's put it this way. You get two sections. But with 12 people, you get one section that's smaller because one candy bar is divided into 12 sections.

Although Nancy's response seems humorous, it is typical of the responses of a small number of children who fail to apply mathematics to these situations at all. For example, in answering the Poker Chips problem, some children will note that no chips were removed from the table, that in both arrangements there were red and blue chips, and that you kept the blue chips on the top and the red chips on the bottom, that the red chips were translucent and the blue ones were opaque, and so on.

Missy and Paul did not attend to all of the relevant information in the given situation, and although their thinking was confused, both at least recognized the inverse relationship between the size of the pieces and the number of cuts. Paul's solution was seriously hampered by his trouble with basic fractions. Because prior knowledge was limited, neither student reached the point where the desired objective was discussed.

Kari was reluctant to completely solve the problem on her own. However, with some coaxing, she was able to discover that the students would get the same amount of candy either way. As revealed by her response to the follow-up question, she had not recognized the invariant relationship of two candy bars to three people. It is likely that additional experience with this type of situation would produce the needed insight. Ted's thinking was the most sophisticated. He realized the need to compare number of people to number of candy bars and recognized the invariant relationship.

One possible scoring rubric for this question might be the following:

> 0 = Student fails to focus on the mathematical nature of the situation.
>
> 1 = Student fails to incorporate all of the relevant information.
>
> 2 = Student is hampered in attempts to reach the desired goal because of insufficient prior knowledge.
>
> 3 = Student reaches a correct answer in this situation, but has not actually built a model of the relationships.
>
> 4 = Student understands the invariant nature of the relationships between pairs of changing quantities.

A Simple Reporting Scheme

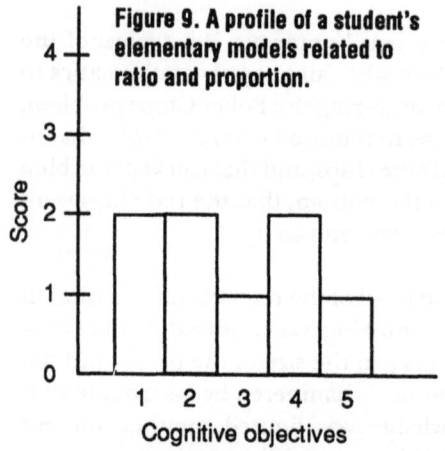

Figure 9. A profile of a student's elementary models related to ratio and proportion.

The preceding problems were written to elicit information about whether students have constructed specific elementary models that are thought to be important to proportional reasoning. When a student has been assessed in regard to all five of those goals, the teacher has a profile of that student's mathematical preparation, information concerning sites where the student is most likely to show development, and information concerning the contexts that are more likely to elicit the desired models from that particular student. As shown in Figure 9, the vector score assigned to that student's performance on the five cognitive objectives may be used to produce a profile of the student's knowledge in the domain.

As instruction proceeds, the entry profile can be updated to record the student's progress as shown in Figure 10. The teacher may also wish to keep a note card on each student, to record information

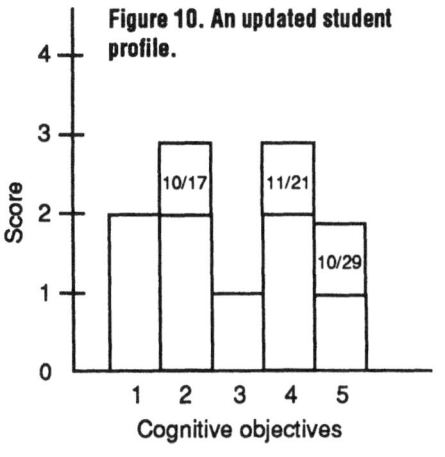

Figure 10. An updated student profile.

concerning problems or contexts that are more conducive to eliciting models, unanticipated problems, assumed background knowledge the student may not have, connections the student has made to other mathematical topics, unusually creative solutions, and so on. Some teachers may prefer to include another score in the student profile, one which indicates an overall level of understanding of proportional reasoning. The two problems that follow are particularly good for capturing various levels of maturity in the domain.

Other Problems With Multiple Levels of Responses

In Figures 11 and 12 are two problems with multiple levels of responses, but unlike the previous examples that were aimed at specific cognitive goals, these are designed to give a more general picture of a student's development in the domain of ratio and proportion. Both problems elicit a wide range of inappropriate solutions including guessing and employing irrelevant information, various insightful preproportional solutions, and true proportional reasoning.

The first problem (Figure 11) is a "classic" ratio and proportion task developed by Karplus (Karplus and Peterson, 1970; Karplus, Karplus, and Wollman, 1972), called "Mr. Short and Mr. Tall." In the most common incorrect solution, the student adds 2 to the number of paper clips to match the difference in heights measured in buttons. Higher-level solutions involve drawing buttons next to the paper clips to discover the relationship that a button is one and one-half paper clips, or finding the scale factor directly from the data.

Figure 11. Mr. Short and Mr. Tall: A proportional reasoning task.

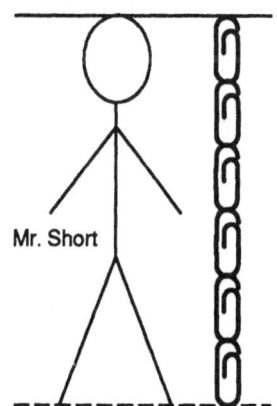

Mr. Short

Here is the height of Mr. Short measured with paper clips. Mr. Short has a friend named Mr. Tall. When we measure their heights with buttons, Mr. Short's height is 4 buttons and Mr. Tall's height is 6 buttons. How many paper clips are needed for Mr. Tall's height?

Figure 12. The Tower: A problem designed to elicit multiple solution strategies.

Why don't you climb the tower? It is only 10 feet!

No! I know I am 6 feet tall. It must be 30 feet.

The people in this picture each give a different answer for the height of the tower. How tall do you think it really is?

The second problem (Figure 12) is an adaptation of a problem posed by Hans Freudenthal (1983). At the very lowest level, students tend to make true/false or right/wrong judgments based purely on a perceptual basis. For example, "The tower doesn't look like it could be ten feet high." Some students use a measurement strategy. This usually involves stacking man on top of man to see how many sets of six feet it takes to get to the top of the tower. Higher-level solutions equate the man's height with eight rungs on the ladder and multiply the man's height by three because there are three sets of eight rungs.

WHY PROBLEMS WITH MULTIPLE LEVELS AND TYPES OF SOLUTIONS?

Especially in complex mathematical domains, it appears that the key to improving learning and instruction will result from dual research perspectives. The first is a top-to-bottom approach in which mathematical phenomena are analyzed to identify the cognitive processes that contribute to competence in the domain. These key processes or cognitive goals provide the content for instruction/assessment activities. Secondly, a bottom-to-top approach involving the analysis of children's thinking is needed to discover initial competencies, to elucidate the manner in which natural strategies develop, to determine when key cognitive processes are acquired, and to identify those contexts most useful in eliciting the processes we would like children to adopt. Problems with multiple types and levels of solutions are the interface between these distinct but complementary endeavors.

According to the National Council of Teachers of Mathematics' *Curriculum and Evaluation Standards for School Mathematics* (1989), one of the tests of the value of assessment should be how well it integrates with and informs instructional decision making and ultimately, how it propels learning. When a teacher uses problems that elicit multiple levels or types of correct answers, instruction and assessment become a seamless process. Problems with multiple levels of responses provide the opportunity to document a student's initial level of performance while allowing or encouraging the student to adopt more mature perspectives. At the same time, the teacher receives a wide range of responses from the class, which, when ordered by sophistication, provide a picture of the manner in which the students' knowledge develops, and gives the teacher a basis for making instructional decisions.

Finally, the kinds of problems we develop for classroom use and the manner in which we interpret student thinking communicate what we believe about learning mathematics with understanding, about individual differences, about what constitutes good teaching, about the nature of mathematics, and about the individual construction of knowledge. Providing students the flexibility to move around in a mathematical territory and teachers the flexibility to interpret student thinking from a variety of perspectives will provide a more realistic conception of mathematical ability.

REFERENCES

Bransford, J.D., Franks, J.J., Vye, N.J., and Sherwood, R.D. (1986, June). New approaches to instruction: Because wisdom can't be told. Paper presented at a Conference on Similarity and Analogy, University of Illinois.

Case, R., and Sandieson, R. (1988). A developmental approach to the identification and teaching of central conceptual structures in mathematics and science in the middle grades. In J. Hiebert and M. Behr (Eds.), *Number concepts and operations in the middle grades*. Reston, VA: National Council of Teachers of Mathematics and Lawrence Erlbaum Associates; pp. 236-259.

Freudenthal, H. (1983). *Didactical phenomenology of mathematical structures*. Dordrecht, Holland: D. Reidel.

Hiebert, J., and Behr, M. (1988). Introduction: Capturing the major themes. In J. Hiebert and M. Behr (Eds.), *Number concepts and operations in the middle grades*. Reston, VA: National Council of Teachers of Mathematics and Lawrence Erlbaum Associates; pp. 1-18.

Hiebert, J. and Wearne, D. (1991). Methodologies for studying learning to inform teaching. In E. Fennema, T.P. Carpenter, and S.J. Lamon (Eds.), *Integrating research on teaching and learning mathematics*. New York: SUNY Press; pp. 153-176.

Karplus, R., and Peterson, R.W. (1970). *Intellectual development beyond elementary school IV: Ratio, a survey*. Berkeley, CA: University of California, Lawrence Hall of Science.

Karplus, R., Karplus, E., and Wollman, W. (1972). *Intellectual development beyond elementary school IV: Ratio, the influence of cognitive style.* Berkeley, CA: University of California, Lawrence Hall of Science.

Karplus, K., Karplus, E., Formisano, M., and Paulsen, A.C. (1979). Proportional reasoning and control of variables in seven countries. In J. Lochhead and J. Clement (Eds.), *Cognitive process instruction: Research on teaching thinking skills.* Philadelphia, PA: Franklin Institute Press; pp. 47-103.

Lamon, S.J. (1992). Ratio and proportion: A framework for connecting content and children's thinking. *Journal for Research in Mathematics Education, 23*(5).

Lesh, R., and S.J. Lamon. (1989). National Council of Teachers of Mathematics. *Curriculum and evaluation standards for school mathematics.* Reston, VA: NCTM

Polya, G. (1945). *How to solve it.* Princeton, NJ: Princeton University Press.

Post, T.R. (1986). The learning and assessment of proportional reasoning abilities. In G. Lappan, and R. Even (Eds.), *Proceedings of the eighth annual meeting of the North American Chapter of the International Group for the Psychology of Mathematics Education.* East Lansing, MI; pp. 349-353.

Stenmark, J. (1989). *Assessment alternatives in mathematics: An overview of assessment techniques that promote learning.* Berkeley, CA: EQUALS and the California Mathematics Council.

Vergnaud, G. (1983). Multiplicative structures. In R. Lesh and M. Landau (Eds.),

14  Using Learning Progress Maps to Improve Instructional Decision Making

Richard Lesh, Susan J. Lamon, Brian Gong, and Thomas R. Post

INTRODUCTION

This chapter considers the following questions: Which decisions are top priorities for educators to address? Which kinds of information do they need to make informed decisions? Which way should that information be reported? Which types of information and data sources should be summarized by reports? And how can such reports be computer-based, interactive, multidimensional, decision-specific, and easy to understand?

The preceding questions reflect a shift in emphasis away from testing, toward informed educational decision-making. The goals are: (i) to intrude as little as possible on instruction, either by using information from instructional activities or by increases in the instructional value of assessment-focused activities; (ii) to provide high-fidelity portraits or multidimensional descriptions of students, teachers, or programs that are as rich as possible; (iii) to reflect information from as many sources as possible in which students naturally exhibit their achievements and abilities; (iv) to facilitate well-informed decision making while avoiding value judgments that artificially and needlessly limit decision-making responsibilities of relevant professionals.

To achieve the preceding range of goals, recent policy statements from a number of relevant professional and governmental organizations have made significant progress to clarify: (i) the nature of the most important *broader, deeper, and higher-order objectives* that will be needed to provide adequate mathematical foundations for knowledge in the 21st century, and (ii) the nature of the most important *new types of tests and items* that will be needed to assess priority-level understandings and processes. However, precisely because significant progress has been made at the level of individual problems and isolated objectives, we believe it is timely to discuss necessary next steps—and that these next steps may need to expand to include the following range of issues.

- New instructional objectives should go beyond behavioral objectives (and basic facts and skills of the past), toward cognitive objectives (and deeper and higher-order understandings and processes for the future).

- New types of problems (or authentic performance activities) should not be mere surrogates for activities we care about; they should directly involve authentic work samples taken from activities that are important in themselves.

- New examination formats should contribute as much to instruction as to assessment, and they should provide as many opportunities as possible for students to demonstrate their knowledge and abilities in the context of familiar and realistic problem-solving and decision making situations, rather than forcing students to prove their capabilities in the context of a few artificial, high-pressure tests.

- New scoring procedures should involve constructed responses, open-ended responses, or graphics-based responses, students' project portfolios, and other responses with multiple levels and types of correctness.

- New data analyses should capitalize on recent advances in computer technology (such as graphical displays and rapid turnaround times), and they should avoid out-dated assumptions about the nature of mathematics, of mathematical talent, and of realistic situations in which mathematics is used, or about the nature of priority educational decision-making issues.

- New reports should go beyond one-dimensional numeric scores that compare strong and weak students, to focus on n-dimensional reports that identify the strengths and weaknesses of individual students.

For continued progress to be made it is important for mathematics educators to deal with issues beyond the level of individual items and objectives. New statements of objectives must go beyond simple, unweighted lists of definitions, facts, and skills that students are imagined to master one at a time. It is not enough to simply replace multiple choice items with problems and scoring procedures that still allow only a single type and level of correct answer. And when conceptually rich and instructionally relevant assessment information is gathered using portfolios, cooperative group projects, or teacher classroom observations, the results should not be collapsed into a single numerical score or left in an uninterpreted form that fails to meet the needs of educational decision-makers.

One way to optimize the responsible use of assessment results is to generate reports that highlight trends and other patterns. Ideally, such reports should filter, simplify, organize, and interpret information in a form relevant to local conditions and to particular decision-making issues. When reports display information from a variety of sources, they should also help their users to make more valid interpretive inferences and adopt more socially responsible courses of action.

Addressing these goals will shift attention beyond traditional bottom-up approaches to assessment (objectives to items to tests to reports). This chapter will take a top-down approach which begins by trying to identify the types of reports and information that educators need to optimize the mathematical experience of every student.

THREE UNDERLYING ASSUMPTIONS

The recommendations that will be made in this chapter are based on the following three assumptions. First, for most of the decisions that are priorities for educators to address, it is inappropriate and simplistic to characterize students (or teachers, or programs) using only a single number. Second, regardless of whether one is interested in evaluating automobiles, or students, or teachers, or programs, it is inappropriate to attach labels or scores without specifying purposes, assumptions, and conditions. For example, to say that a given automobile is "good" clearly begs the question: Good for what purpose? Good under what conditions?

Questions about validity have as much to do with the way assessment information is reported as they do with the nature of the problems and objectives that were used to generate the data. Moreover, the validity of reports is especially relevant to curriculum reform, because the validity of an assessment program depends on the decisions it is intended to inform. Tests and reports that may once have been valid, reliable, or economical may become invalid, counterproductive, and costly when goals and conditions shift. For example:

- Tests and reports based on a narrow conception of talent and designed "weed out" students often create significant barriers to discovering and nurturing diverse kinds of talent (Stenmark, 1989).

- Tests and reports designed to diagnose the learning difficulties of individual students may be costly and inappropriate accountability measures of the cost-effectiveness of complex programs (American Council on Education, 1988; Kellaghan, Madaus, and Airasian, 1980).

- Tests designed to assess behavioral objectives based on basic facts and skills are often counterproductive when used to assess the effectiveness of innovative programs aimed at cognitive objectives and deeper and higher-order understandings and processes (Steen, 1981; Kulm, 1990).

- Pencil-and-paper tests, once considered inexpensive, are increasingly viewed as costly encroachments on valuable instructional time, especially when (i) computer equipment is already available in the school and assessment capabilities can be thought of as "value added" ways to use existing resources, and (ii) the test is known to have negative impacts on both what is taught and how it is taught (Romberg, Zarinnia, and Williams, 1989).

- Increasing pressures for accountability result in more resources devoted to documentation of specified outcomes of instruction; often, this means taking resources away from instruction. U.S. Department of Education data show that a typical school district today devotes more than 50 percent of its staff to nonteaching duties, a percentage that has doubled during the past two decades, in part because of sharp increases in the number of educational middlemen who are not directly involved in the instruction of students.

In the past, when high-stakes tests used exceedingly narrow conceptions of talent the goal was to screen a small-but-adequate number of students for admissions into a few elite programs. But today, even at selective colleges, alarming numbers of students are admitted who must be assigned to remedial courses; and, far too few students are pursuing careers related to mathematics or the sciences that preserve our country's economic prosperity. Therefore attention has shifted toward supporting more diverse conceptions of talent and toward maximizing opportunities for all students.

The third assumption underlying this chapter is that the practice of labeling students with only a single number (which is presumed to be condition free, invariant across all contexts, unmodifiable through experience, and equally applicable for all purposes) is simply *out of date!* It is a throwback to the days when primitive computers led to conditions were very different than they are today. For example:

- *Graphics capabilities were primitive.* Therefore, the main way to simplify a welter of data was to reduce it to a single number – or to a small table of data. Then, numeric reports were restricted even further by the need to reduce computational complexity (e.g., De Soete, 1986). (note: Gleick's popular book, *Chaos*, shows how similar computational restrictions biased the development of mathematical models in a variety of fields outside of education.)

- *Turnaround times were extremely slow.* On-site computers were seldom available to facilitate data interpretation and analysis; and mail-based and telephone-based communication with centralized "mainframes" was also slow and primitive. Therefore, because reports could not be produced in a frequent or timely manner, they were seldom able to be used effectively to inform ongoing instruction aimed at *improving* programs – or at *helping* students ,as opposed to screening them (Clarke, 1988; Popham, 1987).

- *It was not viewed as practical to gather "work samples" from a representative sample of authentic performances in realistic situations that we really care about.* Therefore, brief artificial contexts were created as surrogates for reality; and, the (incorrect) assumption was made that measuring the decomposed parts of a complex system was equivalent to measuring the complex tasks itself. That is, complex tasks were treated as nothing more than strings of isolated simple tasks; and, decontextualized problem situations were used so that no student would have an "unfair" advantage of having the problems fit their specialized experiences and interests (with the exception, of course, of those students whose culture emphasizes and rewards success at dealing with decontextualized problems (Council of Chief State School Officers, 1988).

- *Paper-based tests could seldom be customized or modularized to meet local needs.* For their reports to be valid, they generally had to be used on an all-or-nothing basis. Therefore, students, teachers, and programs were often evaluat-ed using tests that were based on objectives that they had never attempted to address.—Today, because of technological advances, even publications that are expensive to produce (such as Sears catalogues) are often "localized" to address the needs and interests of particular sites (Hunt, 1986; Messick, 1990)

- *Data analysis was based on single-formula models.* Therefore, all data from all sources was processed in the same way; and, when new information was added to a data set, recalculations were performed on the entire data set.—But today, because of technological advances, graphics-based reports (such as national weather maps) often are based on complex mixes of qualitative and quantitative information taken from multiple sources; and, clusters of "local" object-oriented computational models are used so that complex composite models can be created in which "global maps" can be updated locally without regenerating the entire map (Flury & Riedwyl, 1981).

- *Manipulatable and interactive reports were not available.* Even though many of the most obvious misuses of testing information often occur when assess-ment information is used for purposes it was never intended to address, paper-based reports tended to become too complex for decision-makers to use when attempts were made to go beyond simply *presenting* information to also provide guidance about *what the information means* in the context of specific decision-making issues.—Today, because of technological advances, expert systems are often available to produce inquiry-oriented information systems that facilitate decision-making in many contexts (e.g., medicine, weather) that are at least as complex and risk laden as those that occur in education.

This chapter will explore assessment possibilities associated with the fact that modern computer capabilities are able to produce reports that are: (i) *graphics-based* to clarify trends and other patterns beneath the surface of things, (ii) *inter-active* to respond to on-site user inquiries, (iii) *intelligent* to filter, simplify, organize, and interpret information in a form that is especially relevant to local conditions and to particular decision-making issues, and (iv) *multi-dimensional* to integrate information from a variety of data sources, and to produce vector-valued interpretations (or n-dimensional graphics-based interpretations for situations that do not lend themselves to simple numeric quantification).

VISUALIZATION TECHNOLOGY

Scientific visualization refers to the use of computer capabilities to create and manipulate sophisticated graphic renderings of large, complex data sets. Techniques such as contour and vector mapping, three-dimensional plotting, and color raster imaging (a process that assigns a color to each data point) are used to minimize the mental/visual processing required to draw conclusions about complex multivariable data in mathematics, science and engineering. Computer graphics can clarify relationships that are obscure or impossible to grasp in more conventional representations.

Many colleges and universities have visualization laboratories in which students and researchers can create, manipulate, and animate three-dimensional images. Fields as diverse as bioengineering, foreign languages, medicine, and psychology are finding increased uses for these techniques. And because of its growing use for educational purposes, the emphasis in development of visualization techniques is shifting from general purpose tools to discipline-specific software (Publix Information, 1991). We argue in this chapter that visualization technology can drastically change the nature of instruction, assessment, and reporting, using visualization tools now available in the public domain for use on personal computers. We expect the proliferation of visualization tools from special visualization laboratories at universities into elementary schools, and we anticipate profound changes in the mathematics curriculum as visualization technology brings hands-on, experiential learning into the classroom. Likewise, as manipulatable graphic representations become available to most classroom teachers, changes will occur in the reporting of student and class assessment data. (In the appendix, we describe some simple maze image processing techniques and show the progressive transformation of one image into another.)

As new visualization technology enters the classroom, it could be used to aggregate, display, and manipulate complex assessment data to increase the power of reports. The next section of this chapter describes one type of report that exploits this tool and achieves some ambitious objectives for handling complex assessment data. Later sections will then give additional details about how these reports were designed and about how several new types of data sources can be taken into account, such as those based on students' project portfolios, on teachers' classroom observations, or on other authentic performances of students in realistic situations. (Burstall, 1986; de Lange, 1987).

LEARNING PROGRESS MAPS

Figure 1 shows two views of a type of interactive graphics-based report that is being developed in a current NSF-funded project investigating the nature of students' understandings in middle school mathematics (Post, Behr, Lesh, and Harel, 1990). Although the pictures shown here cannot be enlarged or colored to show details, their general character is apparent. We refer to such reports as "learning progress maps" because they resemble three-dimensional topographical maps in historical atlases, where straightforward graphic techniques are used to illustrate periods during the rise and fall of various empires. Some regions are conquered and stable; others are occupied but unstable; and still others are terra incognita.

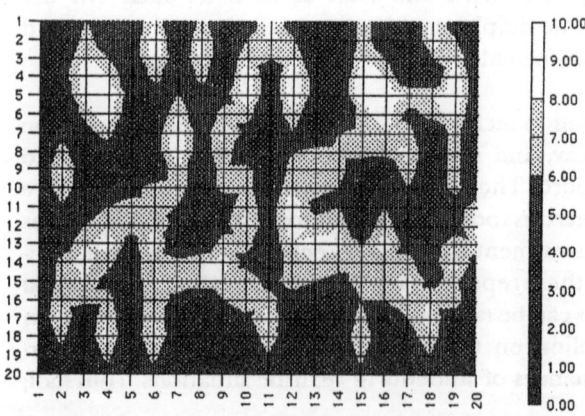

Figure 1A

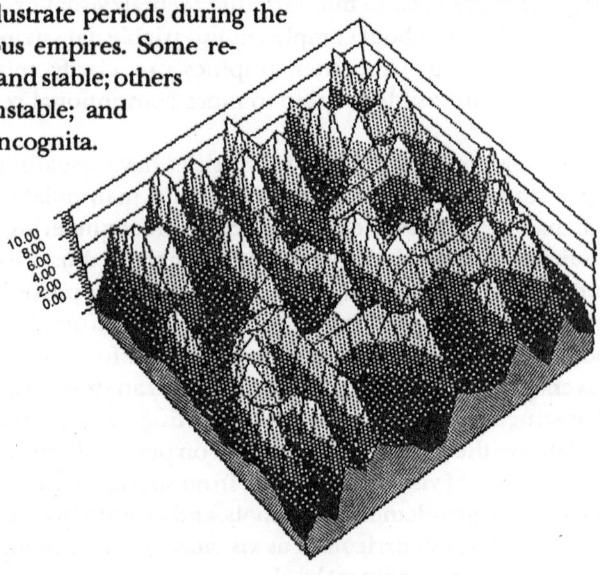

Figure 1B

Learning progress maps can provide easy-to-interpret models of potential student knowledge (Kieren, 1988; Lesh and Kelly, 1990). They can show the conceptual terrain that all students are encouraged to explore in a targeted curriculum; and, they can show n-dimensional profiles of development for individual students in various conceptual regions. A map, in its initial state, is much like a plain (or plane) in which we expect students to "grow" mountains. The growth of a mountain is a physical analogy for the student's internal construction of the major models in the mathematics curriculum. (See discussion of models in Chapter 2.) An individual student's progress in each conceptual region is recorded by altering the map, thus creating a graphic representation of his or her development.

Neighborhoods

As shown in Figure 2, the locations of the anticipated mountains are indicated as neighborhoods on the plane. Each neighborhood contains a conceptual field or a collection of activities whose underlying ideas, concepts, models, procedures, and/or representations are narrowly interconnected (Vergnaud, 1983). Within each neighborhood, the locations of activities depend on the structural complexity of the conceptual systems underlying the activities. The activities at the boundary of the neighborhood are based on elementary ideas needed for understanding the field. The activities near the center are model-eliciting activities (See Chapter 2); these are conceptually demanding activities related to holistic, well-integrated systems of knowledge (elements, relations, and operations).

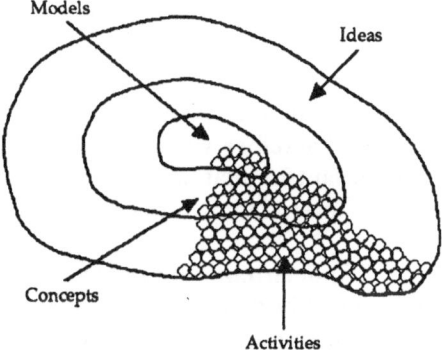

Figure 2. A neighborhood on a learning progress map.

The outer region of a neighborhood corresponds to a collection of instructional/assessment situations in which students deal with elementary ideas that are critical to the development of a model. For example, for proportional reasoning concepts, the outer region of the neighborhood would contain activities that help students do some of the conceptual reorganizing necessary to understand proportions. Critical understandings include (i) distinguishing absolute comparisons and relative comparisons and (ii) recognizing the differences between ratios and fractions. (See Chapter 13 for a discussion of ratio and proportion.) Although activities to facilitate these understandings do not evoke complete models, they provide students with entry points to the model-building process. These activities are analogous to foothills that must be crossed before one can climb the mountain.

The middle section of the neighborhood corresponds to activities that help students integrate ideas and form concepts. For example, in the case of ratio and proportion, the concept of an equivalence class involves the integration of two primitive ideas: (i) the covariation of two terms within the same measure space, and (ii) the relation that yields an invariant function between terms of two measure spaces. Thus, the concept of an equivalence class does not constitute a model but is more than an elementary idea; it is an example of an intermediate conceptual entity.

The center of the neighborhood contains activities involving mathematical models that underlie the most important cognitive objectives students are expected to construct in the targeted mathematics curriculum. During the past decade, some of the most productive areas of mathematics education research have focused on cognitive analyses. This research has identified many important conceptual models that children use to make mathematical judgments in content areas such as whole number arithmetic (Carpenter, Moser, and Romberg, 1982; Fuson, 1988; Steffe, 1988); rational numbers and proportional reasoning (see, for example, Hiebert and Behr, 1988); and geometry and algebra (Confrey, 1990; Thompson, 1990; Tall, 1992; Kaput, 1989; Wagner and Kieren, 1989). Using this research as a point of departure, our own research suggests that, at a given grade level (or within a given course), few curricula deal with more than 10 to 15 distinct foundation-level models (Post, 1987). Yet these models are seldom clearly highlighted in mathematics instruction and assessment. Furthermore, even though a great deal of research has investigated development within these conceptual fields, little is known about developmental relationships among these areas. Therefore, we are designing maps to help educators sort out the most important foundation-level models from lower-order objectives and to emphasize interrelationships among conceptual models.

To determine the relative locations of a collection of activities within a neighborhood, it is necessary to conduct thorough cognitive analyses of the underlying cognitive systems that students use to interpret learning and problem-solving situations (Greeno, 1988b). For example, within the strand of whole number arithmetic, Steffe (1988) and others have shown that conceptual models based on simple counting units provide cognitive foundations for higher-level models based on composite units, or other units of units. Lamon (1992) has shown how conceptual models based on composite units provide cognitive foundations for higher-level systems in which the units consist of (i) relations among lower-level units (such as ratios), (ii) operations with lower-level units (such as functions), or (iii) transformations among lower-level units (such as operators).

Across a variety of mathematical strands, one important way that lower-level systems develop into higher-level systems resembles the way written language proceeds from letters to words to phrases to sentences to paragraphs. Lower-level models based on simple units develop into higher-level models based on units-of-units, or relations among units, or operations on units. The power and complexity of student activities increase as units gradually evolve beyond objects that can be perceived by direct experience, and toward objects based on patterns and regularities beneath the surface of things (Howson and Wilson, 1986).

Progress from the boundary of a neighborhood toward the center is similar to Piaget's notion of progress from concrete operational structures to formal operational structure (Piaget and Beth, 1966). But, Piaget focused on concepts that develop naturally, and on periods in which global conceptual reorganizations occur (such as at the transition periods between the periods of concrete and formal operational reasoning); whereas mathematics educators have focused on (i) concepts that seldom develop beyond primitive levels unless artificial (mathematically rich) learning environments are provided and (ii) detailed transitional stages between periods of global reorganization. Therefore, our maps focus on those dimensions of conceptual development which are needed to achieve a deep understanding of an elementary topic area (Lesh, 1990).

Combining conceptual systems

Different neighborhoods contain activities related to different conceptual systems; and a key criterion for conceptual systems to be different is that they are based on different primitive objects. For example, at the level of high school mathematics, the Mathematical Sciences Education Board identifies strands based on the following primitive types of objects: (i) patterns, (ii) quantities, (iii) shapes, (iv) uncertainties, (v) dimensions, and (vi) changes (Steen, 1990). Or at the middle school level, models which tend to be emphasized are based on primitive objects such as (i) fractions, (ii) ratios, (iii) rates, (iv) percents, (v) proportions, (vi) probabilities, (vii) rules (or fractions), or (viii) coordinates (see, for example, Hiebert and Behr, 1988; Post, 1987; Steen 1986). At the primary school level, models that are emphasized involve primitive objects such as (i) counts, (ii) composite units, (iii) measures, (iv) fractions, (v) locations, and (vi) shapes (see, for example, Steffe, Cobb, and von Glasersfeld, 1988).

In some ways, the preceeding strands and themes are similar to the columns in scope and sequence charts for textbooks or to the content categories in content-by-process matrices for standardized tests. But, whereas traditional types of charts and matrices have tended to fragment and compartmentalize the curriculum, recent curriculum guides have focused on unifying themes and on basic patterns and regularities that cross a variety of conceptual models (Lesh, 1990).

Because our "mountains" emphasize to the most important conceptual models and reasoning patterns that students are encouraged to construct at a given grade level, the location and names of the neighborhoods vary from one curriculum to another, as well as from one grade level to another within a given curriculum; and, progress through a given K-12 curriculum can be thought of as a cluster of tectonic plates that move, rise, sink, expand, and contract from one grade level to another. For example, in a given sixth-grade curriculum, some of the most important conceptual neighborhoods are based on systems of fractions, ratios, rates, or quotients, yet at the seventh-grade level, all of these systems may be integrated into a single powerful and inclusive model based on rational numbers or proportional reasoning (Lesh, Post, and Behr, 1988).

Intersecting neighborhoods on the maps indicate that the underlying conceptual systems are structurally overlapping. That is, spatial closeness indicates conceptual relatedness. Of course, it is not possible to portray all of the logical relationships that exist among systems, but because our maps use more than one or two dimensions, they can capture more of the relatedness of mathematical ideas than the organizational schemes in most scope and sequence charts in textbooks or most content-by-process matrices for tests.

Comparisons with expert knowledge

Learning progress maps are not intended to portray a single "true" picture of expert knowledge. In fact, we reject the notion that there exists a single, static, and correct organization. For example, in mathematics, it is known that experts tend to define, organize, and weight their knowledge in different ways depending on their own personal goals and experiences, and for a given individual—expert or novice—the best way continually evolves as contexts, conditions, and purposes change. Real human cognition tends to be far more situated, dynamic, and purpose-oriented than we can capture in any single map (Greeno, 1988a; Mestre, 1987). Therefore, our maps are considerably different than the usual semantic networks that cognitive psychologists have tended to use to portray states of knowledge.

Another way that our maps are different than descriptions about the state of knowledge of an expert is that the expert's quest for economy, parsimony, and power eventually lead to a well-compiled form of knowledge in which a number of conceptual nuances that are important at lower levels of development become invisible—unless a large amount of psychological unpacking is provoked. For instance, in the case of rational numbers and proportional reasoning, a number of basic constructs take on distinct personalities, depending on whether students encounter them in the form of fractions, ratios, rates, indicated quotients, decimals, or percents (Kieren, 1988). Therefore, for purposes of instruction, it is often important for students to explore similarities and differences among these subconstructs even though mathematicians might treat them all as special cases of a single system.

Local school systems or individual teachers should be able to modify the default values that we assign to the weights, locations, and attributes of various neighborhoods or mountains. Even if these modifications override some of the wisdom that we believe we built into the organization of our maps, local educators should be able to modify parameters to reflect the goals, objectives, and instructional sequences in their own local curriculum.

Capabilities that are granted to local teachers should also be given to those who must make decisions based on the information teachers enter into the database or model. For example, if a school or a teacher insists on modifying default values in a way that emphasizes lower-order facts and skills (perhaps for the purpose of addressing state-level pressures for accountability), then funding agencies or college admissions officers may choose to reset the original default values to emphasize deeper and higher-order understandings, or other preferences.

Creating N-Dimensional Student Profiles

Student profiles are created by modifying the learning progress map through a variety of image processing techniques (similar to those used in the example involving the Mendelbrot Set in the appendix). In addition to obtaining an overall impression of a student's development within a particular neighborhood and across the curriculum, the map should provide information on other relevant questions. For example, while working on a particular activity, did the student make important connections to other topics not located in the same neighborhood? Is the student able to represent a situation using only one mode of representation, or more? Where (and when) has the student's most recent activity occurred? Does the student understand a concept embedded in only one context or in many different contexts?

Visual dimensions of student variables

Using learning progress maps, a number of visual dimensions are available to help provide information about student variables. Height above the plane, colors and various intensities of colors, shading, and a host of local symbols may be used to highlight relevant aspects of a student's development.

Height above the plane. For each activity in which a student has participated, we might represent a student's overall level of development as a surface in 3-space by converting a teacher-assigned score to height above the activity's location in the plane. A student's level of development will usually be judged using a rubric in which development is implicitly defined along one or more of the following dimensions: concrete to abstract (Dienes, 1960), global to analytic to deductive (van Hiele, 1959), preoperational to operational (Piaget and Beth, 1966), concepts to rules to problem solving (Gagne, 1985), enactive to symbolic (Bruner, 1973), external to internal (Vygotsky, 1978), situated to decontextualized (Cole and Griffen, 1987; Greeno, 1988b), facts and skills to applications to analysis/synthesis/evaluation (Bloom, 1956), naive interpretations (based on superficial characteristics of events) to scientific models (focused on underlying patterns and regularities) (Steen, 1988). A rubric highlighting relevant dimensions from this list is developed by analyzing a range of students' responses as they interact with an activity. (See Chapter 13 for a discussion of the scoring of problems with several types of correct answers.) The teacher uses the rubric to assign a score that corresponds to the depth of the student's understanding, and that depth of understanding is represented visually as height above the plane.

Over time, a student's profile should develop from spikes to ridges to mountains. If a student's map continues to show isolated spikes located primarily in the outer regions of various neighborhoods, the student has skipped from neighborhood to neighborhood, working only on basic ideas and avoiding the activities designed to elicit deeper thinking and higher-level reasoning. Alternatively, a ridge in a particular neighborhood would indicate that a student has attacked several related activities and achieved some degree of understanding of the underlying ideas.

A cliff or very sharp transition on the map would suggest potential conceptual trouble spots. For example, if a student developed very high spikes in the outer region of a neighborhood with no scores from the middle region of the neighborhood to help round them into ridges, the student may be having some difficulty in accomplishing reorganizations to connect significant conceptual discontinuities. For example, in progressing from whole numbers to fractions, students must reject a naive model of multiplication in order to develop more powerful models; it is no longer adequate to think of multiplication as repeated addition.

Color, intensity, and shading. On computer screens, color, intensity, and shading are useful to suggest additional dimensions of conceptual development, such as: concrete to abstract (Dienes, 1960), global to analytic to deductive (van Hiele, 1959), pre-operational to operational (Piaget & Beth, 1966), concepts to rules to problem-solving (Gagne, 1985), enactive to symbolic (Bruner, 73), external to internal (Vygotsky, 1978), situated to decontextualized (Cole & Griffen, 1987; Greeno, 1988b), or facts & skills to applications to analysis/synthesis/evaluation (Bloom, 1956). For example, one color could indicate connections students have made between topics in different neighborhoods. Color intensities might indicate breadth of a student's understanding; that is, a light color might show that a student understands the same idea in two different contexts, with darker shades showing the ability to generalize to other contexts. For example, in ratio and proportion, students may think relatively in problems involving speed and distance but not in problems involving scale drawings. Hatching or crossed hatching might indicate activities attempted by the student for which no score was assigned.

Local symbols. Local symbols commonly used on maps can suggest a variety of instructionally significant aspects of a student's activities and achievements in various conceptual regions. For example, rivers, highways, or bridges can depict links established between logically distant conceptual neighborhoods, based on problem-solving experiences that involve more than a single topic. Clouds and weather systems can indicate the general location of a student's most recent learning or problem-solving activities. In addition, various intensities of one color might indicate how recently that activity has occurred.

Interactive Capabilities of Maps

Learning progress maps are designed to capitalize on the dynamic, interactive, and intelligent nature of computer-based graphics; and they are designed to rapidly update reports on a wide range of issues such as (i) the strengths and weaknesses of a given student, (ii) typical problem-solving activities that might cause difficulties in a given conceptual neighborhood, (iii) instructional activities that address a given student's profile of strengths and weaknesses, (iv) which students need special help in a given conceptual region, (v) which students need individual attention, (vi) which topics address the needs of a large group of students, and (vii) which students share a particular common need. By making graphics-based reports both intelligent and interactive they should be able to generate information that is detailed and relatively complete and, at the same time, simple and easy to interpret. A goal is to display details only when they are needed or directly requested, and not to use a single display to serve all purposes (Flurry, 1980; McDonald and Ayers, 1978). For example:

- Teachers should be able to view static snap shots showing the status of a given student (or group of students); or, they should also be able to see animated graphics showing learning progress over a designated (long or short) period of time and for specially targeted (small or large) sets of objectives. And they should be able to factor in or out information from multiple data sources, including traditional assessment activities, homework, classroom observations, and students' project portfolios. For example, a teacher might want to factor out scores on homework, or a funding agency (or a local administrator) might want to factor out data sources that might be viewed as lacking objectivity.

- Teachers also should be able to use straightforward graphics tools to modify the color, shading, and local graphics in a given region based on their own classroom observations, and if they wish, such information should be treated just like information from any other source. And teachers should be able to zoom and scan to view details about a given student, or about a particular conceptual neighborhood. For example, using a standard mouse, teachers should be able to select particular points on a map to see displays of typical problems or instructional activities of the chosen conceptual neighborhoods.

- Teachers should be able to view maps through various windows, each of which calls upon resident computer software to organize, aggregate, and display information in a form that is appropriate to address a specific set of decision-making issues. Rather than simply generating statistics, the maps (and other software) should help decision makers (i) interpret the meaning of available information, (ii) recognize possible decision alternatives, and (iii) identify available resources (textbooks, software, and so on). The goal is not simply to furnish information; it is to help students, teachers, and others make well-informed decisions.

Maps with the kinds of capabilities just described must be linked to (i) a richly cross-indexed library of problems for homework, classroom activities, and tests for specially targeted purposes; (ii) research and instructional information on types and levels of sophistication in students' responses to problems (and the capability for teachers to amend or annotate such information); (iii) objectives-based instructional resources ranging from textbooks to educational software; and (iv) printed reports based on state or local curriculum guides, specific textbook scope-and-sequence charts, or national objectives frameworks such as those published by the National Council of Teachers of Mathematics (1989).

To provide the guidance just described, it is not necessary for the communications links to be as fine-grained as many artificial intelligence researchers have supposed (Brown and Burton, 1978; Brown and VanLehn, 1980). For example, our research suggests that it is seldom necessary or desirable to prescribe instruction at a level of detail below whole lessons; and there rarely exists one—and only one—appropriate assignment that can meet the needs of a particular student (Lesh and Kelly, 1990). Furthermore, because a number of equally beneficial activities usually are possible for a given student, there tends to be considerable merit in allowing students to participate in the selection of their own assignments.

In designing learning progress maps capable of addressing the preceding kinds of goals, the main difficulties are not technical or technological, even though they involve some sophisticated programming and software capabilities. The key difficulties are conceptual. Above all, designing instructionally relevant learning progress maps depends on defining a theory-based objectives framework capable of dealing in an integrated way with both basic facts and skills and higher-order understandings and processes, and also capable of being linked to the best available instructional materials and curriculum guides.

Programming Principles

To clarify the kind of information that is needed for programming learning progress maps, it is useful to think of the maps as fancy smoothed and colored versions of a three-dimensional bar graph similar to the one shown in Figure 3A. In fact, for the purposes of this chapter, the main differences between Figure 3A and Figure 3B are that (i)

3A has only a single value (to specify the height of the bar at each point), whereas 3B has a vector of values assigned to each point (to specify height, color, shading, and the presence/absence of a variety of local symbols), and (ii) 3B has many two-way rules between cells, whereas 3A does not.

Figure 3A

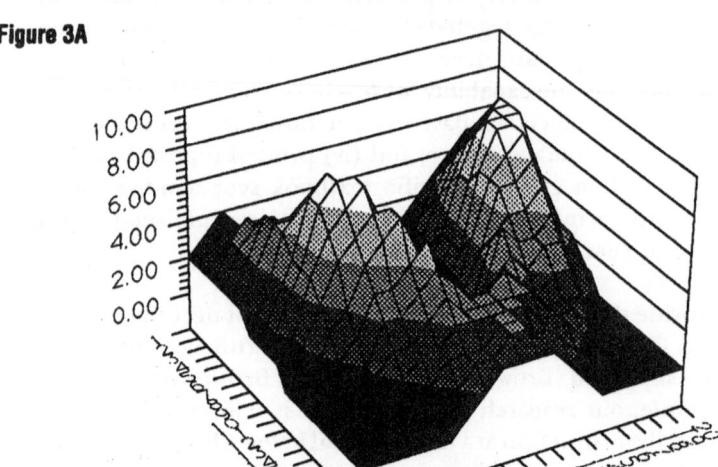

Figure 3B

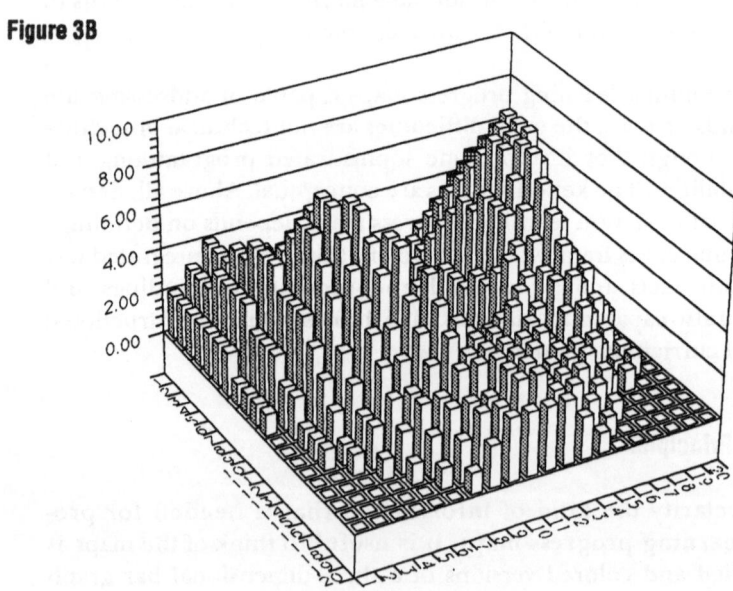

Figure 3 is based on a standard spreadsheet (WINGZ) in which the value in each cell is based on either data taken from some external source or rules based on variables involving values from other cells. This means that neighborhoods of the map inform one another as well as being informed by external data sources such as tests or instructional activities. To serve as a simple learning progress map, the spreadsheet underlying Figure 3B needs one additional feature. That feature is a set of meta-rules to govern how and how often information is changed by the operation of the spreadsheet-level rules. Both the spreadsheet-level rules and the meta-rules must be based on (i) theory-based models describing relationships among understandings represented in various neighborhoods of the map, and (ii) empirical information based on the experiences of individual students (for example, when a given student's problem-solving experiences forge links between logically distant conceptual neighborhoods). Such off-the-shelf meta-rule systems already exist for graphics-based spreadsheets (for example, Axcelis's EVOLVER, 1990), so the programming underlying our learning progress maps can be thought of as involving simply a vector-valued spreadsheet with meta-rules, and with vector-valued links to a library of problems and instructional activities. The key difficulties do not have to do with programming; they have to do with finding appropriate ways to define the vector-valued rules.

Defining vector-valued rules

Four important considerations have influenced the design of our learning progress maps:

- The maps must shift attention away from comparing students on one-dimensional scales and toward instructionally useful, multidimensional profiles of conceptual strengths and weaknesses of individual students.

- The maps must avoid the kind of fragmented objectives frameworks that have characterized past curriculum reform efforts and must focus on systems in which objectives are organized and weighted so as to increase the power and usefulness of the system as a whole.

- The maps must emphasize depth and breadth of understanding rather than simple mastery/nonmastery in which one perceived goal is simply to finish as rapidly as possible. Objectives frameworks must focus on the most powerful and the most useful conceptual models and reasoning patterns that we want students to construct. The terrains of knowledge they depict should encourage students to explore depth and breadth of meaning and abandon the notion of simply checking off mastered isolated activities.

- The maps must not simply crank out statistics and leave students and teachers to interpret on their own what the information means (and doesn't mean) in the context of specific decision-making issues.

In the next section, we describe several reasons why it is especially important to use computer-based reports that are interactive, inquiry-oriented, and decision-focused.

USING INTERACTIVE DECISION-FOCUSED REPORTS

Frederiksen and Collins describe a systemically valid test (or item, or report) as one that "induces in the education system curricular and instructional changes that foster the development of the cognitive traits that the test is designed to measure. Evidence for systemic validity would be an improvement in those traits after the test had been in place within the educational system for a period of time" (1989).

Negative influences of standardized testing have been thoroughly documented in mathematics. Therefore, we are sympathetic to the pro-active policy of Frederiksen and Collins stated at the start of this section. Furthermore, we strongly support the performance assessment movement in which school systems are refusing to use tests that do not involve a representative sample of material that they really want students to learn. Nevertheless, we are concerned about the following kinds of misuses of such policies.

The policy stated by Frederiksen and Collins focuses on tests as leverage points to induce changes in our education systems; and it further suggests that those who implement the policy already know which kinds of changes are positive. We favor a perspective which recognizes that past curriculum reformers (such as those in the new math movements of the 1960s and 1970s) often tried to induce changes in directions that, in retrospect, proved to be at least partly wrong-minded.

Whereas reform-minded educators often support policies that remove decision making from the hands of local educators, we prefer to focus on policies that empower local educators, while doing our best to ensure that their decisions are well-informed concerning relevant conditions, available options, and probable results. Our position is that tests and reports should be used to inform—not preclude—educational decision making.

Using test results as leverage points to clarify objectives and induce curriculum changes focuses on one set of priority decision-making issues, but other issues should not be neglected. For example, a fundamental shortcoming of many curriculum reform movements is their focus on radical change while ignoring the difficult tasks that deal with orderly and evolutionary steps to get from "where we are" to "where we want to be."

Suppose it were possible to instantly change all college admissions tests to conform to the kinds of future-oriented goals cited in recent publications from the National Council of Teachers of Mathematics (NCTM), Mathematical Sciences Education Board (MSEB), American Association for the Advancement of Science (AAAS), and National Science Teachers Association (NSTA). Care must still be taken to ensure fair treatment for students who were indoctrinated through twelve years in an old system that emphasized obsolete views of basic facts and skills (Ford Foundation, 1986; Malcom, 1984).

Our position is that it is not acceptable for testing programs to use students as pawns to induce curriculum changes, especially if the students most likely to suffer are those that have been treated unjustly by the system we are trying to change.

Because their rhetoric is often aimed at rallying political action, curriculum reformers who argue in favor of one point often implicitly suggest that other points should be ignored. For example, because teacher decision making should be a priority, it does not follow that the information needs of other decision makers should be ignored. Because performance assessment items and students' project portfolios should be given greater attention, it does not follow that other sources of information should be ignored.

We take the position that no single decision maker or decision-making issue should blind us to all others, and no single source of information is likely to be appropriate for all purposes and audiences. One of the worst things that could happen to promising new forms of assessment is that they too become used for issues and audiences that they were not intended to address.

Decision-specific Appropriateness of Reports

To describe the main kinds of issues that must be taken into account so that appropriate capabilities can be incorporated into our learning progress maps, it is useful to focus on teacher decision making. For example, consider the following ways in which maps might help inform teacher decision making.

- Teachers assign instructional activities. For each student, a goal may be to identify conceptual strengths and weaknesses so the student won't waste time and effort on activities that are too hard, too easy, or irrelevant to priority goals and needs.

- Teachers group students. For example, when a given conceptual neighborhood is specified, the goal may be to identify students who might profit most from activities in that targeted area.

- Teachers select or screen students for priority access to scarce resources. When a limited resource such as a software program, a tutor, or a field trip experience is available, the goal may be to identify a limited number of students who should be given access.

- Teachers certify learning accomplishments. For example, when achievement expectations have been specified, the goal may be to generate individualized learning reports showing both the status and recent achievements of individual students and their peers.

- Teachers identify students who need special attention, whether a student seems to be experiencing unusual difficulties or to be ready for some significant insight.

- Teachers evaluate programs. They may need to produce accountability measures for kill-or-keep decisions about particular programs, or to identify changes, deletions, or extensions that should be made in the future.

Two important observations about the preceding kinds of decisions are that none can be neglected, and that the data sources and report characteristics relevant for one issue are not necessarily appropriate for others. For example, a test that diagnoses learning difficulties for individual students may be inappropriate if the goal is to prove that a large and complex program works. Or if appropriate instructional activities are needed for a given student, then fast turn-around may be far more important than high precision or high reliability because negative consequences are not associated with errors. Yet for tests that focus on gate-keeping functions, decisions to screen students out of short-term opportunities can result in permanently limiting future educational choices. In this later case, errors may lead to an enormous human losses.

The main point of these observations is that useful and instruction-relevant learning progress maps must be designed to reflect information from a variety of sources, including teachers' classroom observations, students' project portfolios, and performance assessment tests focusing on realistic types of problem solving. Our maps are also being designed to emphasize the strengths and needs of individual students, based on capabilities that we know can be changed through instruction, and to emphasize positive planning and optimizing functions aimed at ensuring maximum success (Shavelson, 1991).

Reports that do not help students are not the kind we are interested in making available. Therefore, our learning progress maps are designed to emphasize a form of assessment that gives students lots of opportunities to demonstrate their knowledge and capabilities through sampling over long periods of time and a variety of areas that explicitly fit students' individual interests and experience. In this way, records are not as easily influenced by factors such as cheating, sickness, or disinterest, and issues related to validity, reliability, and generalizability are addressed in the most straightforward ways possible (Gardner, 1991; Shavelson, 1989).

SUMMARY

Assessment reports. The computer-based learning progress maps described in this chapter are graphic, dynamic, interactive, intelligent, decision-focused reports designed to produce n-dimensional student profiles that are linked in meaningful ways to future-oriented objectives frameworks, such as the NCTM's 1989 *Curriculum and Evaluation Standards for School Mathematics.* They are capable of reporting assessment-relevant information from many types of data sources; and, their primary aims are to identify strengths and weaknesses of individual students for the purpose of optimizing educational opportunities.

REFERENCES

Alexander, L., and James, H.T. (1987). *The nation's report card: Improving the assessment of student achievement.* Washington, DC: National Academy of Education.

American Association for the Advancement of Science. (1989). *Project 2061: Science for All Americans.* Washington, DC: AAAS.

American Council on Education. (1988). *One third of a nation.* Report of the Commission on Minority Participation in Education and American Life. Washington, DC: ACE.

Archibald, D.A., and Newman, F.M. (1988). *Beyond standardized testing: Assessing authentic academic achievement in the secondary school.* Reston, VA: National Association of Secondary School Principals.

Axcelis. (1990). *EVOLVER.* (Computer Program). Axcelis Corp., Seattle, WA, 98101.

Bloom, B.S. (1956). *Taxonomy of educational objectives: The classification of education goals. Handbook 1: Cognitive domain.* New York: Longmans, Green and Co.

Bloom, B.S. (1985). *The development of talent in young people.* New York: Balentine.

Blumberg, F., Epstein, M., MacDonald, W., and Mullis, I. (1986). *A pilot study of higher-order thinking skills assessment techniques in science and mathematics.<* Princeton, NJ: National Assessment of Educational Progress.

Bridger, M. (1988). Looking at the Mendelbrot Set. *The College Mathematics Journal, 19*(4), 353-363.

Brown, J.S., and Burton, R.R. (1978). Diagnostic models for procedural bugs in basic mathematical skills. *Cognitive Science, 2,* 155-192.

Brown, J.S., and VanLehn, K. (1980). Repair theory: A generative theory of bugs in procedural skills. *Cognitive Science, 4,* 379-426.

Bruner, J. (1973). *Beyond the information given.* J. Anglin (Ed.). New York: Norton and Co.

Burstall, C. (1986). Alternative forms of assessment: A United Kingdom perspective. *Educational Measurement: Issues and Practice, 5*(1), 17-22.

Business Week, Human capital: The decline of America's work force. Special Report, September 19, 1988, 100-141.

Carpenter, T.P., Corbitt, M.K., Kepner, H.S., Jr., Lindquist, M. M., and Reys, R.B. (1981). *Results from the second mathematics assessment of the national assessment of educational progress.* Reston, VA: National Council of Teachers of Mathematics.

Carpenter, T. P., Moser, J., and Romberg, T.A. (Eds). (1982). *Addition and subtraction: A cognitive perspective.* Hillsdale, NJ: Lawrence Erlbaum Associates.

Center for the Assessment of Educational Progress, Educational Testing Service. (1987). *The subtle danger: Reflections on the literacy abilities of America's young adults.* Princeton, NJ: Educational Testing Service.

Clarke, D. (1988). *Assessment alternatives in mathematics: Mathematics curriculum and teaching programs.* Canberra: Curriculum Development Centre.

Clement, J. (1982). Students' preconceptions in introductory mechanics. *American Journal of Physics, 50,* 66-71.

Cole, J., and Griffen, P. (Eds.). (1987). *Contextual factors in education: Improving science and math education for minorities and women.* Madison, WI: Wisconsin Center for Educational Research.

Collis, K.F., and Romberg, T.A. (1989). *Assessment of mathematical performance: An analysis of open-ended test items.* National Center for Research in Mathematical Sciences Education, Wisconsin Center for Education Research, School of Education, Madison, WI: University of Wisconsin.

Conference Board of the Mathematical Sciences. (1983). The mathematical sciences curriculum K-12: What is still fundamental and what is not. In *Educating Americans for the 21st century: Source materials*. National Science Board Commission on Precollege Education in Mathematics, Science, and Technology. Washington, DC: National Science Foundation, 1-23.

Confrey, J. (1990, April). Origins, units, and rates: The construction of a splitting structure. A paper presented at the annual meeting of the American Educational Research Association. Boston.

Council of Chief State School Officers. (1988). *Assessing mathematics in 1990 by the National Assessment of Educational Progress*. Washington DC: State Education Assessment Center.

Council of Chief State School Officers. (1987). *Equity and excellence: A dual thrust in mathematics and science education: Model state education agency efforts*. Washington, DC: Council of Chief State School Officers.

de Kleer, J., and Brown, J.S. (1983). Assumptions and ambiguities in mechanistic mental models. In D. Gentner and A.L. Stevens (Eds.), *Mental models*. Hillsdale, NJ: Lawrence Erlbaum Associates.

de Lange, J. (1987). *Mathematics, insight and meaning: Teaching, learning and testing of mathematics for the life and social sciences*, (dissertation). Utrecht, Netherland: Rijksuniversiteit Utrecht.

Denham, W.F., and O'Malley, E.T. (Eds.). (1985). *Mathematics framework for California public schools, kindergarten through grade twelve*. Sacramento, CA: California State Department of Education.

De Soete, G. (1986). A perceptual study of the Flury-Riedwyl faces for graphically displaying multivariate data. *Int. J. Man-Machine Studies, 25* (1986), 549-555. London: Academic Press.

di Sessa, A. (1982, January-March). Unlearning Aristotelian physics: A study of knowledge-based learning. *Cognitive Science, 6*(1), 37-75.

di Sessa, A. (1989). Knowledge in pieces. In G. Gorman and P. Pufall (Eds.) *Constructivism in the computer age*. Hillsdale, NJ: Lawrence Erlbaum Associates.

Dienes, Z. (1960). *Building up mathematics*. London: Hutchinson Ltd.

Donlon, T.F. (Ed.). (1984a). *The College Board technical handbook for the Scholastic Aptitude Test and Achievement Tests*. New York: College Entrance Examination Board.

Dossey, J.A., Mullis, I.V.S., Lindquist, M.M., and Chambers, D.L. (1988). *The mathematics report card: Are we measuring up?* Princeton, NJ: Educational Testing Service.

Edgerton, H.A. (1985). *Identifying high school seniors talented in science*. Westinghouse Science Talent Search. Washington, DC: Science Service.

Flury, B. (1980). *Construction of an asymmetrical face to represent multivariate data graphically*. Technical Report No. 3, University of Berne, Department of Statistics.

Flury, B., and Riedwyl, H. (1981, December). Graphical representation of multivariate data by means of asymmetrical faces. *Journal of the American Statistical Association, 76*(376), Applications Section.

Ford Foundation. (1986). *Minorities and mathematics*. New York: Ford Foundation.

Frederiksen, J.R., and Collins, A. (1989). A systems approach to educational testing. *Educational Researcher, 18*, 27-32.

Frederiksen, N. (1984). The real test bias: Influence of testing on teaching and learning. *American Psychologist, 39*, 193-202.

Frederiksen, N., Mislevy, R.J., and Bejar, I.I., (Eds.). (1992). *Test theory for a new generation of tests.* Hillsdale, NJ: Lawrence Erlbaum Associates.

Fuson, K. C. (1988). *Children's counting and concepts of number.* New York: Springer-Verlag.

Gagne, R. (1985). *The conditions of learning and theory of instruction* (4th ed.). New York: Holt, Rinehart and Winston.

Gardner, H. (1985). *Frames of mind: The theory of multiple intelligences.* New York: Basic Books.

Gardner, H. (1991). Assessment in context: The alternative to standardized testing. In B.R. Gifford and M.C. O'Connor (Eds.), *Future assessments: Changing views of aptitude, achievement, and instruction.* Boston, MA: Kluwer Academic Publishers.

Greeno, J. (1988a). The situated activities of learning and knowing mathematics. In M. Behr, C. Lacampagne and M. Wheeler (Eds.), *Proceedings of the tenth annual meeting of the Psychology of Mathematics Education.* DeKalb, Illinois.

Greeno, J. G. (1988b). For the study of mathematical epistemology. In R. Charles and E. Silver (Eds.), *Research agenda for mathematics education: Teaching and assessing mathematical problem solving.* Reston, VA: National Council of Teachers of Mathematics.

Hiebert, J., and Behr, M. (Eds.). (1988). *Number concepts and operations in the middle grades, 2.* Reston, VA: National Council of Teachers of Mathematics.

Holland, J. L., and Richards, J. M., Jr. (1965). Academic and nonacademic accomplishment: Correlated or uncorrelated? *Journal of Education Psychology, 56*, 165-174.

Howson, G., and Wilson, B. (Eds.). (1986). *School mathematics in the 1990s.* International Commission of Mathematical Instruction Study Series. Cambridge: Cambridge University Press.

Howson, G., Kahane, J.P., Lauginie, P., and de Turckheim, E. (1988). *Mathematics as a service subject.* International Commission of Mathematical Instruction Study Series. Cambridge: Cambridge University Press.

Hoyt, D.P. (1966, Winter). College grades and adult accomplishment: A review of research. *The Educational Record, 47*(1).

Hunt, E. (1986). Cognitive research and future test design. In *The redesign of testing for the 21st century.* Princeton, NJ: Educational Testing Service.

Kaput, J. (1989). Linking representations in the symbol system of algebra. In C. Kieren and S. Wagner (Eds.), *A research agenda for the learning and teaching of algebra.* Reston, VA: National Council of Teachers of Mathematics.

Kellaghan, T., Madaus, G.F., and Airasian, P.W. (1980). *The effects of standardized testing.* Dublin, Ireland/Boston, MA: St. Patrick's College/Boston College.

Kieren, T.E. (1988). Personal knowledge of rational numbers: Its intuitive and formal development. In J. Hiebert and M. Behr (Eds.), *Number concepts and operations in the middle grades, 2.* Reston, VA: The National Council of Teachers of Mathematics; pp. 162-181.

Kirsch, I.S. (1987, September 28). Measuring adult literacy. Paper prepared for the symposium Towards Defining Literacy, sponsored by the National Advisory Council on Adult Education and held at the Literacy Research Center, University of Pennsylvania.

Kirsch, I.S., and Jungeblut, A. (1986). *Literacy profiles of America's young adults.* Princeton, NJ: Educational Testing Service.

Krutetskii, V. (1976). *The psychology of mathematical abilities in school children.* Chicago: University of Chicago Press.

Kulm, G. (1990). *Assessing higher-order thinking in mathematics.* Washington, DC: American Association for the Advancement of Science.

Lamon, S.J. (1992). Ratio and proportion: A framework for connecting content and children's thinking. *Journal for Research in Mathematics Education, 23*(5).

Lapointe, A.E., Mead, N.A., and Phillips, G.W. (1989). *A world of differences: An international assessment of science and mathematics.* Princeton, NJ: Educational Testing Service.

Leinhardt, G., and Seewald, A.M. (1981). Overlap: What's tested, what's taught? *Journal of Educational Measure, 18*(2), 85-96.

Lesh, R. (1990). Computer-based assessment of higher-order understandings and processes in elementary mathematics. In G. Kulm (Ed.), *Assessing higher-order thinking in mathematics.* Washington, DC: American Association for the advancement of Sciences.

Lesh, R., and Kelly, A. (1990). A modeling theory of computer-based tutoring. In J.M. Laborde (Ed.) *Modeling student knowledge in geometry.* IMAG: Grenoble, France.

Lesh, R., and Lamon, S.J. (in press). Assessments of authentic performance. Princeton, NJ: Educational Testing Service.

Lesh, R., Post, T., and Behr, M. (1988). Proportional reasoning. In J. Hiebert and M. Behr (Eds.), *Number concepts and operations in the middle grades, 2.* Reston, VA: The National Council of Teachers of Mathematics; pp. 93-118.

Malcom, S.M. (1984). *Equity and excellence: Compatible goals.* Washington, DC: American Association for the Advancement of Science.

Mathematical Sciences Education Board. (1990). *Reshaping school mathematics: A philosophy and framework for curriculum.* National Research Council. Washington, DC: National Academy Press.

McDonald, G.C., and Ayers, J.A. (1978). *Some applications of Chernoff faces.* In P.C.C. Wang (Ed.), *Graphical representation of multivariate data.* New York: Academic Press.

McKnight, C.C., Crosswhite, F.J., Dossey, J.A., Kifer, E., Swafford, J.O., Travers, K.J., and Cooney, T.J. (1987). *The underachieving curriculum: Assessing U.S. school mathematics from an international perspective.* Champaign, IL: Stipes Publishing Co.

Mestre, J. (1987). Why should mathematics and science teachers be interested in cognitive research findings? *Academic Connections,* The College Board, 3-5, 8-11.

Minstrell, J. (1982). Conceptual development research in the natural setting of a secondary school science classroom. In M.B. Rowe (Ed.), *Education for the 80's: Science*. Washington, DC: National Education Association.

Mislevy, R.J. (1992). *Foundations of a new test theory. In N. Frederiksen*, R.J. Mislevy, and I.I. Bejar (Eds.), *Test theory for a new generation of tests*. Hillsdale, NJ: Lawrence Erlbaum Associates.

Mosenthal, P., and Kirsch, I. (1989). Understanding documents. *Journal of Reading*. October 1989; pp. 58-60.

Mosenthal, P., and Kirsch, I. (1990). *Megatrends 2000: Ten new directions for the 1990s*. New York: William Morrow and Co.

National Assessment of Educational Progress. (1981). Mathematics objectives: 1981-82 assessment, No. 13-MA-10). Princeton, NJ: Educational Testing Service.

National Council of Teachers of Mathematics. (1989). *Curriculum and evaluation standards for school mathematics*. Reston, VA: NCTM.

National Research Council. (1990). *Renewing U.S. mathematics: A plan for the 1990s*. Executive Summary: Committee on the Mathematical Sciences: Status and Future Directions, Board on Mathematical Sciences, Commission on Physical Sciences, Mathematics, and Applications. Washington, DC: National Academy Press.

National Science Foundation. (1988). *Women and minorities in science and engineering*. Washington, DC: National Science Foundation.

National Science Teachers Association. (1990). *Essential Changes in Secondary School Science: Scope, sequence, and coordination*. Washington, DC: NSTA.

Pandey, T. (1990). Power items and the alignment of curriculum and assessment. In G. Kulm (Ed.), *Assessing higher-order thinking in mathematics*. Washington, DC: American Association for the Advancement of Science.

Piaget, J., and Beth, E. (1966). *Mathematical epistemology and psychology*. Dordrecht, Netherlands: D. Reidel.

Pollak, H. (1987). Notes from a talk given at the Mathematical Sciences Education Board. Frameworks Conference, May 1987, at Minneapolis, MN.

Popham, W.J. (1987). The merits of measurement driven instruction. *Phi Delta Kappan, 68*(9), 679-682.

Post, T. (1987). *Teaching mathematics in grades K-8: Research-based methods*. Boston: Allyn and Bacon.

Post, T., Behr, M., Lesh, R., and Harel, G. (1990). *Research and development in middle school mathematics*. (NSF Grant No. MDR-8955346). Washington, DC: National Science Foundation.

Publix Information. (1991, Summer). *Syllabus* (17). Sunnyvale, CA.

Resnick, D.P., and Resnick, L.B. (1991). Varieties of literacy. In A.E. Barnes and P.N. Stearnes (Eds.), *Social history and issues in human consciousness: Interdisciplinary connections*. New York: New York University Press.

Resnick, L.B. (1987a). *Education and learning to think*. Washington, DC: National Academy Press.

Resnick, L.B. (1987b). Learning in school and out. *Educational Researcher, 16*(9), 13-20.

Resnick, L.B. (1989). *Tests as standards of achievement in schools.* Learning Research and Development Center. Pittsburgh, PA: University of Pittsburgh.

Resnick, L.B., and Resnick, D.P. (1989). Assessing the thinking curriculum: New tools for educational reform. In B.R. Gifford and M.C. O'Connor (Eds.), *Future assessments: Changing views of aptitude, achievement and instruction.* Boston: Kluwer Academic Publishers.

Romberg, T.A., Zarinnia, E.A., and Williams, S. (1989). *The influence of mandated testing on mathematics instruction: Grade 8 teachers' perceptions.* Madison, WI: University of Wisconsin-Madison, National Center for Research in Mathematical Science Education.

Romberg, T.A., Zarinnia, E.S., and Collis, K.F. (1990). A new worldview of assessment of mathematics. In G. Kulm (Ed.), *Assessing higher order thinking in mathematics.* Washington, DC: American Association for the Advancement of Science.

Schank, R. (1991). *Tell me a story: A new look at real and artificial memory.* New York: Schribner.

Schoenfeld, A. (1985). *Mathematical problem solving.* New York: Academic Press.

Shavelson, R.J. (1989, June). *Performance assessment: Technical considerations.* Presentation at the seminar on Authentic Assessment, Berkeley, CA.

Shavelson, R.J. (1991). Can indicator systems improve the effectiveness of mathematics and science education? The case of the U.S. Evaluation and Research in Education. In C.T. Fitz-Gibbon (Ed.), *Evaluation and Research in Education.* Vol.4(2); pp. 51-50.

Shavelson, R.J., Carey, N.B., and Webb, N.M. (1990, May). Indicators of science achievement: Options for a powerful policy instrument. *Phi Delta Kappan 8,* 692-697.

Shavelson, R.J., Baxter, G.P., Pine, J., Yure, J., Goldman, S.R., and Smith, B. (1991). Alternative Technologies for Large Scale Instruments of Education Reform. *School Effectiveness & School Improvement and International Journal of Education Policy and Policies.* Vol. 4(2); pp. 51-60.

Steen, L.A. (1981). *Mathematics tomorrow.* New York: Springer-Verlag; pp. 73-82.

Steen, L.A. (1986). A time of transition: Mathematics for the middle grades. In R. Lodholz (Ed.), *A change in emphasis.* Parkway, MO: Parkway School District, 1986, 1-9.

Steen, L.A. (1987, July). Mathematics education: A predictor of scientific competitiveness. *Science, 237,* 251-252, 302.

Steen, L.A. (1988, April). The science of patterns. *Science, 240,* 611-616.

Steen, L.A. (Ed.). (1988a). *Calculus for a new century: A pump, not a filter.* Washington, DC: Mathematical Association of America.

Steen, L.A. (1990). *On the shoulders of giants: New approaches to numeracy.* National Research Council. Washington, DC: National Academy Press.

Steffe, L.P. (1988). Children's construction of number sequences and multiplying schemes. In J. Hiebert and M. Behr (Eds.), *Number concepts and operations in the middle grades*, 2, 119-140. Reston, VA: The National Council of Teachers of Mathematics.

Steffe, L.P., Cobb, P., and von Glasersfeld, E. (1988). *Construction of arithmetical meanings and strategies.* New York: Springer-Verlag.

Stenmark, J.K. (1989). *Assessment alternatives in mathematics: An overview of assessment of techniques that promote learning.* EQUALS staff of the Assessment committee of the California Mathematics Council Campaign for Mathematics. Berkeley, CA: Lawrence Hall of Science, University of California.

Tall, D. (1992). The transition to advanced mathematical thinking: Functions, limits, infinity, and proof. In D. Grouw (Ed.), *Handbook of research on mathematics teaching and learning.* New York: Macmillan.

Tatsuoka, K. (1990). *Boolean algebra applied to determination of universal set of knowledge states.* ETS Technical Report ONR-3. Princeton, NJ: Educational Testing Service.

Thompson, P. (1990). A theoretical model of quantity-based reasoning in arithmetic and algebra. Paper presented at the annual meeting of AERA, San Francisco, CA. Available from author, San Diego State University, Department of Mathematics.

Turner, N., and Rains, D. (1986). *Careers of mathematically talented students: A 27-year study of top-rankers in the 1958-1960 AHSME.* U.S. School Mathematics from an International Perspective. Champaign, IL: Stipes Publishing Co.

van Hiele, P.M. (1959, June). La pens'ee de l'enfant it la geometrie. Bulletin de l'Association des Professeurs Mathematiques de l'Ensignement Public 6.

Vergnaud, G. (1983). Multiplicative structures. In R. Lesh and M. Landau (Eds.), *Acquisition of mathematics concepts and processes.* Orlando, FL: Academic Press; pp. 127-174.

Vygotsky, L. (1978). *Mind in society: The development of the higher psychological processes.* Cambridge, MA: Harvard University Press.

Wagner, S., and Kieran, C. (Eds.). (1989). *Research issues in the learning and teaching of algebra.* Reston, VA: National Council of Teachers of Mathematics.

Wallace. M.A., and Wing, C.W. (1969). *The talented student.* New York: Holt, Rinehart and Winston.

Willingham, W.W., Lewis, C., Morgan, R., and Ramist, L. (1990). *Predicting college grades: An analysis of instructional trends over two decades.* Princeton, NJ: Educational Testing Service.

APPENDIX: SOME SIMPLE IMAGE PROCESSING TECHNIQUES

The following example uses fractal geometry to show some of the intermediate steps in the progressive transformation of one image to another (Bridger, 1988). For those who are uninitiated to the world of visualization, it provides a brief introduction to simple image processing techniques and a demonstration of the power of image processing to change our perspective on things.

The Mendelbrot Set is defined in terms of a sequence whose values for each complex number Q stay inside a circle of radius

R = 2. Color may be assigned to each point Q in the complex plane in such a way that Color (Q) = 0 if Q is in the Mendelbrot Set, or else a color is assigned to the first index N for which a term of the sequence ZN goes outside the circle. If the computer plots a grid of points in the complex plane, each of which is assigned a color, a high resolution monitor capable of displaying hundreds of colors, produces a picture of rare beauty. Since colors are not available here, Figure A-1 shows the Mendelbrot Set on a grid in the complex plane with the color black assigned to points in the set and white to all other points.

Figure A-1. The classical Mendelbrot set M (Bridger, 1988).

When colors are not available, another option is to graph the set in 3-space by letting the index N represent a height above the complex plane, instead of a color. The result is shown in Figure A-2.

Figure A-2. The Mendelbrot set in 3 dimensions (Bridger, 1988).

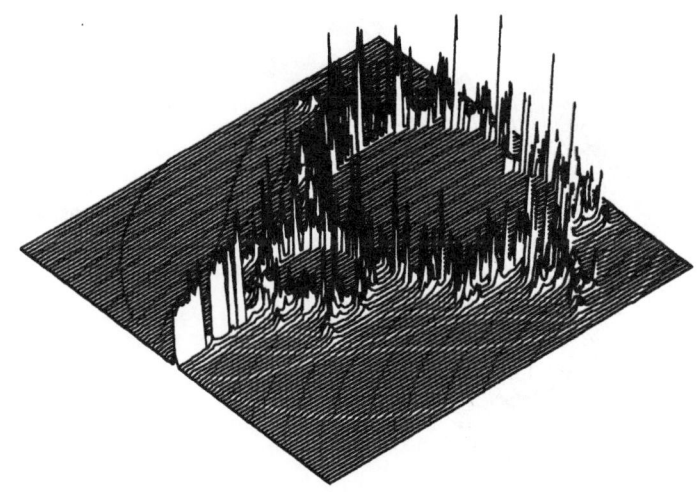

Next, a filtering technique may be used to smooth the spiky nature of the image. In this case, the weighted average of eight points near a given point (with the closer points having the greater weight), was used to determine the height at each point. This smoothing process may be applied as many times as desired. Figures A-3, A-4, and A-5 show the results of applying one smoothing, three smoothings, and six smoothings, respectively.

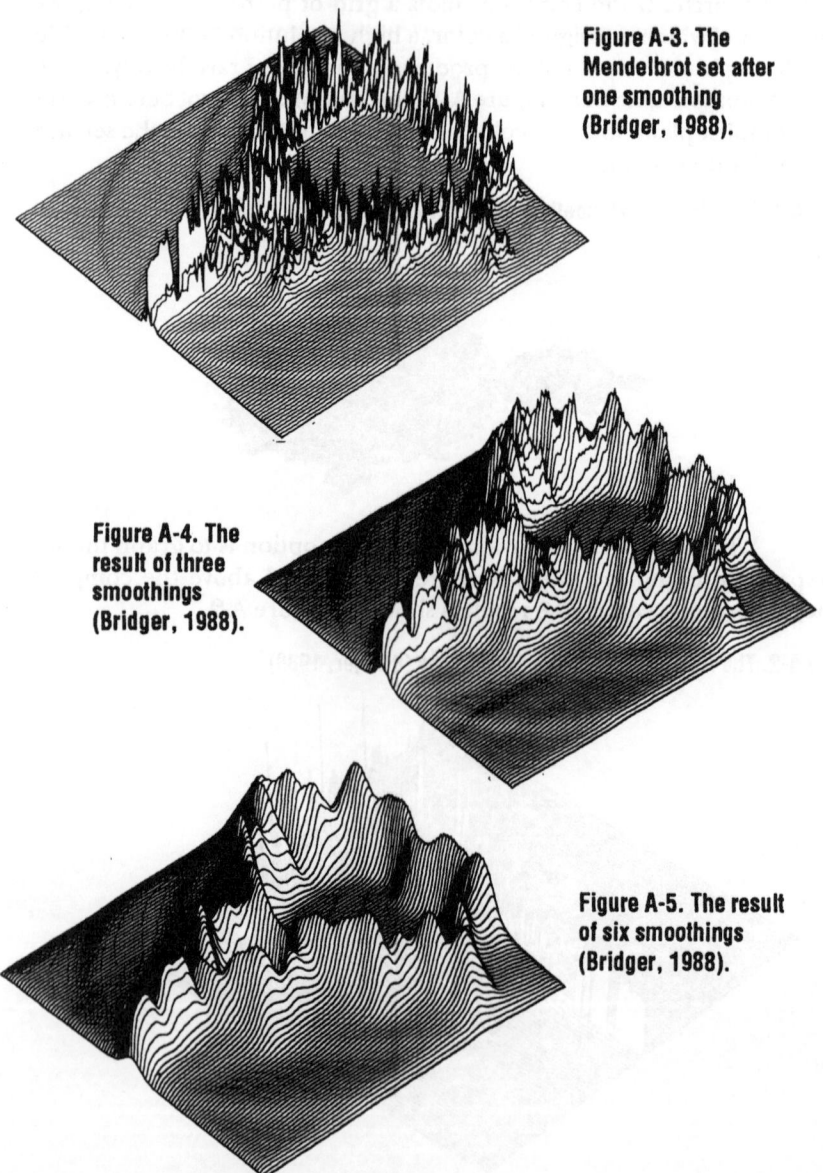

Figure A-3. The Mendelbrot set after one smoothing (Bridger, 1988).

Figure A-4. The result of three smoothings (Bridger, 1988).

Figure A-5. The result of six smoothings (Bridger, 1988).

Since averaging tends to blur the basic shape of the mountain, showing only its well-weathered main ridge, to recover more of its basic shape, an algorithm is applied to detect and accentuate ridges oriented in various directions. Figure A-6 shows the image after applying a ridge-finding algorithm in four directions.

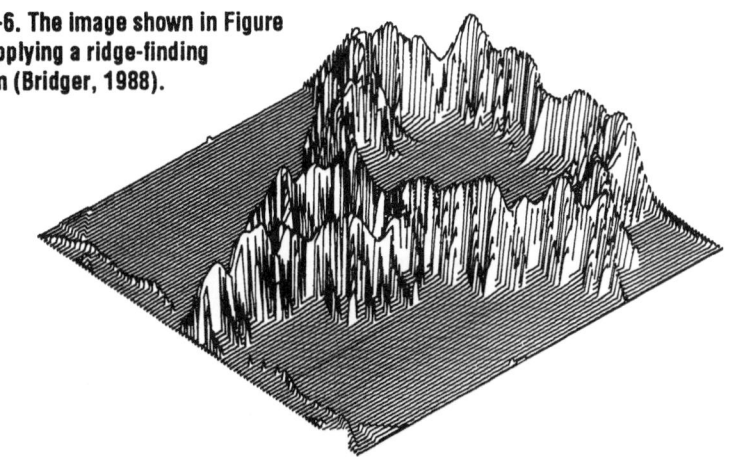

Figure A-6. The image shown in Figure 5 after applying a ridge-finding algorithm (Bridger, 1988).

In this example, a few simple image processing techniques dramatically changed the image of the original black-and-white set of points, and made available information that was not included in the original image. For example, we gained a sense about how long it takes to eliminate a point from the Mendelbrot Set. And today's visualization tools are far more powerful than this example shows!

They can be used to produce reports that are (i) graphics-based to clarify trends and other patterns beneath the surface of things, (ii) interactive to respond to on-site user inquiries, (iii) intelligent to filter, simplify, organize, and interpret information in a form that is especially relevant to local conditions and to particular decision-making issues, and (iv) multidimensional to integrate information from a variety of data sources, and to produce vector-valued interpretations (or n-dimensional graphics-based interpretations for situations that do not lend themselves to simple numeric quantification).

PART V: Difficulties, Opportunities, and Future Directions

15 Future Directions for Mathematics Assessment

*Richard Lesh, Susan J. Lamon,
Merlyn Behr, and Frank Lester*

INTRODUCTION

This chapter consists of three parts. The first part examines the assumptions underlying traditional types of standardized testing compared with assumptions underlying innovative types of performance assessment. The second part focuses on directions for the future; and, in particular, it focuses on three pervasive themes that shaped the perspectives of most of the chapters in this book, even though they are themes that were addressed only indirectly. These themes are *equity, technology,* and *teacher education.* The third part gives examples from three closely related projects which were designed to find practical ways to implement recommendations that were made in other chapters of this book. All three projects emphasize performance assessment activities that focus on: (i) deeper and higher-order understandings of elementary mathematics, (ii) realistic problem-solving situations, and (iii) diverse types of mathematical abilities.

A COMPARISON OF STANDARDIZED TESTING AND PERFORMANCE ASSESSMENTS

The table that follows summarizes some of the most important differences between traditional *standardized tests* and the kind of *performance assessment activities* that were emphasized throughout this book. For example, in general, traditional types of standardized tests have served the information needs of only a narrow range of decision makers and decision-making issues; and, they have tended to be based on exceedingly outdated conceptions of mathematics and mathematical problem solving. On the other hand, the alternative assessment movement has often focused on using tests as a leverage point for curriculum reform, and has tended to give relatively little attention to issues such as fairness and reliability in scoring, the usefulness and credibility of results (for decision makers who are not close to the students or instructional settings that are being assessed), and the scope and representativeness of the constructs that are measured (when attention shifts beyond the quality of isolated tasks to the quality of collections of tasks).

The main purpose of the table is not to point out shortcomings of the two views of assessment. Rather, the table is intended to clarify why the types of alternative assessment materials that have been emphasized in this book are not simply new ways to address old assessment goals. They reflect a complete *paradigm shift* that involves new decision makers, new decision-making issues, new sources of assessment information and new understandings about the nature of mathematics, mathematics instruction, and mathematics learning and problem solving. Consequently, it also involves new ways to think about traditional assessment issues such as reliability, validity, and generalizability.

The table is laid out in a two column form, where the left-hand column describes assumptions on which traditional forms of assessment tend to be based, and where the right-hand column describes corresponding assumptions for alternative forms of assessment. Such a two-column format is useful for clarifying key issues that distinguish contrasting "world views" about assessment. However, such black and white comparisons are also bound to be too simplistic. For example, the table is not intended to suggest that it is necessary for performance assessment programs to adopt *all* of the perspectives described in the right-hand column, nor that it is impossible for such programs to subscribe to *any* of the perspectives described in the left-hand column. The goal is simply to indicate why the transition from traditional to alternative forms of assessment tends to involve a complete paradigm shift and to describe some of the most important issues that distinguish the old and new world views.

To organize the table, five categories of fundamental issues are identified that serve to highlight the nature and extent of the paradigm shift that separates traditional forms of assessment from newer alternatives. These categories are: (i) underlying assumptions about the nature and purpose of assessment, learning, and teaching; (ii) issues associated with the process of assessment; (iii) the importance of alignment of assessment with curriculum emphases and instructional practices; (iv) perceived constraints on assessment practices; and (v) criteria for judging the quality of assessment instruments.

	Traditional Standardized Tests	**Performance Assessments**
UNDERLYING ASSUMPTIONS		
Assumptions about the Nature of Mathematics See chapters 2,3,4,6,8	Mathematical knowledge is characterized as (nothing more than) a list of mechanistic condition-action rules.	According to the Mathematical Sciences Education Board (1990): *Two outdated assumptions are that: (i) mathematics is a fixed and unchanging body of facts and procedures, and (ii) to do mathematics is to calculate answers to set problems using a specific catalogue of rehearsed techniques.* (p. 4). ... *As biology is a science of living organisms and physics is a science of matter and energy, so mathematics is a science of patterns. ... Facts, formulas, and information have value only to the extent that they support effective mathematical activity* (p. 12)
	The most important cognitive objectives of mathematics instruction are models (or patterns, or structures) which can be explored for their own sakes, or which can be used to construct, describe, explain, predict, manipulate, or control structurally interesting systems in real or possible worlds–with refinements, extensions, and adaptations being made to these models when necessary.	
Assumptions about the Nature of Mathematical Problem Solving See chapters 2,3,4,6,8	Problem solving is characterized as getting from givens to goals when the path is not obvious. It is assumed to involve (nothing more than) answering questions which are posed by others, within situations that are described by others, to get from givens to goals which are specified by others, using strings of facts and rules which generally need to be restricted in ways that are often artificial and unrealistic ... so that solutions and solution paths will be rejected if they fail to match the one expected by the authority figure.	Realistic problem solving situations often involve constructing useful ways to "think about" (e.g. describe, explain, manipulate, predict) patterns and regularities governing the behavior of structurally interesting systems. Therefore, solution processes often involve several modeling cycles in which givens and goals are interpreted in a variety of alternative ways. ... The mathematical results which are produced are not restricted to simple counts and measures; they often involve mathematical "objects" such as graphs, equations, coordinate systems, and other types of mathematical models.

A *mystery* is a phenomenon that people don't know how to think about yet. In real life situations in which mathematics is useful, many "problems" are more like mysteries than they are like traditional types of school mathematics problems. For example, consider the case of a mountain climber who wants to scale a cliff. The essence of the problem is to find a productive way to think about the terrain; that is, once the terrain is understood, the task of getting from the bottom of the cliff to the top is simply a strenuous exercise, not a problem. Similar situations occur when business people use spreadsheets to create productive ways to describe cost/benefit trends, or when youngsters use pocket calculators to combine information from situations in which quantitative values are assigned qualitative judgements. Consequently, the kind of problem solving experiences that are emphasized tend to involve the construction of new ideas, rather than simply the ability to use old ideas.

Assumptions about the Nature of Mathematics Ability See chapters 2,3,4,6,8	Experts are considered to be people who are good at remembering facts and following rules specified by others, and who are generally clever about finding ways to assemble these facts and rules to get from givens to goals within situations described by others.	Students (and teachers) are considered to be decision-makers, situation conceptualizers, system builders, problem formulators, and rule generators. Experts are people who have constructed a rich and varied collection of powerful models for making sense of complex systems.
Assumptions about the Nature of Mathematics Learning See chapters 2,3	Humans are characterized as information processors; and, learning is viewed as a cumulative process of gradually adding, deleting, and de-bugging mechanistic condition-action rules (definitions, facts, rules, or skills).	Humans are viewed as model builders, theory builders, and system builders. The models they construct are used to describe, explain, create, modify, adapt, predict, and control complex systems (in real or possible worlds). In other words, the emphasis is on structuring experience at least as much as on processing information, and the models that are constructed develop along dimensions such as concrete-to-abstract, intuitive-to-analytic-to-axiomatic, particular-to-general, global or undifferentiated-to-refined, fragmented-to-integrated, enactive-to-symbolic, and situated-to-decontextualized.
	In general, model construction processes involve the extension, refinement integration, and differentiation of existing models; deeper and higher-order understandings of these models develop as students go beyond thinking with the model to also think about it, or as the model is gradually embedded in more powerful and economical notation systems, or as students go beyond automatic responses to also think about thinking. Consequently, cognitive development often involves discontinuities and major conceptual reorganizations; and, some of the most useful activities to facilitate and document development are the kind of project-sized activities that have been emphasized throughout this book.	

Assumptions about the Nature of Mathematics Teaching See chapters 2,3	Instruction consists mainly of: (i) demonstrating relevant facts, rules, skills, and processes, (ii) monitoring activities in which students repeat and practice the preceding items, and (iii) correcting errors that occur.	For both instruction and assessment, some of the most useful experiences are problem solving activities that are similar to the Case Studies which have been used for years in professional schools in fields such as business and engineering (where there is a long history recognizing that many of the most important goals of instruction should focus on models for constructing and exploring complex systems). In such activities, students not only solve complex and realistic problems, their solutions also involve the creation of conceptual models which can be used to interpret similar problems in other situations.
Assumptions about the Nature of Individual Differences See chapters 3,4	Students are branded with labels such as *impulsive* (versus *reflective*), or *field dependent* (versus *field independent*) –that are supposed to be: (i) invariant across all (or most) topics and contexts, and (ii) difficult or impossible to change through hard work, instruction, experience, or development. Sometimes, these labels involve a continuum (e.g., in the case of IQ tests); but, more often, the characteristics involve only a few values, one with negative connotations, and one with positive connotations. The assumption is that teachers should use some sort of APTITUDExTREATMENT approach to instruction– where type #N students are matched with type #N instruction.	Research in mathematics education has furnished overwhelming evidence that there are many alternative types of mathematical talents and styles. Further, students who are most successful tend to be flexible at modifying their styles and approaches to different types of problems and alternative stages in the problem solving process. Also, many different kinds of personalities, knowledge, and capabilities can lead to success (e.g., Krutetski, 1976); and, many different types of success are possible (Begle, 1979). Both students and professionals exhibit irregular profiles of expertise, with strengths in some areas and weaknesses in others. In general, in classroom settings, instruction which has proven to be best for individual students is the same type that is best for all students; it is instruction that encourages conceptual and procedural flexibility, and that gives students access to a variety of approaches.

ALIGNMENT WITH CURRICULUM STANDARDS		
Assumptions about the Impact of Assessment on Instruction See chapters 1,7,8,14,15	Measurement instruments are treated as if they were neutral indicators (similar to unobtrusive thermometers) so that the act of measurement does nothing to induce changes in the system that is measured. Yet, in mathematics, it is a well documented fact that "high stakes" standardized tests tend to have strong (and often negative) influences on both <u>what</u> is taught and <u>how</u> it is taught. In fact, too often, curriculum and instructional practices have been driven by standardized testing, rather than the other way around.	Assessment activities are selected mainly because they are important from the point of view of instruction; and, care is taken to ensure that the testing program helps to influence the education system in positive ways by ensuring that the goals being measured are aligned with priority goals of instruction. The constructs being assessed must go beyond simply being correlated with desirable outcomes. The performances that are assessed should reflect appropriate scope, breadth, and depth of understanding and abilities, without emphasizing constructs that are irrelevant to (or conflicting with) the important understandings and capabilities.
Assumptions about the Instructional Value of Assessment Activities See chapters 2,6,9	In general, tasks used in traditional forms of assessment are not intended to have positive instructional value. In fact, it is often considered to be desirable for scores to remain invariant even if equivalent forms of the test are repeated again and again. Also, because severe and unrealistic constraints tend to be placed on time, tools, and other resources, the tasks are only surrogates for those that are really considered to be important. Consequently, because such tasks generally have characteristics that their real counterparts do not, they often reinforce misleading beliefs about: (i) the nature of mathematics, (ii) the ways mathematics is used in an *age of information*, and (iii) the kinds of abilities that are productive.	To document learning progress, it is not necessary to interrupt students from their most important instructional activities. The goal is to focus on activities that contribute to both learning and assessment. That is, students should be able to simultaneously <u>learn</u> and <u>document</u> what they are learning, while solving problems which might reasonably occur in "real life" situations.

Assumptions about Alignment with National Standards See chapters 1,2	Until 1989, there was no national consensus about what should be taught in the school mathematics curriculum. Curriculum guidelines varied from state to state, and from textbook to textbook. Therefore, tests which were focused on national student populations generally had to adopt the following strategies: (i) focus on the "basics" which are common to all mathematics curriculum materials, and/or (ii) focus on general understandings or capabilities which are independent of any specific curriculum materials. The result was a practice in which not much attention was paid to what was actually being tested (as long as the tasks seemed to be generally mathematical, and as long as they passed sensitivity reviews, statistical reviews, and other reviews related to nonsubstantive issues). This situation persists today in most subject matter areas except mathematics. In mathematics education the situation has changed dramatically. In 1989, the relevant professional and governmental organizations all reached a formal consensus about national *Curriculum and Evaluation Standards for School Mathematics* (NCTM, 1989).	
	In a series of studies comparing the alignment of existing standardized tests with endorsed *Curriculum and Evaluation Standards for School Mathematics*, Romberg, Wilson, and Khaketla (1991) formed the following conclusion. *These tests are based on different views of what knowing and learning mathematics means. ... These tests are not appropriate instruments for assessing the content, process, and levels of thinking called for in the STANDARDS.* (p. 3)	A primary goal is to develop assessment activities that are closely aligned with the NCTM's *Curriculum and Evaluation Standards*, emphasizing • mathematical structure, • mathematics as communication, • mathematics as connections, • mathematics as reasoning, and • mathematics as problem solving (with special attention to number sense and estimation)
Assumptions about Instructional Objectives that should be Emphasized See chapters 2,3,4,6,9	In the case of aptitude tests, a single construct (g) is measured which is supposed to be deep, significant, and difficult to modify through instruction. Yet, in alignment studies such as those conducted by Romberg and his colleagues (Romberg, Wilson, and Khaketla, 1991), when the content of nationally significant standardized tests have been compared with nationally endorsed *Curriculum & Evaluation Standards for School Mathematics*, results have shown that:	Both conceptual and procedural knowledge are addressed, and special attention is given to (i) <u>cognitive objectives</u> (which emphasize the models that students must use to describe, explain, construct, refine, manipulate, predict, and control complex systems) and, (ii) <u>deeper and higher-order understandings</u> of such cognitive models, as specified in the new national consensus about *Curriculum and Evaluation Standards for School Mathematics*.

	(i) most items focus on procedural knowledge rather than on conceptual knowledge. (ii) the procedural knowledge tends to focus mainly on low-level behavioral objectives or global process objectives which are not connected to any particular mathematical knowledge, (iii) the topic areas that are emphasized then to be ancient topic areas such as Aristotelian Logic, Egyptian Number Theory, and Euclidean Geometry, (iv) little attention is given to new types of understandings and processes that have become important because of the prevalence in our society of technological tools, and (v) the "applied" problems tend to be only "pure math" problems with the names of everyday objects substituted for abstract symbols.	Special attention is also given to the kind of problem-solving and decision-making situations that occur in "real life" situations. A goal is to encourage students to interpret the situations mathematically without having to "turn off" their everyday knowledge and experience. Another goal is to recognize that technological tools are not just new ways to perform old skills; their ubiquitous presence in our society has resulted in radical changes concerning the mathematical knowledge and abilities needed by most people.
THE ASSESSMENT PROCESS		
Assumptions about Priority Decisions and Decision Makers See chapters 12, 14	Administrators are treated as the most important decision makers; and, the decisions which are considered to have the highest priority usually involve (i) program accountability, or (ii) accept or reject decisions about programs, teachers, or students.	The most important decision makers are students, teachers, parents, and others who are interested in optimizng students' achievements.

The NCTM's *Curriculum and Evaluation Standards* are clear about the following points.

- Because teacher decision-making should be a priority, it does not follow that the information needs of other decision-makers should be ignored.

- Because *performance assessment items* and *students' project portfolios* should be given greater attention, it does not follow that other sources of information should be ignored.

No single decision-maker or decision-making issue should preclude all others, and no single source of information is appropriate for all purposes and audiences. One of the worst things that could happen to promising new forms of assessment is that they (too) become used for issues and audiences that they were not intended to address.

Assumptions about Sources of Assessment Information See chapters 1,2,5,10,11	Students must prove their knowledge and abilities within the context of a small number of brief "high pressure, low interest" testing situations in which severe and unrealistic constraints must be placed on time, tools, and other resources. Also, when responses are restricted to those that have simple "right-wrong" answers, all tasks must be eliminated in which the goals involve descriptions, explanations, predictions, or other results in which several alternative levels and types of "correct" responses are nearly always possible.	Many "low pressure, high interest" opportunities are provided in which students can simultaneously develop and document their increasing knowledge and abilities. Assessment-relevant information is taken from a variety of contexts which include not only tests but also teachers' clinical interviews and classroom observations, and students' extended projects and project portfolios.
Assumptions about the Purpose of Assessment See chapters 1,7,8,14,15	Descriptions of standardized tests often suggest that they were intended to be used primarily to generate positive predictions about the probability of future success for students (or teachers, or programs). In fact, however, "high stakes" tests tend to be used to generate: (i) absolute cut-off scores which are used to screen students (so that some are never allowed to try the predicted activities), (ii) labels which claim to reflect static traits which are difficult or impossible to change through hard work, instruction, experience, or development, and (iii) operational definitions for <u>what</u> should be taught and <u>how</u> it should be taught.	The goals are: (i) to describe past accomplishments rather than to predict future accomplishments, and (ii) to facilitate learning progress. Therefore, it is important to identify strengths and needs of individual students. Also, because the goal is to <u>document progress</u>, the characteristics that are emphasized are assumed to develop and change over time. Therefore, it is important to monitor performance over a long period so that trends become apparent. In mathematics, just as in most fields, the best predictor of future success depends on being interested and productive over a long period of time (Bloom, 1985; Krutetski, 1976).
	Whether or not a .5 correlation is "good" depends on: (i) the type of decisions the measure is actually used to inform, (ii) the extent to which the fraction of the variance that is not taken into account corresponds to random error, or to important characteristics that are simply ignored. If we were predicting winners at horse races, a .5 correlation might be viewed as very good. However, if we were deciding which horses to send to the glue factory, it might not be good at all (especially from the perspective of the horses). Is it good to be accounting for 50% of the desirable characteristics? Or is it bad to be ignoring 50% of the desirable characteristics? Are the characteristics that are being measured having unproductive influences on breeding and training practices?	

Assumptions about Locus of Responsibility See chapters 1,2,9	Testing instruments assume responsibility for documenting whether or not students (or teachers, or programs) are competent or to what extent they have achieved specified goals.	Students (and teachers, or programs) must assume greater responsibility to produce credible documentation about their own abilities and achievements. Furthermore, schools (programs and districts) must also help to address credibility, fairness, and other issues which cannot be handled adequately by individuals who are being assessed.
	When individuals assume greater responsibility for documenting their own abilities and achievements, the situation resembles the way things work in universities, businesses, or industries. For example, in such situations, individuals usually develop resumes which provide brief overviews of achievements, and the credibility of entries depends on a variety factors. Entries similar to publications are evaluated by referees from relevant professional organizations. For entries that are similar to courses completed (or programs completed, or projects completed), the quality of results is certified by relevant institutions, or "blue ribbon" references are cited who can describe the quality of entries. In other cases, a portfolio of actual products is available for inspection. In all of these cases, the individual assumes a large part of the responsibility for documenting achievements and abilities, even though various institutions, organizations, or individuals help to verify the quality of items that are listed.	
Assumptions about Assessment Information that should be Generated See chapters 12,14	Single-number quality ratings are assigned to students (or teachers, or programs). The emphasis is on comparing students to students along a one-dimensional "good-bad" scale.	Multidimensional profiles are generated (for students, or teachers, or programs). These profiles are aimed at high-lighting: (i) strengths and weaknesses of individual students (or groups), (ii) progress that has been made over specified periods of time, and (iii) conditions under which achievements occur.
	Rather than comparing students to students based on (for example) one-dimensional conceptions of *general intelligence*, mathematics educators tend to be more concerned about comparing students' capabilities with priority instructional objectives. Whereas standardized tests have emphasized the power of *general intelligence*, mathematicians and scientists tend to be more impressed with the power of having a pocket full of "capability amplifiers" which: (i) are based on elementary-but-deep mathematical models (e.g., involving rational numbers, signed numbers, vectors, coordinates, functions, graphs, etc), (ii) it took our society centuries to invent, but which can be learned by average ability middle schoolers, (iii) are designed help	

students generate insightful interpretations, explanations, and predictions about sophisticated problem solving situations, (iv) enable students' profiles of capabilities to vary dramatically, depending on which models they have and have not constructed, and (iv) which would enable average ability middle schoolers to perform like "geniuses" if they could somehow be transported back to a time before these conceptual tools had been invented.

Assumptions about Responsibility for Assigning Quality Ratings See chapters 3,7,9,10,11, 13,14	Value judgements are assigned to students by testing programs based on very small samples of performance. These value judgements often convey the impression that "good" individuals have no weaknesses, and that "weak" individuals have no strengths. Or, if profiles of strengths and weaknesses are recognized, the "weights" assigned to various components are assumed to always be the same regardless of the purpose of the evaluation. Thus, the aggregated quality rating typically turns out to be the same regardless of factors that influence most decision-making endeavors.	The goal of the assessment program is to generate simple and yet high-fidelity and multi-dimensional "portraits" to describe students, teachers, or programs. The purpose of these descriptions is to facilitate decision-making by informed professionals. Therefore, reports should avoid unnecessary value judgements which artificially and needlessly limit decision-making responsibilities by relevant professionals. If sound and reliable information is provided, the acts of judging, valuing, comparing, and ranking should be reserved for qualified people who are closest to the students, or closest to the decision-making issues that need to be confronted.
Assumptions about Interpretions of Students' Responses. See chapters 2,3,4,10,12,13	Responses to individual items are typically machine scorable, and are assigned simple right-wrong evaluations. In general, scores for pools of problems are assigned by simply calculating sums of scores on individual items.	Students' cognitive models cannot simply be sorted into categories labeled mastered or not mastered. Levels of development are significant; and, these levels tend to be reflected in patterns of responses, not simply in aggregated counts of correct and incorrect answers.
	The difficulty of a problem depends on how it is interpreted by the student, not just on objective task variables. Therefore, a problem does not have a single difficulty level that is independent of the way students interpret it. And, this fact is especially applicable to performance activities in which appropriate responses involve descriptions, constructions, explanations, or predictions, where there is always more than a single level or type of "correct" response. Yet, in spite of the fact that several levels and types of responses are acceptable, objective criteria are usually available so that students themselves can judge the quality of their own work. There should be no need for students to depend exclusively on the judgement of external authorities.	

Assumptions about Accountability: Documenting Achievement over Time See chapters 7, 14	To document achievement over time, pretest/posttest designs tend to be used where the test measures <u>states</u> of knowledge at the beginning and end of instruction, and deficiencies are compared with some ideal model. Then, pretest scores are subtracted from posttest scores to infer that improvement has occurred. One of the dangers associated with this approach is that, even when the primary goal is to prove that progress has been made, the pretests and posttests that are used often help to create conditions under which the possibility of success is minimized.	Activities can be used that encourage development in directions that are increasingly "better" without using pretests and posttests as operational definitions of "best." Furthermore, these same activities can also be used to produce documentation that development has occurred by producing a trace of progress that has been made. In this way, it is possible to focus directly on the <u>process</u> of change without simply making comparisons involving <u>static states</u>; and, it is possible to document progress without taking students and teachers away from valuable instructional activities.
Assumptions about the Role of Teachers See chapters 3, 10, 11, 14	Often, in the past, the goal was to produce "teacher proof" tests, textbooks, and programs of instruction.	Input from teachers is essential because it is important to interpret the meaning of students' responses and assessment results, and because it is important for teachers to contribute relevant information about conditions in which students' achievements occurred.
PERCEIVED CONSTRAINTS ON ASSESSMENT		
Assumptions about Constraints involving Cost-Effectiveness See chapters 1, 5, 8, 12, 14	Little attention tends to be given to the benefit side of cost/benefit ratios, even though single-number scores and letter grades provide very little information to inform most decisions and decision-makers. The main goals are to minimize costs and time commitments because (i) the tasks are expensive to develop and calibrate, and (ii) the tests are non-beneficial intrusions on instructional time (for students, teachers, and instructional programs). Negative effects on instruction are often simply ignored.	Attention is focused on increasing effectiveness and on increasing the richness and quality of the information that is generated. Because the assessment activities that are emphasized are those that are important from the point of view of both assessment and instruction, they are not intrusions on instruction, and they can occur often. Also, the resources that are available for instruction can be put in the service of assessment, and vice versa.

Assumptions about Time Constraints See chapters 2,5,8,9	Questions must typically be answered at a rate of approximately one per minute. Therefore, categories of problems that are "out of bounds" include all of those in which 60 second solutions would be inappropriate.	Response times vary, as in real situations, but many solutions require at least an hour to construct, with constraints on time, tools, and other resources being only those that might occur in "real life" situations.
Assumptions about Constraints on Tools & Resources See chapters 2,5,8,9	In general, students are prohibited from using tools, including the kind of pocket calculators or notebook computers that are common in "real life" situations. In fact, even if tools such as calculators are allowed, efforts tend to be made to avoid tasks that are not tool neutral.	In general, realistic tools and resources are available, including "how to" manuals, colleagues, and consultants.
Assumptions about Constraints involving the Simplicity of Reports See chapters 1,7,14	It is assumed that the only way for (paper-based) reports to be sufficiently simple is to reduce all available information to single-number "scores." And, in spite of the fact that decision-makers consistently misuse these scores for pur--poses that they were never intended to address, efforts to avoid such misuses tend to be even less conscientious that the kind of labels on cigarette cartons, which say "cigarettes (tests) can be hazardous to your health (education system)."	To highlight patterns of understandings, abilities, and achievements, information is simplified using computer-based reports which are dynamic, graphic, interactive, and easily modified to focus on a variety of alternative decision-making issues. Therefore, complexities are only highlighted when they are requested for a specific purpose.
CRITERIA FOR JUDGING THE QUALITY OF ASSESSMENT INSTRUMENTS		
Assumptions about Fairness See chapters 2,4,6,8,9	Decontextualized "vanilla" problems are treated as though they didn't favor privileged students who are schooled in the art of answering such questions (where both the questions and the answers would often be nonsense in the "real life" experiences of less privileged students). In general, no attention is given to the conditions under which students develop the knowledge and abilities presumably being evaluated.	Special attention is given to precisely those tasks in which targeted groups of students have special interests and experience. Yet, students are not forced to demonstrate their knowledge and capabilities within situations in which they have no interests or experience because a variety of options are available for every major construct that is assessed. Also, care is taken to describe relevant conditions that influence achievement opportunities.

	note: In principle, using carefully decontextualized problems means that all activities should be eliminated if some subgroup of students might consider them to be especially meaningful or interesting (because special technical or conceptual tools might have been developed to deal with them).	**note:** Because decontextualized questions must be answered without knowing the purpose of the question, there are often no cues to help a student make reasonable judgements about appropriate levels of precision, specificity, or other issues that govern the quality of responses to "real life" problems.
Assumptions about Rationale for Selecting and Equating of Tasks See chapters 2,3,6,8,12,13	In aptitude tests, all tasks are interpreted as measuring the same construct ("g"). Therefore, tasks are selected, rejected, and equated mainly on the basis of statistical properties. In traditional types of criterion referenced achievement tests, each item focuses on an isolated behavioral objective; the assumption is generally made that complex behaviors consist of chains of basic facts and skills.	Tasks are selected to reflect a representative sample of problem types, knowledge, and abilities which "cover" the field being assessed. Problems no longer have a single level of difficulty or a single level and type of "correct" response. For example, the difficulty of a task depends on the way individuals interpret the problem, and on the student's prior experiences. Therefore, two tasks are only equivalent if they are interpreted in the same way and if they elicit the same level and type of response.
Assumptions about Reliability and Response Interpretation See chapters 3,12,13,14	A score on an test (or item) is reliable if: (i) students would get the same score if they did the item again, and (ii) experts consistently give the same score. The interpretation of a response only depends on task variables and on whether the student got the correct answer.	Because the tasks (and tests) that are emphasized are those that contribute to learning as well as to assessment, students who do them repeatedly would be expected to improve. Also, because complex performances are involved, it is expected that a given expert might assign different quality ratings to a given performance, e.g., depending on the purpose of the evaluation, or depending on the weights that are assigned to various attributes or sub-components of performance.
	Suppose the goal were to assess carpenters, and the tasks involved building houses (of a variety of types in a variety of specific settings). It would be sensible to assume that quality ratings would improve if the activity were repeated multiple times; it would be sensible to assume that quality ratings would vary due to a variety of factors. For example: Was the house energy-efficient but ugly? Was it innovative but too costly? Was it sturdy but out of place in the given location? Various "experts" might not agree. Value judgements might be conditional, depending on the weights that are assigned to various attributes, perspectives, and purposes.	

Assumptions about Justifications for Generalizations See chapters 2,3,6,8,12,13	The only information that is taken into account is based on a small number of brief tests; these generally involve a restricted and unrepresentative class of problems, knowledge, and abilities. Sophisticated statistical techniques tend to be used to make inferences based on underlying subtle patterns of behavior in the restricted situations included in the tests. Attention tends to shift toward hypotheses about students themselves (with the assumption that these hypotheses apply to a more representative class of situations).	First, the tasks are not just correlated with important real life tasks, the goal is for them to actually include a representative sample of such tasks. Second, like spreadsheets which provide frameworks for dealing with data for a variety of situations, the responses that students generate are not just specific answers to isolated questions, they are explanations, descriptions, and predictions that generate useful information in a whole class of specific situations. Third, the documentation of students abilities is gathered over a long period of time and in a variety of situations. Therefore, generalizations that are stated about a given student are summaries about actual past performance. Conjectures about future performance are left to decisionmakers who are close to the students that are involved and who are close to the situations in which the hypothesized capabilities should be relevant.
Assumptions about Criteria for Judging Validity See chapters 1,2,7,12,13,15	In the past, nationally influential standardized tests had to reflect the lack of consensus about what should be taught and tested in the mathematics curriculum. Therefore, rather than focus on instructional goals that would vary from one curriculum to another, standardized testing programs tended to emphasize general abilities and aptitudes which were assumed to be curriculum-independent and/or common to all relevant programs and materials. Validity was interpreted as *predictive validity*. That is, the goal was for the test to be correlated with: (i) <u>past tests</u> (which tend to be obsolete), or (ii) <u>global performances</u> in introductory courses (which tend to be the most in need of curriculum reform, and the least aligned with current national standards).	A national consensus has been reached about *Curriculum and Evaluation Standards for School Mathematics*. Therefore, "validity" is interpreted in terms of *content validity* and *systemic validity*; and, *content validity* is measured in terms of alignment with national *Standards*. (i) Does the test emphasize constructs that are priorities for instruction? (ii) Does the test avoid emphasizing constructs that are narrow, shallow, untypical, obsolete, and/or counterproductive in terms of instruction? *Systemic validity* refers to the fact that assessment instruments should help to promote (and at least not subvert) positive changes in the students, teachers, and programs whose performances they describe.

Using tests as leverage to clarify objectives and induce curriculum changes focuses on <u>one</u> set of decision-making issues that are priorities to address, but there are also <u>other</u> issues that cannot be neglected. For example, suppose it were possible to instantly change all "high stakes" screening and accountability-oriented tests to conform to the kinds of future-oriented goals cited in recent publications from the NCTM, MSEB, AAAS, and NSTA. How could we ensure fair treatment for students who were indoctrinated with more than twelve years of an old system that emphasized obsolete views of mathematics, problem solving, teaching, and learning?

Whether attention is focused on standardized tests or on alternative sources of assessment information, responsible people in both areas recognize that it is not acceptable for assessment programs to use students as pawns to induce curriculum changes, especially if the students most likely to suffer are those who have been treated unjustly by the system we are trying to change. Still, questions remain. What should be the relative emphasis on revolution versus evolution? How much should assessment be permitted to lead or lag curriculum and instruction?

Assumptions about Secrecy for Problem Banks See chapter 15	Because high stakes consequences are associated with performances on small items and pools of items, and because item calibration is expensive, security tends to be carefully guarded. Consequently, the only descriptions of tasks that are available tend to be global content-by-process matrices and examples of released items from past tests.	The activities that are emphasized tend to involve complex constructed responses and these constructions tend to focus on the cognitive objectives (or models) that are the most important goals of instruction. Therefore, performing the task is equivalent to demonstrating the desired competence. So it is possible to be extremely straightforward and open about the nature of assessment tasks.

In discussions between proponents of traditional testing and their counterparts in performance assessment, it tends to become clear that these two perspectives are based on completely different sets of assumptions. Therefore, it is often difficult to initiate a dialogue because, within each of the two columns of the table, the cells are all closely interconnected and need to be considered together as a whole. For example, in the table at the end of this chapter, if a particular cell is considered in either the left or right column, it is usually difficult to isolate the issues addressed in this cell from issues addressed in other cells within the same column. That is, if a person insists on clinging to the points of view expressed in most of the other cells in the same column, then it often doesn't make sense to abandon the principles expressed in the remaining cell. Piecemeal transitions often do not make sense.

When mathematicians and mathematics educators call for assessment instruments to focus on *authentic mathematics, realistic problems,* and *genuine mathematical abilities,* they are not simply demanding that tests should focus on new types of items and testing formats. New conceptions need to be adopted about the nature of mathematics, problem solving, teaching, and learning. For example, concerning the psychological foundations of traditional test theory, Mislevy has stated:

> *The test theory that dominates educational measurement today might be described as the application of twentieth century statistics to nineteenth century psychology.* (Mislevy, 1991, p. 234).

> *The essential problem is that the view of human learning that underlies standard test theory is not compatible with the view rapidly emerging from cognitive and educational psychology.* (Mislevy, in Introduction, Chapter 12, above).

Similar views have been expressed in virtually every recent report from organizations such as the National Council of Teachers of Mathematics, the Mathematics Association of America, the American Association for the Advancement of Science, the National Academy of Sciences, and the Mathematical Sciences Education Board. There is overwhelming agreement that the kind of standardized testing programs that exist today generally emphasize "industrial age" views that are inconsistent with the needs and realities of an *age of information,* and that they emphasize views that are counterproductive to badly needed curriculum reforms in mathematics education (Romberg, Zarinnia, & Williams, 1989). For example:

- The National Research Council writes: *The most important components of mathematical talent cannot be addressed: (i) using timed tests with large numbers of small decontextualized questions, or (ii) when artificial restrictions are placed on the resources that are available. ... Most of the tests used for mathematics assessment have too narrow a focus. They do not measure the wide range of mathematical skills and abilities that educators and business leaders believe is needed for a population to live and work in a world increasingly shaped by mathematics, science, and technology.* (NRC, 1990. p. 21).

- Leading mathematicians and mathematics educators write: *Current tests ... force students to answer artificial questions under artificial circumstances; they impose severe and artificial time constraints; they encourage the false view that mathematics can be separated out into tiny water-tight compartments; they teach the perverted doctrine that mathematical problems have a single right answer and that all other answers are equally wrong; and, they fail completely to take account of mathematical process, concentrating exclusively on the "answer."* (Romberg, Zarinnia, & Collis, 1990, p. 23).

- Leaders in cognitive science write: *The educational system we have inherited was not, by and large, designed to prepare people for adaptive functioning in a technically complex environment. ... Its goals for students did not include the ability to interpret unfamiliar texts, construct convincing arguments, understand complex systems, develop approaches to problems, or negotiate problems resolutions in a group.* (Resnick and Resnick, 1989, p. 37).

According to the preceding perspectives, the standardized testing industry is badly out of step with the past quarter of a century of research in cognitive science; the constructs it measures are inconsistent with modern advances in mathematics and technology; and, the problems and abilities that are emphasized are completely inadequate to reflect the needs of citizens and workers in an *age of information*.

- Romberg, Zarinnia, and Collis (1989) have singled out the following two assumptions that are particularly important sources of outdated beliefs. (i) *Reductionism* assumes that, if you have something you want to explain, you take it apart until you identify its simplest parts. In this way, mathematics is partitioned into fragmented lists of behavioral objectives that are each treated as an end in itself. (ii) *Mechanism* assumes that once you break something apart, you build it up again based on simple chains of condition-action rules. That is, mathematics is considered to be nothing more than a list of factual and procedural rules, and systems of mathematical knowledge are considered to be nothing more than the sum of their parts, with the meaning of isolated rules being the same regardless of conceptual structures in which they are embedded.

- Resnick and Tucker (1991) have used similar language to describe some of the most important sources of outdated beliefs. (i) *Decomposability* likens thought to a simple machine; first, learn isolated facts and skills in isolation, then simply link them together to learn higher-order concepts and principles or to solve more complex problems. (ii) *Decontextualization* assumes that teaching (or testing) a skill out of context is the same as teaching (or testing) it within a realistic and meaningful situation; it assumes that each component of a complex skill is fixed and that it will take the same form regardless of the context in which it is used.

According to the preceding points of view, the shortcomings of current standardized testing are not superficial; they are deep; and, they will not be ameliorated using simplistic techniques such as converting multiple-choice questions to their fill-in-the-blank counterparts. In fact, when such strategies are used, the tests that result often become worse rather than better in terms of their alignment with nationally endorsed *Curriculum and Evaluation Standards for School Mathematics* (NCTM, 1989). Why? The central reason is that, to improve the alignment of a multiple-choice test, two main shortcomings must be addressed. (i) Multiple-choice questions are only able to focus on a small and nonrepresentative sample of problems that occur in "real life" situations. (ii) Multiple-choice questions tend to reinforce misleading notions about the nature of mathematics, problem solving, and problem-solving abilities. Therefore, if no attempt is made to deal with larger and more representative classes of problems, there is no way for the alignment of a test to improve.

In a recent review of Roger Schank's book, *Tell Me A Story*, the New York Times Book Review gave the following succinct summary of the shortcomings of psychometric theory.

> *Psychometrics has fallen into disrepute ... because tests: (a) at best measure only a very narrow part of intelligence, (b) may predict academic performance to some degree, but predict outside the school only poorly, (c) are biased in favor of middle- to upper-social-class individuals who received the kind of education that prepares them for these rather trivial kinds of test items, or (d) all of the above.* (New York Times Book Review, September, 15, 1991, p. 3).

In spite of the preceding kinds of concerns, however, large numbers of students continue to be evaluated, screened, and labeled using tasks which ignore modern conceptions about mathematics, problem solving, teaching, and learning. This is why agencies such as the National Academy of Sciences, the National Research Council, and the Mathematical Sciences Education Board have been issuing the following kinds of policy guidelines to schools and universities:

> *Align institutional admissions and placement testing practices with contemporary standards for school mathematics. ... Discontinue use of standardized tests that are misaligned with national standards for curriculum.* (*Counting on You*, p. 21).

As a result of the preceding kinds of pressure, changes in testing practices are being explored in states school districts throughout the nation. But, change is difficult within complex systems, where solutions to one problem often introduce difficulties in others. Consequently, at the same time that innovative materials are being developed, it is also important to broaden and strengthen the knowledge base on which these initiatives are based.

CONTENT QUALITY, TECHNOLOGY, EQUITY AND TEACHER EDUCATION: FUTURE DIRECTIONS

Three of the main reasons to focus on alternative forms of assessment are (i) to emphasis broader and more realistic conceptions about mathematics, mathematical problem solving, and mathematical abilities, because the goal is to prepare students for productive participation in a technology-based society, (ii) to identify talented students and to give special attention to targeted groups of minority students and women whose abilities have not been recognized, cultivated, or rewarded by traditional textbooks, teaching, or tests, and (iii) to help optimize all students' opportunities for success by facilitating informed decision making, by encouraging the development of solid conceptual foundations for future success, and by providing as many "low pressure, high interest" opportunities as possible for students to simultaneously develop and document their increasing knowledge and capabilities. In other words, some of the main reasons to focus on alternative assessment have to do with *content quality*, *equity*, *technology*, and *teacher decision making* (or *teacher education*).

To begin, it is useful to notice that most of the chapters in this book were concerned with content quality even though the authors might ultimately have been interested in other issues such as equity, technology, or teacher education. In the matter of equity, for example, students will not be treated fairly on a given test if the constructs being measured do not correctly reflect the nature of mathematics and mathematical problem solving. When screening applicants for admission into instructional programs, for example, construct validity without predictive validity is (by definition) impossible; whereas if selection procedures are based on only modest (.5) levels of predictive validity, and if the authenticity (or alignment with instructional goals) is low, then test results are likely to encourage discrimination since decision making may be based on inappropriate criteria. As a result, students may be discouraged from studying mathematics, not because they are incapable, but because of misguided beliefs about the nature of mathematics and real-life situations in which mathematics is useful.

Research Issues Related to Content Quality

In recent years, as instruction has become more individualized, and as more diverse types of understandings and abilities have been recognized and encouraged, the predictive validity of most nationally influential standardized tests has steadily decreased (Donlon, 1984; Willingham et al, 1990; Edgerton, 1985; Holland and Richards, 1965; Hoyt, 1966; Wallach and Wing, 1969; Shavelson, 1989, 1991). In fact, even with all of the subjectivity, inconsistency, and unfairness that sometimes goes into teachers' grades, the predictive validity of students' grade point averages still tends to be at least as high as standardized tests. Yet, even when test scores and coursework are combined to predict performance in other areas, only about half of the variance is taken into account, and correlations tend to be even lower when attempts are made to predict performance on more realistic kinds of problem-solving activities.

Studies by Edgerton and Shavelson are instructive. Edgerton (1985) studied students being considered for scholarships and awards in the Westinghouse National Science Talent Search. Applicant evaluations by leading scientists were based on academic transcripts, standardized tests, and actual performance on research projects analogous to the work of adult scientists. The results showed that only about one-third of the students who were chosen on the basis of realistic scientific performance would have been chosen if decisions had been based on only academic transcripts and/or scores on standardized multiple-choice tests.

Shavelson conducted a series of studies of the predictive validity of standardized multiple-choice tests relative to performance measures involving tasks or projects with high construct validity (Shavelson, Carey and Webb, 1990; Shavelson, Baxter, Pine, Yure, Goldman and Smith, 1990). He and his colleagues found that, in general, the higher the construct validity of performance measures, the lower the predictive validity of the standardized tests.

Both Edgerton and Shavelson concluded that the notion of talent that is reflected in most standardized tests and/or academic transcripts is far too narrow, and that serious questions must be raised about the appropriateness of using such predictors for purposes such college admissions and scholarships. Other researchers have formed similar conclusions (see, for example, Collis and Romberg, 1989; Conference Board of the Mathematical Sciences, 1983; *Business Week*, 1988; Hawson et al, 1988). In fact, even in the assessment of general intelligence (g), researchers such as Gardner (1985) have found that a more pluralistic conception of talent must be developed.

Resnick (1987a, b) cites the following reasons why traditional textbooks, teaching, and tests have been inconsistent with real-life problem solving and decision making:

- School learning emphasizes individual cognition, while learning in everyday contexts tends to be a cooperative enterprise.

- School learning stresses "pure thought," while the outside world makes heavy use of tool-aided learning.

- School learning emphasizes manipulation of abstract symbols, while nonschool reasoning is heavily involved with objects and events.

- School learning tends to be generalized, while the learning required for on-the-job competency tends to be situation specific.

Resnick and Resnick conclude that ". . . school work draws on only a limited aspect of intelligence, ignoring many of the intelligences needed for vocational success, especially in the more prestigious vocations" (1989, p. 21).

This claim has been verified by ethnographic studies focusing on ordinary people solving problems in real-life situations (see, for example, Lave, in press; Saxe, 1991; Carraher et al, 1985). For example, a prototype of such studies involves following expert grocery shoppers while they are shopping. What researchers typically see is that experts do a lot of decision making that depends on elementary arithmetic. As long as experts are engaged in these real activities, their arithmetic abilities seem nearly flawless. Yet if the same experts are given a paper-and-pencil arithmetic test with standard word problems that are seemingly isomorphic to the real problems, success rates are typically low. How can this be? In fact, given real needs, people invent a variety of clever and reliably accurate procedures. Consider how most people, even those who lack confidence in their ability to answer school-related percent problems, perform routine real-world tasks such as calculating 15 percent tips. One way involves calculating 10 percent of a given amount, then finding half of this result, and adding to get 15 percent. Another way involves remembering that $1.50 is 15 percent of $10, and that $15 is 15 percent of $100; other amounts can be estimated using proportional reasoning.

Although ethnographic studies have been conducted in realistic situations, they have generally been based on rather impoverished notions about the kinds of situations in which mathematics is useful. In on-the-job problem solving, most studies have looked at entry-level positions, and even for these, studies have seldom looked at situations in which teams of people work together and use technology-based tools. It seems likely that Resnick's claims will be supported even more strongly when less restricted types of tasks, knowledge, and abilities are studied.

Although recent standards and policy-directing statements have made great progress in clarifying the kinds of mathematical knowledge and abilities that citizens and workers in the twenty-first century will need, research is needed to answer such questions as these: If a cognitive objective is a mathematical system (or model) for describing real or possible worlds, then what are the ten to twenty most important cognitive models that students should construct during a given course (or year of schooling)? What does it mean to understand such objectives? And in particular, what does it mean to have a deeper or higher-order understanding of these objectives?

Advances in technology and cognitive psychology and dramatic changes in demands for a competitive work force will almost certainly accelerate. Thus, as first-round goals of the NCTM's *Standards* begin to be achieved, new second-round goals will become necessary. There will be a continuing need to modify and refine existing standards, to go beyond stating desirable goals for instruction, and to develop operational definitions (procedures and criteria) for measuring the extent to which these goals for programs, teaching, textbooks, and tests are being met.

Research Issues Related to Advances in Technology

Technology-based tools have created some of the most important driving forces behind (i) changes in the types of problem-solving situations in which mathematics is useful, (ii) changes in the types of mathematical knowledge and abilities that are useful in newly generated situations, and (iii) changes in the types of jobs and professions in which mathematics is used on a day-to-day basis. As a result, when new conceptual and procedural tools are used to address new types of problems, past conceptions of mathematical ability tend to be far too narrow, low-level, and restricted for productive participation in an age of information. For example, the Mathematical Sciences Education Board stated:

> Communication has created a world economy in which working smarter is more important than working harder. Jobs that contribute to this economy require workers who are mentally fit—workers who are prepared to absorb new ideas, to adapt to change, to cope with ambiguity, to perceive patterns, and to solve unconventional problems. It is these needs, not just the need for calculation ... that

make mathematics a prerequisite to so many jobs. More than ever before, Americans need to think for a living; and, more than ever before, they need to think mathematically. (1989, p. 3).

In a subsequent report, the board continued:

Today's worldwide computer-driven competitive economy demands workers with thinking skills, workers who can deal with computer terminals, automated equipment, and visual data displays; who can make estimates and solve problems; who have a mental "toolkit" of number-managing techniques. . . . Workers now frequently form groups or teams, collaborating rather than working alone. More and more occupations require the ability to understand, communicate, use, and explain concepts and procedures based on mathematical thinking. . . . The rudimentary skills that satisfied the needs of the workplace in the past no longer suffice. . . . Workers are less and less expected to carry out mindless repetitive chores. Instead, they are engaged actively in team problem-solving, talking with their co-workers, and seeking mutually acceptable solutions . . . . They define problems, collect data, establish facts, and draw valid conclusions. (1990, p. 3).

Unlike earlier periods when people demonstrated their knowledge and abilities mainly by showing how well they could remember facts and follow rules, employers today increasingly emphasize abilities such as (i) making sense of complex systems and experiences, (ii) formulating problems in ways that lend themselves to useful solutions, and (iii) learning and adapting to rapidly changing challenges and circumstances. In a technology-based society, new tools do much more than provide new ways to process information from an external world that is "out there" (independent of human influence). The essence of an age of information is that many of the most important influences on peoples' daily lives are businesses, communication networks, and other systems created by humans, often as a direct result of using powerful new technologies.

To create such systems, or to make sense of existing systems, the tools that are needed are not conceptually neutral. Such tools make it easy to think about the world in some ways but not in others, to create certain kinds of systems but not others, and to develop certain kinds of understandings and abilities while also increasing the likelihood that new types of higher-order misconceptions and disabilities will occur. On the one hand, students can use certain tools to develop ways of thinking that are not accessible to students without such tools. With these tools, for example, students can see patterns and regularities that are otherwise difficult (or impossible) to find. On the other hand, "When students are given a hammer, lots of things begin to look like nails!" That is, there is a tendency to use a tool because it's there, not because it's appropriate.

Actually, in mathematics, such tools as pocket calculators or notebook computers with modeling capabilities function in much the same way as conceptual tools such as Cartesian coordinate systems, or proportional reasoning, or the Calculus. Both kinds of tools enable students to produce results that would have been difficult (or impossible) to produce otherwise; and both enable students to think differently (and often better), not just faster. That is, both types of tools have distinctive conceptual frameworks as well as efficient techniques and procedures associated with them. In fact, since both embody powerful conceptual/procedural systems, both might be called conceptual technologies.

In spite of these facts, modern technology-based tools tend to be treated as though they are radically different from conceptual technologies of the past. For the latter, it seemed obvious that both instruction and assessment should focus on the conceptual/procedural systems that had the greatest power and usefulness (whether they were real-world applications or pure math explorations). Yet this is not the case for modern conceptual technologies. Many of the most powerful systems are treated as though they involve cheating—partly, perhaps, because too few people are familiar with the new knowledge and abilities that these tools presuppose. Too many people assume that (i) students who are familiar with such tools are still conceptually equivalent to students who are not, (ii) the tools are just new ways of doing old things, (iii) both groups of students are relying on the same mathematical ideas, and (iv) the technology-enhanced group is simply able to work faster (but not differently, and not better) than their peers.

If school mathematics becomes more sensitive to the needs of our rapidly changing society, these assumptions will not be valid. In the area of technology and assessment, then some important research issues include the following:

- How can assessment instruments go beyond simply being correlated with success in real-life situations to directly involve authentic kinds of problems, knowledge, and abilities?

- What new types of understandings and abilities should be emphasized for problem solving that involves groups of specialists working with powerful technology-based tools?

- What new types of understandings and abilities will prepare students to continually learn and adapt to conceptual/procedural systems associated with new technologies?

Research Issues Related to Equity

Research in mathematics education has produced overwhelming evidence that different types of mathematical talent and personalities, knowledge, and capabilities can all lead to success and that many different types of success are possible. Furthermore, most people have quite irregular profiles of strengths and weaknesses (Begel, 1979). Consider the following everyday examples:

- In business and industry, people who are good at working alone to answer other people's clearly formulated questions are not necessarily good at thinking about fuzzy situations, or dividing complex problems into subcomponents, or adapting to new tools and resources, or working in teams coordinating the efforts of people with diverse talents and expertise.

- In universities, mathematics professors are not necessarily good accountants or at following the rules for filling out income tax forms. And people who are good at answering other people's questions about tax forms (in typical situations) are not necessarily good at knowing which questions to ask (in nonstandard situations).

- In computer programming, people who are good at working within the constraints of a given language (or programming environment) are not necessarily good at developing modifications to existing languages or at developing new programming environments to fit changing needs.

- In mathematics, specialists in algebraic/analytic/written-symbolic forms of thought are not necessarily good at geometric or graphic forms. People in departments of pure mathematics tend to have quite different personalities that those in applied departments such as statistics, computer science, econometrics, or psychometrics.

Similarly, students who are good at working alone, using pencil and paper to quickly and flawlessly solve traditional word problems, are not necessarily good at realistic problems, where there is no single correct answer, where sense-making is as important as answer-giving, and where rapidly generated, single-rule solutions tend to be inadequate. In school, being caught thinking too often means that you were caught not knowing; taking more than three seconds to respond means you'll probably be passed over by the teacher; getting an answer that isn't a whole number is a clue that you're doing something wrong; and using more than a single rule means that you aren't doing something the right way. In fact, many of the skills and beliefs that contribute to success in school tend to be extremely counterproductive in real-life situations where inappropriate responses are often associated with quickly generated answers and single-rule "canned" solution procedures.

In spite of these facts, most high-stakes testing programs continue to rely on psychometric models that assume (i) that mathematics questions nearly always have one correct answer, (ii) that mathematical abilities and talents can be collapsed into a single trait (g, general intelligence), and (iii) that correlation with performance in other areas is enough, even if the sample of problems is only a small and atypical subset of those that are occur in realistic situations.

In the past, if a narrow conception of talent was correlated with a more representative conception, testing specialists tended to treat modest (.5) correlations as though they were sufficient—that is, sufficient to select small-but-adequate numbers of students for access to scarce-but-adequate resources. But today, national assessment priorities have changed. At a national level, our foremost problem is not to screen talent but rather to identify and nurture capable students. The pool of students receiving adequate preparation in mathematics is no longer adequate; far too many capable students are being shut out or turned off by textbooks, teaching, and tests that give excessive attention to constructs that are narrow, shallow, obsolete, and often fundamentally inconsistent with national standards that have been endorsed by all relevant professional and governmental organizations. This is our greatest challenge related to equity.

Traditionally, mathematics courses and mathematics tests have been popular devices for screening students for entry into programs, colleges, courses, and professions, regardless of whether the mathematics that was taught (and tested) matched the actual requirements for success in the programs or on the job. But today, when more than nineteenth century shopkeeper arithmetic really is used on a day-to-day basis, mathematical deficiencies have far more debilitating consequences than in the past, and mathematics is a key to success in matters related to equity. Fortunately, there is evidence that there is a large supply of potentially capable students available. For example, the last section of this chapter describes a set of projects demonstrating that, when problem-solving and decision-making activities encourage students to make sense of situations based on their own personal knowledge and experiences, even students labeled average or below average can emerge as exceptionally capable. Many are able to routinely invent (or significantly extend, modify, or refine) mathematical models that go far beyond those that their teachers believed they could be taught. However, simply introducing new types of problems (or new levels and types of mathematical ideas) will not necessarily allow a wider variety of students to demonstrate their capabilities. In fact, if issues of fairness are not also taken into account, the privileged simply tend to get more privileges.

To address these concerns, important research priorities are (i) clarifying the kind of knowledge and abilities that are actually needed in a representative sample of realistic jobs and settings, (ii) developing new types of problems for use in high-stakes examinations, and (iii) exploring new sources of information (beyond tests) to inform high-stakes decision-making issues.

Research Issues Related to Teacher Education

To improve assessment, two distinct approaches can be taken: One involves *improving the content quality of standardized tests* by focusing on authentic mathematical knowledge, realistic problem-solving situations, and the diverse mathematical abilities that are productive in realistic situations. The other involves *improving the credibility, reliability, and fairness of alternative forms of assessment* including teacher's classroom observations, and students' extended projects. Improvements in teacher education are obviously critical to the second approach, but for the first, it is also necessary for teachers to assume greater responsibility in the assessment process. To see why, it is useful to consider assessment that evaluates programs or teachers.

When evaluating programs, simply labeling them successes or failures tends to be misleading. All programs have profiles of strengths and weaknesses; most are effective for some students (or teachers or situations) but not for others. But other factors may also be important. For example, regardless of quality, a program seldom succeeds when the principal of a school doesn't understand or support the program objectives. Therefore, when programs are evaluated, the roles of key administrators should also be assessed. And, regardless of quality, if program implementation is half-hearted, total success can hardly be expected. Moreover, powerful innovations usually need to be introduced gradually. Therefore, when programs are evaluated, the implementation processes should also be assessed.

Similarly, when evaluating teachers, both their own profiles of strengths and weaknesses and the conditions under which they have been achieved should be identified. No teacher can be expected to be good in bad situations (such as when students do not want to learn, or when there is no support from parents and administrators). Not everything experts do is effective, and not everything novices do is ineffective. Furthermore, no teacher is equally effective across all grade levels (from kindergarten through calculus), with all types of students (from the gifted to those who are challenged with physical or mental handicaps), and in all types of settings (from those dominated by inner-city minorities to those dominated by the rural poor). In fact, characteristics that lead to success in one situation often turn out to be counterproductive in others.

Such observations suggest the need for close connections among assessments of teachers and of programs, program administrators, and program implementations. Similar connections should exist for assessments aimed at students, programs, and teachers. But when we are assessing students, who should describe these profiles and conditions? The responsibility must fall mainly to teachers, because they are usually closest to students, having observed them over long periods of time and in diverse kinds of learning and problem-solving situations. Two distinct types of assessment, one emphasizing *state-focused documentation* and the other emphasizing *progress-focused documentation*, are involved in creating an assessment program where such teaching roles are emphasized.

State-focused documentation aims mainly at (i) evaluating (or assigning values to) states of development, (ii) identifying deficiencies with respect to some standard, and (iii) inferring that progress has been made by comparing one evaluation with another using a subtraction-based model to describe differences. For example, traditional pretest/posttest models are frequently used for program accountability, and here one of the dangers associated with state-focused documentation often becomes clear. While the primary goal may be to prove that a certain program works, the pretests and posttests that are used may promote conditions in which the program will not work. For example, if it is known that school principals are important to the success of most programs, then there is no reason not to try to optimize the chances that school administrators will have positive influences, while, at the same time, documenting the extend to which optimization occurs. (Note that this last tactic is a key characteristic of progress-focused documentation.)

Progress-focused documentation monitors progress directly by focusing on activities that simultaneously encourage and document development in directions that are "better," without necessarily using pretests and posttests that embody a fixed and final definition of "best." (Examples of progress-focused documentation will be described in the next section.) Since such activities tend to contribute to both learning and the documentation of learning, distinctions between instruction and assessment blur and teachers' contributions become particularly important. The quality of teachers' contributions tends to be strongly influenced by their own knowledge; thus they need to develop deep understandings about the content they are teaching.

In matters related to assessment, important challenges for mathematics education research focus on clarifying what it means for teachers to construct deep understandings of elementary mathematical ideas. For a given collection of ideas, it may mean knowing how these ideas developed in the history of our culture, how they typically develop in the minds of children, and how they are developed in curriculum materials. It may also mean being familiar with how these ideas have been influenced by modern technology and how their uses have changed in response to problems in our technology-based society. In any case, these kinds of understandings and abilities are clearly quite different from those traditionally emphasized college-level mathematics courses for teachers, which too often are characterized by superficial treatments of advanced topics, rather than by deep treatments of elementary topics. The effects of this instruction can be particularly negative because teachers tend to teach as they themselves were taught, and because instructors of college-level mathematics courses for nonmath majors have typically not been the best role models for elementary teachers.

EXAMPLE PROJECTS ON TECHNOLOGY, EQUITY, AND TEACHER EDUCATION

To be more explicit about the directions for research that were suggested in the preceding section, this section focuses on three closely related projects, one emphasizing technology issues, one emphasizing equity issues, and one emphasizing teacher education issues. These projects are briefly summarized below; then each is discussed at greater length later in this chapter. These projects emphasize performance assessment activities that focus on (i) deeper and higher-order understandings of elementary mathematics, (ii) realistic problem-solving situations, and (iii) diverse types of mathematical abilities.

The "Math-rich Newspapers Project" (Newspapers) emphasizes the use of technology in real-life situations. Intended to identify key characteristics of high-quality performance assessment activities, it focuses especially on the understandings and abilities that average-ability middle schoolers need to work in teams, on realistic problems, and using a variety of technology-based tools. The project goal is to develop a library of exemplary problems and an interactive computer-based guide to help teachers develop activities that will (i) contribute to both instruction and assessment, (ii) integrate both basic facts and skills and higher-order understandings and processes, and (iii) gather and report assessment information from settings ranging from tests to interviews to open-ended projects.

The Pre-college Mathematics Ability: Discovery, Development, and Documentation project (3-D project) emphasizes issues related to equity. It is aimed at identifying students in targeted populations (minorities or young women) whose abilities have not been recognized, cultivated, or rewarded by traditional textbooks, teaching, or tests. These students are being identified using problems from the Newspaper project that fit the interests and experiences of the targeted groups and which emphasize broader and more realistic conceptions of mathematical abilities.

The Continuously Developing Teachers Project (Teachers) studies the development of teachers' knowledge (about the nature of mathematics, problem solving, learning, and instruction) within the context of a ten-week sequence of activities. The activities encourage the continuous development of either experts or novices and enable teachers to automatically document their development without taking time away from their instructional responsibilities. The approach is to create performance activities for teachers based on teaching activities associated with performance activities for students. Some teachers' activities include writing or evaluating performance assessment activities for students (discussed in chapters 2 and 5 through 9 in this volume), evaluating the scope and depth of clusters of such activities (see chapters 2, 14), interpreting students' responses to the activities (see chapters 4, 13), developing response interpretation rubrics for assessing students' performances (see chapter 13), observing students' behaviors using videotapes of the student performances (see chapter 10), generating insightful hypotheses about their underlying knowledge and capabilities (see chapters 4, 10), and interviewing and tutoring students as follow-ups to the observations (see chapters 4, 11).

These three projects are closely connected: Teachers in the Teachers project are helping to develop exemplary activities for the Newspaper project; problems from the latter are being used to identify students in the 3-D project, and teachers in 3-D classrooms are participating in the Teachers project. The following general operating principles apply to all three projects:

- Do not use one-number characterizations of students, teachers, or programs.

- Do not brand students with labels that are supposed to be unaffected by conditions, instruction, or experience.

- Problems must have more than a single level or type of correct response.

- Do not use decontextualized ("vanilla") problems that discourage students from making sense of situations based on their own everyday knowledge and experience.

- Focus on model construction/exploration/application sequences in which students are (i) encouraged to develop mathematical models that have the greatest power and usefulness in students' current (or anticipated) lives, and (ii) required to explicitly document how they are thinking about problem situations.

The Math-rich Newspapers Project

Each math-rich newspaper consists of one sheet of paper folded to make four pages; each newspaper is modularized (and desktop publishable) so that it can be tailored to match the interests and experiences of targeted groups of students (for example, middle school students considered to be at risk in particular cities, or on-the-job adults working at the level of middle school mathematics). Each newspaper has four types of problem sets associated with it (Table 2). They are (i) test-sized problems, (ii) homework-sized problems, (iii) project-sized problems, and (iv) Math Olympics problems. All are intended to feed assessment information into the kind of learning progress maps that were described in chapter 14.

Table 2. Four types of problems associated with each math-rich newspaper.

	Learning Focused Activities	Evaluation Focused Activities
Smaller Problems	Homework-sized Problems (5-10 min.)	Test-sized Problems (1-3 min.)
Larger Problems	Project-sized Problems (60 min.)	Math Olympics Problems (30 min.)

Several examples of project-sized problems were given in chapter 2. All four levels of problems involve realistic givens, goals, tools, and settings. The project-sized problems and the Math Olympics problems tend to be similar to the kinds of performance activities that are emphasized in portfolio-based forms of assessment, while the test-sized problems and the homework-sized problems are smaller and somewhat more restricted. In many ways, the test-sized problems and the homework-sized problems involve the kind of questions that an expert interviewer might like to ask students before or after they work on the project-sized problems.

The newspaper articles are similar to those that occur in a real newspaper and are organized within traditional sections: editorials; world, national, and local news; sports; entertainment (radio, television, records, movies), travel, home and family (home repairs, cooking), business; politics; weather; comics; and display and classified ads. The articles are selected and/or written to focus on the interests and experiences of specially targeted groups of students; they supply mathematically rich information to encourage students to construct/explore/apply especially important mathematical models, rules or principles; and they are simplified to reduce reading difficulties and to deemphasize mathematically irrelevant factors that contribute to task difficulty. The topics follow guidelines specified in the NCTM's *Curriculum and Evaluation Standards for School Mathematics* and focus on the kinds of models described in chapter 2 of this book.

Test-sized problems are similar in size and format to textbook word problems, but are more authentic and realistic. In particular, (i) they focus on problem-solving situations that might reasonably occur in the students' everyday lives, (ii) they always involve more than a single level and type of correct answer, (iii) pocket computers with graphing capabilities are available, and (iv) sense-making phases of problem solving are emphasized (such as those that focus on generating graphs or other mathematical descriptions of problem situations). They are also intended to be easy and inexpensive to administer and score.

Adequate attention is given to accountability issues, such as how to minimize cheating, unfairness, and subjectivity in scoring. Students generally work alone in monitored environments, entire problem sets can be completed during single class sessions, and individual tasks can usually be completed in less than two or three minutes. Also, to address concerns about the scope of topic coverage, test-sized problem sets focus on large numbers of small problems. Because of these constraints, the test-sized problems tend to focus on students' abilities to use already constructed models and reasoning patterns rather than requiring construction of new models or making significant modifications to familiar models. That is, these problems tend to focus on modeling skills more than on higher-order understandings and processes; they tend to emphasize understandings and abilities at the *exploration and application phases* of model construction/exploration/application sequence.

Homework-sized problems have no time constraints, and a variety of resources are available, including peers, adult consultants, library materials, and tools such as computers and appropriate software. The goal is not so much to test students as it is to provide activities in which students can simultaneously learn and document what they are learning.

Like the test-sized problems, the homework-sized problems focus on the exploration and application phases of the modeling sequences. But because they contribute to instruction as well as to assessment, these problems can occur more often than tests. Each problem set can focus on a few larger problems, can involve more complex responses, and can focus on deeper and higher-order objectives which are difficult to address within artificially constrained test-like situations.

Compared to test-sized problems, homework-sized problems elicit responses that are more complex and difficult to score. But because the emphasis is on instruction and instructional decision making, teachers and students are the primary consumers of the information generated. Thus rapid turn-around times and meaningful feedbacks tend to be more important than reliability and other criteria that are typically emphasized in high-stakes testing. Also, because there is instructional value in both doing the problems and scoring the responses, time and effort associated with these activities is not an intrusion on instruction. In fact, students may be involved in the evaluation process so that they understand why their answers are evaluated in particular ways, and it is important for students to examine and compare alternative solutions to problems. Such activities can occur frequently; the results from individual sessions become relatively low risk; and attention focuses on performance "traces" across many situations and over long periods of time.

Project-sized problems usually require one or more hours for students (or groups of students) to complete. Realistic resources and tools are available, including reference books, relevant data sources, computer-based tools, and simulations. In many ways, the project-sized problems are similar to the kind of activities emphasized in most portfolio-based assessment projects; however, in the Newspaper project, these problems give special attention to eliciting the fifteen to twenty models that are priorities for students to address at a given grade level. Each project is also part of a model construction/exploration/application sequence, as described in chapter 2.

For such projects, student results quite often consist of written articles suitable for publication in our newspapers (in an inserted middle page). The goal is to give students as many opportunities as possible to build a portfolio of newspaper articles summarizing the results of the projects that they have completed. Each project summary (or written article) is evaluated by an editorial panel (a teacher working with a peer group of students). Editorial panels assess the quality of each article, suggest needed revisions, and decide which articles should be published. In this way, the articles in students' portfolios become distilled and evaluated descriptions of the projects that they have completed, and each student builds a resume that can be viewed as an assessment "cover sheet" to their portfolio of projects.

One of the common criticisms of portfolio-based assessment projects has been that they often produce more information than decision-makers are able to use, with the result that the rich information available is ignored or reduced to simplistic generalizations that have the same negative characteristics as single-number test scores. Therefore an important goal of the Newspaper project is to reduce portfolio-based information to a manageable form while still preserving richness and depth. For example, students' newspaper-like summaries of their own projects (and students' resumes and editorial panels) are straightforward attempts to get students themselves involved in summarizing and evaluating the significance of their own work. In as many ways as possible, responsibility is reflected back onto students for documenting their own achievements, and for analyzing, summarizing, and evaluating the quality for their own work (as long as these practices do not compromise necessary assessment needs).

Each student is expected to complete at least twenty such projects over the course of a school year, and they are expected to focus on the ten to twenty models that the teachers consider the most important for a given student to address in a particular course (or at a particular grade level). Most projects are evaluated by local teachers working with students. Then a small sample of these projects is reviewed by a district-wide panel of experts. These blue ribbon panels (for example, state or school district representatives) spot check the quality of the assessment process and consider exemplary projects for special awards. One goal of this process is to gather assessment-relevant information from a large number of sources, and a separate goal is to increase the credibility of the conclusions that are reached.

As in real life, students receive credit for serving on editorial panels. When students work in groups, a proposal has to be written stating the responsibilities that each student will fulfill. One student in each group has to serve as the "manager" who produces a report on the extent to which each student fulfilled his or her role. This enables appropriate credit to be given to each student for results that are produced by a team. In other words, the evaluation of students is done in the same way that businesses evaluate employees based on their work. Students know what they are signing up for when they begin a project with a group, and teachers tutor designated students, for example, when those responsible for data analysis need to know how to use a particular software program.

Mathematics Olympics problems are similar to project-sized problems that students have already solved, but they involve a new context, or new data, or new questions based on the same data. Each Math Olympics lasts two to three hours; during this time, each student completes three to five problems. As in sports or performing arts competitions, participants know in advance that they must perform a specified set of complex tasks, but the tasks are sufficiently rich so that focusing on them (and knowing in advance which ones would be emphasized) still represents a sufficiently broad, deep, and complex sample of behavior to measure expertise in the field.

Both the Math Olympics and the project-sized problems help students document complex capabilities in settings that are familiar, meaningful, and interesting. But because the general nature of each Math Olympics problem is known in advance, it is reasonable to expect students to complete three to five of these complex problems during a two-hour period (even though the problems on which they are based originally required at least one hour per problem to complete). Essentially, the Math Olympics problems are test-focused counterparts of our project-sized problems. Students work independently within a monitored environment, as in traditional types of testing situations; however, the goal in a Math Olympics is to be as straightforward as possible with students by allowing them to chose the prototype problems that they want to use to demonstrate their knowledge and capabilities. The restriction is that the problems must be chosen from a specified collection of project-sized problems.

Why emphasize problems that are based on math-rich newspapers? One of the main reasons is that current mathematics textbooks and tests pay far too little attention to problem solving in realistic situations. Therefore we have found that these math-rich newspapers provide a good way to help teachers develop activities that fit their own students' interests and experiences. For example, to help teachers develop activities that go beyond traditional types of word problems, we begin by writing (or finding) a math-rich newspaper article that involves a situation familiar and interesting to their students. Then, after the newspaper article is written, all four levels of problems should be written for each article. Several different projects and problems are all based on the same newspaper article. When the articles themselves involve graphs and tables of data, it is natural for students to have to sort out information which is relevant or irrelevant. Problems naturally tend to involve both too much and/or not enough information; some of the relevant information may be based on patterns or trends in the data; and a graph or table may over-simplify or distort the underlying information it was intended to describe. Therefore, it is natural for students to have to critically analyze the story, graph, or table and to focus on deeper and higher-order understandings of the type emphasized in this book.

Why are four different levels of problems based on the same newspaper articles? Since each serves some legitimate purpose, the goal is to ensure that none of the purposes will be ignored or eliminated (Resnick and Resnick, 1989). In many districts where teachers do not have the option of ignoring traditional high-stakes standardized tests, successful assessment programs must find ways to deal with the fact that teachers tend to teach to the high-stakes test (Romberg, Zarinnia, and Williams, 1989). The four levels of problems based on the same newspapers are as closely associated as possible so that teachers will see that the sensible thing to do is to teach and test skill-level knowledge and abilities in the context of larger, more realistic, and more meaningful types of problem-solving situations.

Why is it useful for students' projects to result in written articles similar to those in real newspapers? One reason is that, when students' responses are though of as being whole articles, it is natural to emphasize goals that consist of more than producing narrow types of explicitly requested mathematical results. For example, it is natural to focus on justifying or explaining answers, or finding useful ways to describe problem situations. Also, because the written articles are condensed versions of entire projects that students conducted, it is natural for students to think back about their work and to summarize the most significant aspects of their results.

The 3-D Project

When problems from the Newspaper project have been used with students whose mathematical abilities have not been recognized or encouraged by traditional textbooks, teaching, or tests, their regular teachers nearly always make a comment something like the following: "Several of my best students haven't done very on these projects;, but several of my 'worst' students have done very well." Follow-up interviews often verified such observations. Among the students with histories of A's and good scores on standardized tests, several nearly always turn out to have extremely shallow understandings. As a result, they often have unusual difficulties with more realistic problems in which quick-answer/one-rule solutions tend to be superficial or wrong. On the other hand, some students with histories of low grades and poor test scores turn out to be exceptionally able on relatively complex real-life problems. In a class of 30 such students, it is common to find at least two or three who are exceptionally capable... as long as it is clear to them that they will be rewarded for trying to make sense of problems using extensions of their own real- life knowledge and abilities.

In classrooms participating in the 3-D project, we have found that, when a students work on model-eliciting problems selected to fit their interests and experiences, and when the tasks emphasize a broad range of abilities, a majority of the students routinely invent (or extend, or refine) mathematical ideas that are far more sophisticated than their teachers would have guessed they could be taught. Consider the following example, which involves the CD throwing problem discussed in chapter 2 (problem 2).

When college-level mathematicians are given five or ten minutes to work on this problem, they are likely to do the following: First, instead of thinking about where the disk as a whole must be located in order to win the game, they may notice that it is useful to think only about the center of the disk. As Figure 1 shows, for any given square on the gameboard, if the center of the disk lies within the small shaded square, the player is a winner. Therefore, for a given disk, the probability of winning is equal to the ratio of two areas—the area of the shaded square and the area of the squares on the gameboard.

Figure 1. A way to solve the CD throwing problem.

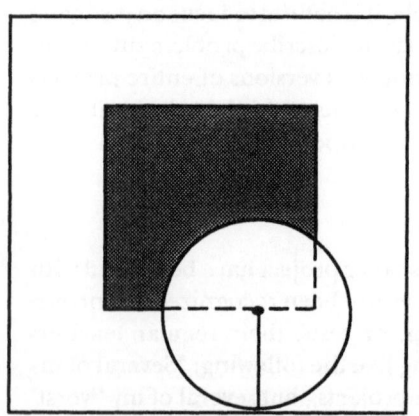

When middle schoolers have at least an hour to work on the problem, a large percentage of them come up with similar (though less formal) solutions. For example, in Newark, New Jersey, a group of three minority seventh graders (who were generally regarded as "goof offs" by their teachers and peers) solved the problem in the following way.

Carl: I played a game like that ... at a carnival with my uncle.

Others: Yah! Me too.

[The boys began by drawing a gameboard on a large sheet of paper. Then they cut out several cardboard disks and (for about fifteen minutes) acted out the game. Because they cut out several disks that were not all the same size, they noticed that the size of the disk made a difference.]

> Jamal: Mine [small] is better. Yours never works [too large].

[Five minutes passed as the boys tried several more disks, gradually making them smaller and smaller (so they could win more often).]

> Bart: This would be easy if this thing [disk] was a dart.
>
> Carl: Yah, but it's not a dart.
>
> Bart: But what if it was. Look! Look at this. [Bart draws lots of little squares inside the original gameboard squares as shown in Figure 2.] I win if my dart lands in here [referring to his disk as a dart, and pointing to the shaded squares in the diagram in Figure 2]. And, I lose if the dart lands out here [pointing to the unshaded squares]. I've got this many chances to win, and this many to lose.
>
> Jamal: Let's count 'em. [He counts, but not very carefully.] There are 300 here [pointing to the unshaded squares]. And 100 here [pointing to the shaded squares].
>
> Bart: So, I've got a 100 to 300 chance to win.
>
> Carl: Yah, but what if you was Michael Jordan? He'd throw it like this [down the center of one row of squares]. He wouldn't throw it like this [from corner to corner].
>
> Jamal: That's right! And, Tina, my sister, she's so dumb, she'd miss the whole thing.

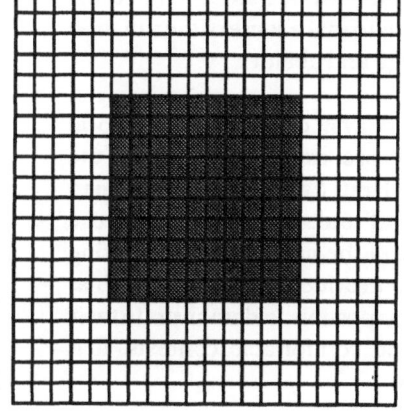

**Figure 2.
A key insight in the boys' solution.**

The solution to the CD throwing problem involves writing a letter to the student council explaining how the game should be set up, and what the chances would be of winning. Excerpts from the boys' letter are shown in Figure 3 (with diagrams omitted).

Figure 3. Excerpts from the boys' letter.

Excerpts from the Boys' Letter

Dear Student Council,
This is a good game. Don't make the squares on the board too big. Make it about this size (a drawing is given). If your like Tina Jackson, your chances to lose are about 100 to 1. If your like Michael Jordan, your chances are about 50 50. Most people will be about 1 to 8.
500 people will come. About half will be like Tina. They won't play much. Only a few guys will be like Michael Jordan. They like to play a lot. So, make them quit after they win once. There are about 20 of these guys. They will all win sometime. Most people who play aren't like Tina or Jordan.
About 200 will play. You should charge 1 dollar to play the game.
Sincerely,
Bart, Carl, and Jamal

The boys' solution used pictures, intuitive language, and "odds of winning" that were based on counting "small squares" (as suggested in the transcript diagram). The boys did an outstanding job of identifying the relevant factors—the relative size of the disks and the gameboard squares, the expected skill levels of the players, the expected number of customers at each skill level, and the cost of the game compared to the cost of the disks. Also, ideas such as conditional probabilities were clearly recognized in the pictures and counts that were made. On the other hand, the answer that the university professors generated seemed to be more like the kind that teachers would expect, but it was based on the assumption (which is generally false) that the throws were random and, at the same time, still good enough to hit the gameboard.

After similar performances on a whole series of realistic problems, it became obvious that Bart, Carl, and Jamal had much more potential in mathematics than their school grades and test scores had suggested. We believe that similarly talented youngsters will emerge in nearly every classroom that allows a broad range of mathematical abilities and working styles to be recognized as productive, and that encourages students to make sense of the situation using extensions of their own real-life knowledge and abilities.

Developing such activities depends heavily on the ability of local teachers to create (or modify) activities that are exactly opposite to the decontextualized problems in most standardized tests (where problems tend to be screened out precisely because some group of students might have special interests or experiences related to them). One of the main purposes of using problems based on math-rich newspapers is to provide every student with as many low-pressure, high-interest opportunities as possible to demonstrate their knowledge and capabilities within familiar and comfortable contexts. It is important to offer every student many options.

In the 3-D project, local teachers began with a library of problem prototypes that needed to be localized to fit the interests of their students. They found the following:

- Problems that appeal to a general audience tend to be far less successful than problems that focus on the current local interests of targeted groups of students. That is, even when exemplary generic problems are provided, local teachers had to localize and particularize problems to match the interests and experiences of their students.

- A collection of exemplary problems is not enough. If innovative types of problems are scored in traditional ways, few improvements are likely to occur in instruction or assessment. Therefore, new types of response interpretation procedures are also needed; these procedures and criteria need to be understood (and used) by local teachers, because the ultimate goal is for these procedures and criteria to be communicated to students (so that they can judge the quality of their own work).

- Assessment-relevant information should be taken from multiple sources–tests, students' extended projects, teachers' classroom tests, interviews, and observations. All of these contexts must reinforce a consistent view about the nature of authentic mathematics, the nature of problem solving in realistic situations, and the nature of the abilities that contribute to success in realistic situations.

Over all, in all of the areas critical to the success of an innovative instruction/ assessment program, factors that have to do with teachers' own understandings about the nature of mathematics, mathematics learning, and mathematics problem solving are found to be critically important.

The Continuously Developing Teachers Project

The Teachers project has three main objectives. The first is to study the development of expert teachers' knowledge about the nature of mathematics, problem solving, learning, and instruction. The second is to create learning environments that do not take teachers away from important instructional activities — that is, learning environments that encourage the development of teachers' knowledge and competencies and that also enable teachers to automatically document their achievements in a form that is impressive to administrators. The third goal is for teachers to contribute exemplary activities to the library of problems being used in the 3-D and Newspaper projects.

Recently, the National Council of Teachers of Mathematics produced *Professional Standards for Teaching Mathematics* (NCTM, 1991). These standards for teaching (and teachers) have received the same kind of national endorsement as the NCTM's earlier *Curriculum and Evaluation Standards for School Mathematics* (NCTM, 1989), which focused on students in the K-12 curriculum. However, if these teaching standards are thought of as describing the mathematics education community's current collective conception of an expert teacher, it is clear that a great distance exists between (i) clear statements of instructional objectives and (ii) operational definitions specifying criteria for measuring acceptable levels of goal achievement.

Although current understandings about the nature of excellent teachers enable us to point to vignettes that exemplify excellence in teaching (and to compare performances and point to directions that represent improvement within a given vignette), we cannot give a measurable definition of an "excellent teacher." What is known, and what the NCTM teaching standards make clear, is that expert teachers cannot be characterized by simple lists of condition-action rules.

Being an outstanding teacher involves much more than being able to flawlessly execute a list of teaching rules (behavioral objectives). In fact, rather than teaching teachers facts and procedures, successful teacher development programs generally focus on helping teachers to refine the ways they think about (i) the nature of children's mathematical knowledge and its development, (ii) the nature of situations in which mathematics is useful, and (iii) the diverse nature of abilities that are productive in such situations.

In the area of assessment for teachers, alternatives to traditional tests can be based on a number of models. For example, in athletics and in the arts, one model is the Olympic gymnastics competitions in which a small number of relatively complex events are prescribed. Expert judges are able to analyze and assess these performances in a reliable manner; then a small number of additional events are defined and executed by each participant. The vignettes in the NCTM *Teaching Standards* provide guidelines that are relevant to both of the preceding factors. But for future progress to be made, more specific prototypes need to be developed and refined, and specific procedures and criteria need to be developed to analyze and assess performances.

Another point that the NCTM teaching standards makes clear is that, when assessing teachers, teaching, or programs, it is important to recognize that continuous growth is needed, even when high levels of excellence have been achieved. There is no fixed and final state of expertise, nor of excellence. In teaching, as in other fields, experts must always continue to develop. In sports, when Michael Jordan develops new capabilities, his environment soon adapts to him, so he must engage in another round of development. The situation is similar in mathematics teaching. The ways that teachers think about mathematics, teaching, learning, and real-life problem solving strongly influence what goes on in their classrooms; but what goes on in their classrooms also requires teachers to develop more powerful and sophisticated understandings about mathematics, teaching, and learning. The cycle is never-ending, and teachers who fail to get better risk being not very good at all.

These principles should be taken into account in assessment programs for teachers. Just as in the assessment for K-12 students, the assessment of teaching should focus on activities that are meaningful and important in their own right. These activities should enable teachers to both develop and document their development without interrupting their instructional activities. And, as in assessments of students or of programs, assessment activities for teachers are more than indicators of progress. They are interventions that can induce either positive or negative changes in the systems and individuals they describe. Therefore, care must be taken to ensure that these influences are positive. This "progress-focused assessment" is aimed at documenting progress in directions that are increasingly "better" without necessarily beginning with a fixed and final definition of "best" and without labeling individuals as good or poor relative to one another.

In the Teachers project, teachers collaboratively write performance assessment problems for their students and analyze students' responses to such items. Then, in weekly meetings over a ten-week period, these teachers produce a library of useful problems and response analysis procedures while refining their collective conceptions about the nature of good problems and good responses. During the process, many participants have developed new insights about the nature of their discipline, of its applications, and of students' understandings and capabilities. And the problems that were written and responses that were analyzed produced a trace of the teachers' own progress that was very impressive to school administrators.

SUMMARY

Enormous progress has been made in clarifying future-oriented instructional objectives in mathematics. Now efforts are being made to create new types of tests and test items consistent with these instructional objectives. Yet, precisely because significant progress has been made at the level of individual problems and isolated objectives, mathematics education researchers should take these next steps:

- *New instructional objectives* should be translated into operational definitions specifying acceptable levels of goal achievement, going beyond behavioral objectives (and basic facts and skills) to assess cognitive objectives (deeper and higher-order understandings and processes).

- *New types of problems* (or authentic performance activities) should be developed that directly involve authentic work samples taken from activities that are important in themselves.

- *New examination formats* should contribute to instruction by providing many opportunities for students to demonstrate their knowledge and abilities in contexts that are meaningful, realistic, and interesting.

- *New response interpretation procedures* must deal with activities in which several levels and types of correct answers are possible.

- *New data analysis models* should deal with patterns of information from multiple sources and be capable of displaying these patterns in a variety of forms to match the needs of various decision makers and decision-making issues.

- *New reports* must be n-dimensional, interactive, and graphics-based, and display the strengths and needs of individual students.

- *New objectives frameworks* must keep pace with continuing advances in technology, cognitive science, and mathematics, and with increasing educational needs for citizens and workers in a technology-based society.

In addition to the traditional bottom-up approaches to assessment (from objectives to items to tests to reports), this book has also emphasized the importance of a top-down approach that begins by determining the priority decision-making issues and the kinds of reports, information, and data sources (testing formats, item types, scoring procedures, aggregation techniques) needed to inform these decisions. And most importantly, we have stressed the special attention that should be given to teacher education and to issues related to equity and the continuing influence of technology in our society.

If mathematics educators neglect issues beyond the level of individual items and isolated objectives, they may find themselves in a situation similar to that of the peasant in the fairy tale who was granted three wishes by the genie in the magic bottle. They will get what they asked for, but not what they wanted.

REFERENCES

Begel, E. (1979). *Critical variables in mathematics education: Findings from a survey of the empirical literature.* Mathematics Association of America. Washington, DC.

Bloom, B.S. (1985). *The development of talent in young people.* New York: Balentine.

Business Week. (1988). Human capital: The decline of America's work force. Special Report, September 19, 100-141.

Carraher, T. N., Carraher, D. W., and Schliemann, A. D. (1985). Mathematics in the streets and in schools. *British Journal of Developmental Psychology,* 3, 21-29.

Collis, K. F., and Romberg, T. A. (1989). Assessment of mathematical performance: An analysis of open-ended test items. National Center for Research in Mathematical Sciences Education, Wisconsin Center for Education Research, School of Education. Madison, WI: University of Wisconsin.

Conference Board of the Mathematical Sciences. (1983). The Mathematical Sciences Curriculum K-12: What is still fundamental and what is not. In *Educating Americans for the 21st century: Source materials.* National Science Board Commission on Precollege Education in Mathematics, Science, and Technology. Washington, DC: National Science Foundation, 1-23.

Donlon, T. F. (Ed.). (1984). *The College Board technical handbook for the Scholastic Aptitude Test and Achievement Tests.* New York: College Entrance Examination Board.

Edgerton, H. A. (1985). *Identifying high school seniors talented in science.* Westinghouse Science Talent Search. Washington, DC: Science Service.

Gardner, H. (1985). *Frames of mind: The theory of multiple intelligences.* New York: Basic Books.

Holland, J. L. and Richards, J. M., Jr. (1965). Academic and nonacademic accomplishment: Correlated or uncorrelated? *Journal of Education Psychology,* 56, 165-174.

Hawson, G., Kahane, J. P., Lauginie, Pl, and de Turckheim, E., Eds. (1988). Mathematics as a service subject. *International Commission on Mathematics Instruction study series.* Cambridge: Cambridge University Press.

Hoyt, D. P. (1966). College grades and adult accomplishment: A review of research. *The Educational Record,* 47, (1).

Krutetskii, V. (1976). *The psychology of mathematical abilities in school children.* Chicago: University of Chicago Press.

Lave, J. (in press). *Cognition on practice: Mind, math, and culture in everyday life.* Cambridge: Cambridge University Press.

Mathematical Sciences Education Board. (1989). *Every one counts: A report to the nation on the future of mathematics education.* Washington, DC: National Academy Press.

Mathematical Sciences Education Board. (1990). *Reshaping school mathematics: A philosophy and framework for curriculum*. National Research Council. Washington, DC: National Academy Press.

Mathematical Sciences Education Board. (1991). *Counting on you*. Washington, DC: National Academy Press.

Mislevy, R. J. (1991). A framework for studying differences between multiple-choice and free-response test items. In R. E. Bennett and W. C. Ward, Eds., *Construction vs. choice in cognitive measurement: Issues in constructed response, performance testing, and portfolio assessment*. Hillsdale, NJ: Lawrence Erlbaum Associates.

Mislevy, R.J. (1992). Foundations of a new test theory. In N. Frederiksen, R.J. Mislevy, and I.I. Bejar, Eds., *Test theory for a new generation of tests*. Hillsdale, NJ: Lawrence Erlbaum Associates.

National Council of Teachers of Mathematics. (1989). *Curriculum and evaluation standards for school mathematics*. Reston, VA: NCTM.

National Council of Teachers of Mathematics. (1991). *Professional Standards for Teaching Mathematics* Reston, VA: NCTM.

National Research Council. (1990a). *Renewing U.S. mathematics: A plan for the 1990s*. Executive summary: Committee on the Mathematical Sciences, Commission on Physical Sciences, Mathematics, and Applications. Washington, DC: National Academy Press.

National Research Council. (1990b). *Renewing U.S. Mathematics: A plan for the 1990s*. Executive summary: Committee on the Mathematical Sciences: Status and Future Directions, Board on Mathematical Sciences, Commission on Physical Sciences, Mathematics, and Applications. Washington, DC: National Academy Press.

Resnick L. B. (1987b). Learning in school and out. *Educational Researcher*, 16 (9), 13-20.

Resnick, L. B. (1987a). *Education and learning to think*. Washington, DC: National Academy Press.

Resnick, L. B., and Resnick, D. P. (1989). Assessing the thinking curriculum: New tolls for educational reform. In B. R. Gifford and M. C. O'Connor, Eds., *Future assessments: Changing views of aptitude, achievement, and instruction*. Boston: Kluwer Academic Publishers.

Resnick, L., and Tucker, M. (1991). The standards project: Creating a national examination system that prepares students for the challenges of the 21st century. (Unpublished overview). Pittsburgh, PA: Learning Research Development Center, University of Pittsburgh.

Romberg, T.A., Wilson, L., and Khaketla, M. (1991). The alignment of six standardized tests with the NCTM standards. Madison, WI: University of Wisconsin.

Romberg, T. A., Zarinnia, E. A., and Williams, S. R. (1989). *The influence of mandated testing on mathematics instruction: Grade 8 teachers' perceptions*. Madison, WI: National Center for Research in Mathematical Science Education University of Wisconsin-Madison.

Romberg, T.A., Zarinnia, E.A., and Collis, K. F. (1989). A new worldview of assessment of mathematics. In G. Kulm, Ed., *Assessing higher order thinking in mathematics*. Washington, DC: American Association for the Advancement of Science.

Saxe, G. B. (1991). *Culture and cognitive development: Studies in mathematical understanding*. Hillsdale, NJ: Lawrence Erlbaum Associates.

Shavelson, R. (1989). Performance assessment: Technical considerations. Presentation at the Seminar on Authentic Assessment, Berkeley, CA.

Shavelson, R. J. (1990). *Can indicator systems improve the effectiveness of mathematics and science education? The case of the U.S. Evaluation and Research in Education*, 4(2), 51-60.

Shavelson, R. J. Baxter, G. P., Pine, J., Yure, J., Goldman, S. R., and Smith, B. (1990). Alternative technology for assessing science achievements. Paper presented at the AERA annual meeting, Austin, TX.

Shavelson, R. J., Carey, N. B., and Webb, N. M. (1990). Indicators of science achievement: Options for a powerful policy instrument. *Phi Delta Kappan*, May 1990, 692-697.

Wallace, M. A., and Wing, C. W. (1969). *The talented student*. New York: Holt, Rinehard and Winston.

Willingham W. W., Lewis, C., Morgan, R., and Ramist, L. (1990). *Predicting college grades: An analysis of institutional trends over two decades*. Princeton, NJ: Educational Testing Service.

16  The Intellectual Prices of Secrecy in Mathematics Assessment

Judah L. Schwartz

*E*ditors' *Note: So far in this book, the authors have focused on examples and descriptions intended to clarify substantive issues influencing the nature of appropriate alternatives to traditional forms of assessment. These issues have dealt with assumptions about the nature of mathematics, of real-life problem solving, or of exemplary mathematics learning and instruction. However, the instructional impact of assessment is also strongly influenced by policy decisions that seem to have nothing to do with assumptions about mathematics, problem solving, or teaching and learning. For example, this chapter summarizes conclusions that were reached by a number of leading scholars and educators about the impact of secrecy, as reported in* The Social, Intellectual, and Psychological Costs of Current Assessment Practices *(Schwartz and Viator, 1990).*

This project, funded by the Ford Foundation, identified a number of policy issues that tend to have particularly negative effects on assessment reform, and among these issues, secrecy in assessment emerged as particularly important. Other issues that were given special attention included equity (Hilliard, Willie), legal and economic issues (Heubert, Barryman), issues in science education and language arts education (Raizen, Chomsky), and issues involving the development of individual students and teachers (Wiggins, Stage, Perrone).

Even though this book has examined assessment mainly from the perspective of people whose main areas of expertise have to do with learning and instruction in mathematics, it is fitting that we close by considering ways that substantive issues in mathematics instruction influence (and are influenced by) a larger context which included policy issues, social issues, and instruction in other curriculum areas—such as those that focus on reading, writing, or the sciences.

INTRODUCTION

My purpose in writing this chapter is to describe what I feel are the several prices we as a society pay for using nonpublicly available instruments for the assessment of the effectiveness of mathematics learning and teaching in our schools. Our society is profoundly undereducated and incapacitated in dealing with public policy matters that have quantitative dimensions, that is, all public policy. Our school systems either do not have the freedom or do not believe they have the freedom to challenge students to think inventively and creatively about mathematics. Finally, teachers and parents feel torn between the desire to educate youngsters richly and imaginatively and the need to prepare them to demonstrate their competence on examinations that are deeply flawed.

All the ills of the present methods of accountability assessment in mathematics are not due to the nonpublic nature of the instruments, but I believe that many of them are. Moreover, I believe that many of the ills that do not result directly from the secrecy of the instruments are nonetheless indirect consequences of the secrecy and are substantially exacerbated by it. Finally, because the field of assessment has seen more than its share of complaints about the ills and evils of educational testing, I shall sketch what I believe to be a viable and pragmatic alternative approach to assessment that does not seem to be flawed in the ways that our present methods of assessment of mathematics teaching and learning are.

HOLDING THE SYSTEM ACCOUNTABLE: HOW WE NOW DO IT?

Most state departments of education and local school boards depend heavily on the results of standardized multiple choice tests to decide how well their systems are educating students mathematically. Even the federal Department of Education must use of the results of such instruments. In fact, much of the current public concern about U.S. students' mathematical incapacity is due to reports in the press and electronic media about poor performance on such instruments.

A remarkable feature of the reports that reach the public's attention is that the public cannot usually see the questions that are asked. The media do not publish the questions; rather, they publish reports about students' performance on the questions. In fact, the media are not able to examine the test instruments either. To some extent, the media are to blame for publishing reports based on instruments they are not allowed to examine, but one must not be too hasty in criticizing the media. The fact is that the general public cannot purchase or even see copies of the tests that are used to report to the public about their schools. Instead, the public is simply told some sort of aggregated score about performance on questions they are not permitted to evaluate themselves. The public also has no access to the scoring criteria and procedures, nor to the methods of aggregating performances on subtests into a single or small group of numbers.[1]

Can we be well served by such procedures? Perhaps, if testing companies produce tests that are free of erroneous questions and if they grade them in error-free ways. We raise the issue of error in the spirit of the ancient adage, "this above all—do no damage." Can the test makers be trusted to introduce no mistaken questions or answers, that is, to do no damage?

Most commercial manufacturers of standardized tests are well respected organizations with long traditions of involvement in education. Many are also publishers of text materials. It would seem only reasonable to assume that they can be trusted to design error-free tests. Certainly they have many subject matter experts on staff; they consult with others in putting together the tests, and with still others who review the tests before they are used.

I do not question the integrity or good intentions of the test manufacturers, but I must point out that in every field of intellectual endeavor there is a publicly available literature in which findings are reported, discussed, and debated. Results that are flawed are, by virtue of open discussion, ultimately exposed and discarded. No journal in the natural sciences, for instance, would accept for publication an article that contained results of measurements made with instruments whose internal structure could not be examined, debated, and evaluated.

The situation with respect to assessment in education does not seem to be parallel. Tests are made and administered without the scrutiny of the community that ultimately depends on the results that the tests report. Occasionally detected errors appear on the front pages of the *New York Times* in stories about the ingenious high school student who slew the Princeton dragon. Do we know how many errors go undetected? Example 1 shows a question from a recent test for high school students that was designed by probably the most prominent American testing organization. The answer the testing organization's experts thought was correct was, in fact, incorrect. The question and the five preset choices for the answer, including the purportedly "correct" answer, had all been extensively reviewed by the organization's internal experts and external consultants.

Example 1.
Two identical coins are placed flat on a table and in contact with one another. One of the coins is held still while the other is rolled without slipping all the way around the circumference of the stationary coin until it returns to its original position. The student is asked how many turns the rolling coin has made and is offered a choice of several answers.

Readers who think they have an answer to the problem should actually try the experiment. This advice would have served the test designers well had they heeded it.

Beyond the price of error in questions and answers, there are other costs to the public in using assessment instruments whose contents they cannot examine and debate. The prices, in my view, are sufficiently high that even if the error problem were otherwise resolved, I would argue there is still adequate basis for insisting on an end to reliance on nonpublic instruments. Stated briefly, precluding error is absolutely necessary and absolutely insufficient.

A continuing, and proper, concern of every teacher as well as every curriculum designer is the level and tone of the instructional materials that they write and present to the students. These materials are influenced by many sources, including the assessment instruments that are used to judge how well the educational system is functioning. This is as it should be. But if the instructional materials are influenced by the assessment instruments, then it behooves us to see to it that this influence is as salutary as possible. And not only level and tone, but taste and judgment are also important. The level can be demanding and the tone good, but the questions may be of little use in helping students develop a sense of, or taste for (or judgment in), mathematics.

Although level and tone are logically distinct from taste and judgment, it is hard to examine questions without attending to both sets of issues. We shall keep both sets of notions in mind as we turn our attention to an examination of the kinds of questions we now use to build our mathematical assessment instruments.

THE TEMPTATION OF PRE-ANSWERED TESTS

If a society has a tradition of using examinations that are secret, it is difficult to resist the temptation to use multiple-choice tests that can be graded automatically. In principle, at least, widespread testing can be done often and economically. Such tests are feasible only if their secrecy can be maintained, since this testing technology requires very large numbers of questions that are expensive and difficult to generate. The source of this difficulty and expense is the need for establishing the validity and reliability of the questions. If, however, the questions can be maintained in item banks that are not made public, then they can be reused, and the cost of generating them can be amortized over many administrations.

Whether the development of the technology of multiple-choice, machine-scorable tests was the cause of, rather than the result of, the imposition of secrecy on the assessment process is probably not important, although it is probably true that the need to keep a test secret even after its administration is a consequence of the economics of standardized testing. For the present analysis, it matters little which is cause and which is effect. Standardized testing and secrecy of items before and after administration currently entail each other. Even though we all know well the form, flavor, and feel of multiple-choice tests, the economics of large scale standardized testing dictates that, except for selected sample items, the actual content of such tests not be made available for public discussion and debate. I shall try to show that not being able to see all the items that are used to test our children has led us to accept mathematics texts and teaching that do a profound disservice to us individually and as a society.

Recently I spoke with an official of the Dutch Ministry of Education who was concerned with assessment of mathematics at the secondary level. He told me that the test instruments they used consisted of about a dozen extended problems, each of which required the student to understand a problem in context, formulate an approach to the problem, use that approach to design a procedure for solving the problem, carry out that solution procedure, and finally, explore the reasonability of the result obtained. Following their administration, the examinations are published and enter the available body of curricular and instructional materials.

Example 2 shows the sort of question I mean, taken from the 1989 end of secondary school examination for students who will not pursue further studies in science or mathematics. The question in Example 2 is reasonably structured and does not present a student with an impossibly wide universe of circumstances to analyze. On the other hand, it does demand that students formulate and quantify such constructs as risk, advantage, and disadvantage; it also requires them to devise a procedure for calculating probabilities and expected outcomes and to carry out those calculations.

Example 2.

The grapes in a certain vineyard are ready to be harvested. The taste of the grapes, and the wine to be made from them, is likely to be better if they are allowed to stay on the vine somewhat longer. On the other hand, the grapes could be badly damaged by heavy rains. The vineyard owner makes two analyses of the situation.

> **I.** Harvest the grapes immediately:
>
> The quality of the grapes is reasonable.
>
> Half the harvest can be sold for direct consumption at a price of $2.00/kilo. The other half can only be used for processing into grape juice. These grapes would bring in $1.30 a kilo.
>
> In this harvesting scheme there is limited risk.
>
> **II.** Harvest the grapes in two weeks time:
>
> The quality of the grapes is now "good."
>
> The entire harvest can be sold for $2.30/kilo. This harvesting scheme involves a greater risk: If it rains on more than two days in the next two weeks, the entire crop of grapes will only be usable for processing into grape juice at $1.30/kilo.
>
> The vineyard owner can count on a crop of 12,000 kilo.

The student is asked to consider how the risk involved in pursuing strategy II compared to the certainty of strategy I, and to quantify the potential advantage and disadvantage of strategy II.

Further, the student is asked to calculate the likelihood of rain on two or more days in the intervening two week period given the datum that the likelihood of rain on any single day in that period is ≈ 15. Initially, the student is asked to calculate expected outcomes for each of the strategies, to choose a strategy, and to present a justification for the choice.

How does the publication of these examinations affect the intellectual quality of what is taught and learned? Do math teachers in Holland teach to the test? In some sense they do, as do teachers the world over. By virtue of the fact that the examinations in Holland contain problems that are rich in structure and that demand that students perform a wide range of mathematical actions, instruction in mathematics tends to emphasize such problems and make such demands on students.

In contrast, in the United States we tend to rely on examinations that make extensive, if not exclusive, use of the multiple choice format. What are the effects of using such tests on the intellectual quality of what is taught and learned? In answering this question, it is important at the outset to note that a multiple-choice item does not ask students to construct a solution to a problem or an answer to a question. Rather, it asks them to recognize a solution or an answer. Recognition and production are fundamentally different abilities—a fact that is well recognized by people with a reading knowledge of a foreign language who find themselves in a restaurant or a shop in a country in which that language is spoken. Do we really want to say to students that being able to recognize an answer to a question is a sufficient level of expertise to attain?

There is a second, and in my view more destructive, intellectual consequence of using multiple-choice tests for serious assessment purposes. The implicit message conveyed to students is that all issues worth discussing and examining can be reduced to a selection among four or five alternatives. This can be presumed to be true no matter how much subtlety or nuance may be involved. As a result, I believe, we have come to be a public that thinks mathematics (and science and history and much else) is an intellectual domain in which in which questions necessarily have answers, and that these answers can be briefly stated. There is also a corollary to this last point. It is that all questions worth asking have correct answers; and one implication of this corollary is that correct answers are unique.

Let us return to the issue of the influence of the assessment instruments used on the level and tone of our instructional materials. If our instruments demand that students recognize answers rather than construct solutions, we will teach students tricks to recognize answers rather than strategies for constructing solutions. If we use instruments that suppress subtlety and nuance, it should not surprise us that our students' analyses tend to be superficial and simplistic.[2]

These consequences are particularly painful for our society. We must have people who can recognize the validity of a quantitative argument offered in support of an important public policy matter, but we need more than that. We need a society of people who are as nimble using quantitative tools of analysis appropriately as they are using the vastly subtler qualitative tools of language. We properly demand, with respect to language, that the people we educate be willing and able to use their production skills of speaking and writing as well as their recognition skills of listening and reading. We can afford no less in mathematics. Assessment, and, by implication, instruction, that asks our students to display only recognition skills and not production skills does not serve us well.

There is yet another painful consequence of current mathematics assessment techniques. Mathematics, and its use in analyzing the quantitative dimensions of the world about us, is not a "right or wrong" kind of enterprise. To be sure, it is possible to ask questions for which there are single correct answers. But one can do this in any domain. And in many domains, we have come to understand that such questions are fundamentally trivial. History is more than dates. Literature is more than names of famous authors. Yet for the most part we haven't gotten beyond this abysmally low level of sophistication in mathematics.

It is relatively easy to see how such a view of mathematics would be appropriate to the application of mathematics to judgmental situations. For example, one could pose a problem of the following sort: "Design the largest doghouse you can using a single 4" x 8" sheet of plywood." But many people will argue that students must first learn the "basics" in mathematics, and that such matters as number facts, multiplication tables, and the like are not really given to interpretation. While it is true that the product of 6 and 9 has only one value, it does not follow that the only way to ascertain whether or not someone knows the basics is to ask questions of the form, "What is the product of 6 and 9?"

For those readers whose education cheated them of the possibility of thinking about mathematics in this way, Examples 3 and 4 are test questions that deal with the same topic, subtraction of whole numbers. It is evident that these two problems assess the same skill. The first one has exactly one correct answer, while the second has many correct answers (and even more incorrect ones). Moreover, the second question offers the possibility of solving the problem by invoking a conceptual understanding of the meaning of subtraction that is independent of the mechanical mastery of the computational procedure. This is important because many computational procedures are learned by rote without a glimmer of conceptual understanding on the part of the student (and often of the teacher).[3]

Example 3.

What is the result of the following subtraction?

$$\begin{array}{r} 7102 \\ -\ 4595 \\ \hline \end{array}$$

a) 3493   b) 3507   c) 3697   d) 3617   e) don't know

Example 4.

Here are two subtraction problems. Make up a subtraction problem whose answer lies between the answers to the two problems that are given.

$$7102 \qquad\qquad 6241$$
$$-\underline{4595} \qquad\qquad -\underline{3976}$$

The general proposition here is that it is possible to pose questions in mathematics that allow for creativity and invention. Moreover, it is possible to do this even for those topics that are generally believed to be the least open to variation. Clearly teachers and students who know that performance will be assessed with such items will engage the subject more richly and deeply than they do now. Further, it is clear that problems of this sort can be made public with no loss of usefulness.

SECRECY: BEFORE AND/OR AFTER THE FACT

To use assessment to influence the teaching and learning of mathematics in a constructive way, I believe that at least two conditions must prevail. First, the assessment instruments must not contradict, either explicitly or implicitly, our pedagogic goals. That is, they must not be mathematically wrong in those areas of mathematics where we really care about students being mathematically right. They must not be simplistic where we want students to discern and deal with complexity. They must not convey, as they do now, an image of mathematics that is at odds with the nature of the discipline.

The second condition for using assessment in a constructive way is that the questions must be (at a minimum) mathematically interesting. I am willing to take as an educational axiom the proposition that questions that have more than one correct answer are inherently more interesting than those that have only single correct answers. Moreover, I believe that any question that has a single right answer can be replaced by a question with a set of correct answers that probes the same mathematical skills, and that at the same time is more interesting and affords greater insight into the diversity of strategies that students employ in solving problems.

Suppose we succeed in altering the nature of the assessment instruments we use so that these two conditions are met. How might we then best make use of the opportunity to influence intellectuality and teaching and learning in the schools? It seems clear that the possible salutary intellectual effects of high-quality assessment items on teaching and learning are maximized by making them widely and easily available. In this way a wide variety of interested audiences, such as teachers, students, parents, school boards, state authorities, colleges and universities, and industrial and commercial organizations interested in hiring young people can all readily see what is expected. Moreover, widespread public availability of assessment instruments makes possible a continuing public discussion of standards by these various interested audiences, a process that can only benefit the educational system.

So far I have been talking about publication of tests after they have been administered. If the tests are good ones, then publishing them can have desirable effects on what is taught and learned in schools. However, dissemination of the tests after administration leaves unanswered the problem of how to avoid errors in the formulation of problems and their solutions. There is little doubt that errors will, in time, be detected after the tests are published. However, that is often too late. Damage may have already been done.

In what follows I suggest a procedure that addresses the problem of error while preserving the potentially useful effects of assessment on instruction. Suppose that past examinations are published for enough time so that a large collection of very good problems become available. Clearly, at some point the collection becomes large enough so that problems that have been used before can be used once again. Note that if we do begin to do this, we move from after-the-fact open examinations to before-the-fact open examinations. Is this workable?

I believe that the widespread availability of small microcomputers and easily manipulated database software for these machines make possible new approaches to the filing, indexing, and retrieving of previously used problems. Publicly available, richly indexed databases of problems and projects can have the kinds of salutary effects on intellectuality discussed above. They also provide the opportunity for scrutiny, discussion, and debate about the quality and correctness of questions and answers. In addition, from a methodological perspective, they alter completely the traditional psychometric questions of reliability and validity.

These new approaches offer the promise of an openness that we have not seen before in education. While such openness is almost certainly not, in and of itself, sufficient to repair the ills of mathematics education in our country, at least it establishes some necessary conditions for reform.

NOTES AND REFERENCES

1. The procedure of making examinations publicly available after administration is not unknown in the United States. The New York State Regents examinations are regularly published in their entirety after they are administered. Generations of high school students and teachers in New York have used these published tests as curricular materials. On the other hand, there are truth-in-testing laws in several states that require that testing companies make available to test-takers, for a fee, the questions and answers on the test they have taken. These laws have not produced an avalanche of interested test-takers eager to see what the testers were and were not asking and what they thought the answers were. I believe that it would be wrong to infer from this experience that the publication, after administration, of a test is of little value. Contrast this case with the Regents' examinations cited above.

2. The most notorious of the rote procedures that are ill understood by students (and teachers) are long division and division of fractions. The computation of logarithms and the procedure for extracting square roots, now almost never taught, were rarely understood.

3. A way to solve this problem without actually carrying out the subtractions is to construct a string of equivalent subtractions. For example: 7102 - 4595; 6102 - 3595; 6202 - 3695; 6242 - 3735; 6241 - 3734; and so on. Thus, the first of the original problems, 7102 - 4595, is equivalent to the problem 6241 - 3734. This is to be compared to the second of the original problems, 6241 - 3976. Even if we limit ourselves to integers, there are more than two hundred problems of the form 6241 - (some number) that can be made up to correctly answer the question. Moreover, a modest amount of reflection will probably persuade the reader that there are an infinite number of incorrect answers to the problem as well.

Schwartz, J.L., and Viator, K.A. (Eds.) (1990). *The social, intellectual, and psychological costs of current assessment practices.* Cambridge, MA: Educational Technology Center, Harvard Graduate School of Education.

Index

A

Abstract concepts. *See also*
 Proportional reasoning
 in experiential learning model, 197
 in mathematical models, 54, 197
 premature introduction of, 195-196
 transition to, 236-237
Accountability
 resources expended on, 346
 secrecy and, 185-186, 428-430
Affect
 assessing, 85-86, 259, 287
 role of, in mathematical understanding, 269-270
Algorithms, 219
American Association for the Advancement of Science, 363, 395
 on nature of real mathematics, 24
Arithmetic
 assessing understanding of, 265-266
 elements of understanding in, 266, 269-270
 knowledge structures in, 266-268
 mathematical models in, 27-30, 352
Assessment. *See also* Assessment reports; Classroom assessment; Testing; Tests
 balance beam tasks in, 302-310
 balanced, principles of designing, 122-124, 181-182
 California Assessment Program, 104, 156, 323
 central limit theorem in, 181
 cognitive models in, 11-12, 21-22, 31, 63-64, 67-71, 82-86, 327
 computers in, 6, 71, 89-91, 100-104, 344, 347-349, 348-349, 356-357, 373-375
 content-specific framework for, 86-87, 162
 cost issues, 180, 183-184, 187-188, 346
 creative challenges in, 123
 curriculum reform and, 121-122, 177-179, 183, 189-192, 340-341, 435
 defining, 7
 elements of, 120, 125-127, 326-328
 fairness issues, 186-187, 398, 404-406, 427
 research project on, 409-410, 415-419
 goals of, 4-6, 65-66, 67, 82

in model-eliciting instruction, 31, 57-58, 327-328
multiple choice questions and, 322-323, 430-435
National Council of Teachers of Mathematics standards, 33
of affective components, 85-86
of arithmetical understanding, 265-266, 270
of extended tasks, 120, 145-146, 150-151, 154-155, 161-162, 163-164, 167, 170-171, 173-175
of geometry inquiry skills, 91-93, 111-114
of higher-order skills, 119-120, 219, 287, 311
of individual understanding
 via probes, 272-273
 via standard test, 270-272, 310
 via structured interview, 273-280, 287
of individuals in groups, 163-166
of learning vs. understanding, 76-82, 433
of multiple levels of response, 339-341
of multiplication understanding, 273-277
of proof strategies, 135-136
of student generalizing ability, 94-101, 137-142
of student representations, 249-250, 261-263
 case examples, 250-252, 258-260
of teachers, 406-407, 420-421
packages, 124, 179-180, 188-189
pilot testing, 180, 181
political issues in, 177, 185-186, 363
portfolio-based, 5, 413
progress-focused documentation, 407-408, 421
public policy in, 177, 433
reform, objectives of, 3, 9-11, 19, 21, 64-65, 319-326, 398
reliability concepts in, 180, 183
scoring considerations in design of, 124
short mathematical tasks in, 127-140
standardized vs. alternatives, 379-398
state-focused documentation, 407
technology in, research project, 408-415
Assessment reports. *See also* Learning progress maps
 applicability of computers to, 348-349

439

Index

cognitive objectives profiled in, 338-339
decision-specific, 5, 363-365
designed for outmoded technology, 347-348
goals of, 4, 11, 13-14, 343-347
graphics in, 6, 338-339, 356-357, 372-375, 422
n-dimensional, 344, 422
secrecy issues, 185-186, 428-430, 435-436
validity of, 5, 345-346
Associativity concepts, 76-82
Australia, assessment in, 192
Authenticity
defining, 17
evaluation criteria for, 17-18
in mathematical problem-solving, 47-52, 344
Autonomy, student, 126, 146

B

Balance beam tasks, 293, 302-310
Behavioral psychology
in testing objectives, 21
influence of, 18-19, 64
Bruner's modes of representation, 236

C

California Assessment Program, 104, 156, 323
California Framework, 121
Cartesian coordinate systems, 26-27
Case example, developing representations
student-student interactions in, 253-258
teacher-student interaction in, 249-252
Case studies, instructional role of, 24, 34
Central limit theorem, 181
Classroom assessment, 13, 65, 66-67, 407-408
microlevel analysis, 249
observational techniques, 284-286
of geometry inquiry skills, 104-111
of student representations, case example, 251, 258-259
screening instruments for, 280-284
tools/skills needed for, 82, 280, 287-288
via multiple choice tests, 286, 322-323
Cognitive models. *See also* Representational systems
assessment and, 11-12, 21-22, 31, 63-64, 67-71, 82-86, 216, 219, 265, 325-326
Cartesian coordinate systems as, 26-27
characteristics of mathematical, 24-26, 31-34, 52-57
conceptual mathematization, 196-197
developing, in mathematics, 34-38, 58-60, 327
experiential learning, 197
experimental learning, 197
heuristics in development of, 215-216

higher-order
characteristics of, 30-31
instructional goals of, 31-34, 68-71
in learning progress maps, 352
non-mathematic research in, 297
perspective shifting in, 223, 230-231
problems in constructing, 296-297
proportional reasoning in construction of, 326
representational systems in, 83-86
role of, 21-24, 37-38
testing performance and, 29-30
Cognitive science
behavioral vs., in mathematics education, 18-19
internal models in, 21-22, 60
research models in, 297
Commutativity concepts, 76-82, 279
Computers. *See also* Technology
applicability to assessment, 6, 348-349
Geometry Supposer programs, 89-91, 100-104
in individualized assessment, 71
role of, 344
visual presentation from, 347, 348-349, 356-357, 372-375
Conceptual fields, 53
Conceptual mathematization, 196-197
Constructivism, in mathematics education, 19, 60, 196
Continuously Developing Teachers Project, 409-410, 419-421
Cost issues, 180, 346, 427
in design of assessment, 187-188
in machine-scorable test design, 431
in standardized testing, 183-184
Covariance/invariance concepts, 334-338
Curriculum
basic vs. problem-solving skills in, 20-21
changing, via test reform, 8-9, 362-363, 380
geometry, 91-93
goals coordinated with assessment, 121-122, 177-179, 183, 340-341
goals distinct from assessment, 68
implementing large-scale change in, 189-192
in transition, 120-121, 127, 178, 191, 363, 398
standards for mathematics (NCTM), 33-34
trends in development of, 218-219

D

Decision-making
assessment reports in, 5-6, 9-10, 362-365
empowering local educators, 362-363
goals of, 10, 343-345
identifying priorities in, 10, 11

learning progress maps in, 363-365
portfolio-based assessment and, 413
standardized tests in, 379-380
Definition of Spontaneous Analogy, 98
Dienes, Z., 59-60
Division, meaning of the remainder in, 230, 234

E

Electromyography, 299
Elementary education
  arithmetic, 27-30
  bridging informal-formal mathematics, 267-268
  conceptual systems in, 353
  evaluating individual understanding
    via probes, 272-273
    via standard test, 270-272
    via structured interview, 273-280, 287
  mathematical model in, 26-30, 31-34, 328, 353
  national testing, 66
  observational assessment techniques in, 284-286
  screening in, 280-284
Equity issues, 30, 186-187, 404-406, 427
  research project on, 409-410, 415-419
Essay questions, 212
Ethnographic research, 400-401
Evaluating, defined, 7
Experiential learning, 197
Extended tasks
  developing, 145, 146
  examples of, and assessment schemes for, 155-173, 188-189
  in applied mathematics, example of, 146-151
  in pure mathematics, example of, 146, 151-155
  role of, 145-146

F

Face validity, 125, 181
Feedback
  in experiential model, 196
  in measurement framework, 312
Fisher-Landau Foundation, 280
Fractal geometry, 372
Fractions, 27-28, 216-217, 221, 237-238, 437
  assessing understanding of, via interview, 278-280
  student interactions while studying, 253-258
  teacher-student interaction while studying, 250-252
Freudenthal, H. F., 196, 223
Freudenthal Institute, 140

G

Generalization processes, 93-94, 137-142
  assessing, via Geometry Supposer, 95-111
Geometry
  assessing inquiry skills via Supposer, 104-111
  assessment of proof strategies in, 135-136
  assessment of symbolization in, 138-139
  conjecture in, 93, 104, 108-110, 111-112
  curricula comparison study, 91-93
  fractal, 372
  generalization processes in, 93-94, 137-142
    assessing, 94-101
  generating hypotheses in, 101-102
  inquiry skills in, 92, 111-114
  perspective shifting strategies in, 223
  practical, exercises in, 168, 207-208
  Supposer programs, 89-91, 100-111
  systematic constraints in, 222-223
Germany, assessment techniques in, 65
Group/team activities
  assessment considerations, 163-166, 176, 280
  in realistic approach, 237
  interactive learning, 198-199
  model-eliciting problems and, 41, 44
  role of, 49-50, 120-121
  student interaction developing representations, case example, 253-258

H

Heuristics, 72, 85, 215-216, 222-223, 224-225, 243, 324
High school education, conceptual systems in, 353

I

Inductive reasoning, 93-94, 105
Instructional activities. *See also* Problem-solving processes; Sample problems
  cognitive objectives in, 344, 422
  Geometry Supposer programs, 90-91, 100-104
  heuristics in, 225, 230-233, 324
  identifying process characteristics in, 220
  in developing mathematical language, 231-233
  in goal description model, 204
  in Math-rich Newspapers Project, 411-415
  meaning of remainder in division, 230, 234
  numeracy skill-building, 159-173, 191-192
  realistic, in fractions, 216-217
  realistic, strategies for, 233-235
  reasoning solutions to doubling problem, 217-218, 225
  short tasks, 127-134

Index

tests as, 9
to encourage thinking strategies, 216-218, 220-230
Interaction
  author-reader, 224
  between students, 198-199
  learning as social activity, 237
  student-student, developing representations, 253-258
Interaction
  teacher-student, studying fractions, 250-252
International Congress on Mathematical Education, 189-190
Interviewing, structured, 273-280, 287
Item response theory, 293, 294
  models based on, 298, 449

J

Job skills
  authentic performance activities and, 39
  higher-order understanding as, 30
  in information age, 23-24
  research needed, 406
  technological advancement and, 401-402

L

Language
  changing level of, in mathematics, 231-233
  competency, 84
  of mathematics, 22
  technical, role of, 30
Learning progress maps
  cognitive objectives in, 352
  combining conceptual systems in, 353-354
  constructing, 351-355, 359-361
  creating n-dimensional student profiles, 355-357
  dynamic nature of, 354-355
  graphic profiling in, 356-357
  interactive capabilities of, 357-359
  model-eliciting activities in, 351
  proportional reasoning in, 326
  role of, 350-351, 361-365
Legal issues, 427
Level of analysis, in mathematics, 53, 56-57
Lewin, K., 197
Log books, 242
Logarithms, 437

M

Math Placement Problem, 40-45, 58
Math-rich Newspapers Project, 408-415
Mathematical Sciences Education Board, 121, 353, 363, 395, 397
  on curriculum reform, 8
  on nature of real mathematics, 24
  on technological advancement, 401-402
Mathematics
  applied, 25, 123, 145
  assessing understanding, 265-266, 287-288
  assessment needs, 63-64
  characteristics of models in, 52-57
  cognitive representational systems in, 83-86
  constructivism in, 19, 196
  diversity of skills in, 404-405
  future-oriented objectives in, 421-423
  heuristics in, 215-216, 222-223, 224-225, 243, 324
  informal forms, 266-268
  instructional goals in, 57, 219-220
  mechanistic view of, 396
  modeling of, 26-30, 64-65, 195, 201-204, 295-297
    higher-order, 30-34, 119-120, 343-344
  numeracy skills, 159-162
  primitive objects in conceptual systems of, 353
  pure, 25, 39, 123, 145
  real, nature of, 24-26
  realistic, testing in, 239-242
  reductionist view of, 396
  task characteristics in, 125-127
  technical integration in, 53, 199-201
Mathematics Association of America, 395
  on nature of real mathematics, 24
Mathematics Olympics, 414
Megatrends 2000, 22
Mendelbrot sets, 373
Middle school education
  model-eliciting activities in, 327
  teachers' higher-order reasoning, 29
Model-eliciting activities
  assessment in, 31, 57-58
  creating, 38-40, 58-60, 324-325
  examples of, 34-37, 40-47, 330-340
  for proportional reasoning, 34-38, 326-328, 330-340
  framework for, 319
  in elementary school, 328
  in learning progress maps, 351
  in middle school, 327
  in non-mathematic disciplines, 23-24, 30
  perspective of, 263
  realistic applications, 47-52
  role of, 37-38, 57-58, 329-330
Multiple choice tests, 9, 180, 182, 205, 214, 286, 322-323, 397
Multiplicative thinking, in children, 273-277, 330-334

N

N-Dimensional student profiles, 355-357
National Academy of Sciences, 395, 397

National Center for Research in Mathematical Sciences Education, 243
National Council of Teachers of Mathematics, 3, 19, 121, 242, 363, 395
  Curriculum and Evaluation Standards for School Mathematics, 33-34,
    45, 50, 340-341, 365, 397, 401, 420
  on curriculum reform, 8
  on nature of real mathematics, 24
  Professional Standards for Teaching Mathematics, 420
  Setting a Research Agenda for Mathematics Education, 8
National Research Council, 395, 397
National Science Teachers Association, 363
Netherlands
  assessment in, 192
  goals of teaching in, 215, 219
  national examinations in, 205-209
  problem-solving task design in, 198, 204
  research in, 243
  testing in, 210-211, 431
  two-stage school tests in, 211-213
Notational systems, 25, 27-28, 32, 85, 231-233
Numeracy, skill-building tasks, 159-173

O

Observational assessment, 284-286

P

Piaget, Jean, 53, 54, 197, 353
Political issues, 177, 185-186, 363
Polya, George, 222-223, 224, 324
Pre-College Mathematics Ability: 3-D project, 409-410, 415-419
Probes, in assessing understanding, 272-273
Problem-solving processes. *See also* Model-eliciting activities;
Representational systems
  assessing multiple levels of response, 339-341
  assessing thinking strategies through, 284-286
  assessing via test probes, 272-273
  assessment of, 67
  basic skills vs., 20-21
  cognitive representation systems in, 83-86, 132
  content-independent, 21
  covariance/invariance concepts, 334-338
  features of, 222
  free productions, 198, 237
  generalizations as, 93-94
  grading conjecture in, 108-110
  heuristic concepts in, 222-225, 230-231, 324
  higher-order goals in, 69-70, 119-120, 205
  in arithmetical understanding, 266-270
  in balance beam tasks, 303-304
  in extended tasks, 150-151
  in geometry, 91-94, 222-223
  in Geometry Supposer, 100-104
  in learning progress maps, 351
  key-word approach to word problems, 74
  nonroutine solutions in, 126, 140-142
  numeracy through, 159-162, 191-192
  open-ended questions, 320-321
  presentation of problem and, 72-82
  realistic, 10, 47-52, 127-132, 204
  role of cognitive models in, 21-23, 24-26
  rules for transfer, 269
  screening instrument in assessing, 282
  single answer tests and, 37
  tacit understandings in, 73-75
  technical integration in, 199-201
  technology-based tools in, 18, 20, 41, 44-45
Progress-focused documentation, 407-408, 421
Proportional reasoning
  assessing, 328-334
  assessing multiple levels of response, 339-341
  assessment framework for, 328-334
  covariance/invariance in, 334-338
  in learning progress maps, 351-352
  in mathematical models, 29
  model-eliciting activities in, 34-38, 326-328
  research on, 297
  role of, 331
Psychological science, role in test theory, 272, 287, 294-295, 297, 395
Psychometric theory, shortcomings of, 121, 397, 405
Public policy
  assessment and, 177, 428
  curriculum reform and, 363
  secrecy in testing, 185-186, 428-430, 436
  test design and, 433

R

Representational systems
  affective, 85-86
  assessing, 132, 249-250, 261-263
  difficulty of building, 260-261
  evaluating, 32
  formal notation in, 85
  heuristic, 85
  identifying, 250
  imagistic, 84-85
  in mathematics, 22
  in tacit, naive cognitive model, 75
  misinterpreting student's, case example, 251, 258-259

Index

modes of, 236
process of understanding, student interaction in, 253-258
student use of, 250
types of, 83-86
verbal/syntactic, 84
Research needs cited, 14
content-specific assessment framework, 87
describing mathematical models, 401
heuristic prototype development, 243
implications of technology, 403-404
in assessment of understanding, 287-288

S

Sample problems. *See also* Extended tasks; Instructional activities
assessing covariance/invariance understanding, 334-338
balance beam tasks, 302-310
CD throwing problem, 416-419
conjecture/generalization test, 95-98, 105-107
essay question, Netherlands, 212
for developing mathematical language, 231-233
fraction monographs, 221
in realistic mathematics testing, 239-242
integrated learning strands, 199-201
model-eliciting, 34-37, 45-46
Netherlands, 206-211
numeracy, 162-175
open assignments, 146-154
realistic instruction in fractions, 216-217, 237-238
reasoning solutions to doubling problem, 217-218, 225
short tasks, 127-142
starters, Shell Centre, 155-159
testing for general goals, 226-230
to develop proportional thinking, 330-334
two-stage tests, 211-213
variations on nontraditional, 79-81
with multiple levels of response, 339-340, 434-435
School administrators, assessing, 406
Schrank, Roger, 397
Scoring
assessment task selection and ease of, 123
cost issues in, 187
extended tasks, 154-155
fluency with representations, 132, 137-139
mathematical thinking in covariance/invariance problems, 338
methods, 182-183
multiple choice, 182
of multiple level responses, 9, 344
of student conjectural ability, 108-110

relative thinking, 334
role in assessment design, 124
statistical moderation in, 184
Screening instruments, 280-284
Secrecy in testing
accountability and, 428-430
development of, 430-431
public policy and, 185-186, 428-430, 436
Self-assessment, 32-33, 82, 85, 113, 164, 269
Set theory, 196
Shell Centre, 122, 140
numeracy modules, 160, 176, 191-192
Situated knowledge, 20
Sociocultural factors, in writing authentic performance activities, 38-39
Standardized testing
ethnographic research and, 400-401
inadequacy of, in information age, 395-398
predictive validity of, 399-400
Standardized tests, 9, 310, 362
National Council of Teachers of Mathematics standards and, 6
of mathematical thinking, 271-272
openness/secrecy issues in, 430-435
performance assessment instruments compared to, 379-398
statistical moderation in, 184
validity vs. reliability in, 183-185
State-focused documentation, 407
Statistical theory, 181, 183-184, 294
latent class model for balance beam tasks, 305-308
probability-based inference networks, 299-302
Structure-Mapping Theory, 98
Subtraction skills, assessing, 434-435

T

Tacit understandings, 73-75
Teachers
assessment of, 406-407, 420-421
assessment role of, 65, 66-67, 183, 192, 287-288
classroom geometry assessment by, 93, 104-110
education needs of, to improve assessment, 406-408
research project on, 409-410, 419-421
higher-order reasoning in, 20, 29, 119
monitoring, 124, 188
open-ended testing design and, 209
qualities of succesful, 420
role in understanding student representations, 261
use of learning progress maps, 364-365
writing model-eliciting problems, 37-47, 324-325

Teaching
  classification of strategies in, 220
  globally formulated goals in, 202, 215
  heuristic methods in, 225, 230-231, 243, 324
  lessons from videotaping, 259
  of algorithms, 219
  outdated concepts in, 396-397
  realistic, goals of, 233-234
  recent reform efforts in, 64-65
  shift in goals of, 219-220
  test-oriented, 63, 209, 286
  time spent, vs. administrative time, 346
  trends in goals of, internationally, 215, 218-219
Technology. *See also* Computers
  in assessment, research project, 408-415
  obsolete, shaping assessment, 347-348
  research issues related to, 401-404
Tell Me A Story (Schrank), 397
Test of Early Mathematical Ability, 271-272
Testing. *See also* Tests
  adaptive, 311, 316-318
  alternatives to, for decision-making, 343
  behavioral objectives in, 21
  classical theory, 293, 294
  cognitive models and, 30
  cognitive objectives in, 21
  defining, 7
  goals of, 205
  heuristics in, 225
  hybrid model of, 298, 311
  in Britain, 178-179
  in Netherlands, 205-213
  in realistic mathematics instruction, 239-242
  item response theory, 293, 294, 311
  latent class models for binary skills, 298, 305-308, 315-318
  mixture models of, 298
  multicomponent models of, 298
  partial credit rating-scale model of, 299
  probability-based inference networks in, 299-302
  rule space analysis in, 298
  saltus models of, 298
  secrecy in, 185-186, 428-430, 436
  student ability to generalize, 94-100, 140-142
  theory development, 293-295, 298-299, 395-398
Tests. *See also* Assessment; Scoring
  assessment reforms and, 7, 66-67
  behavioral objectives in, 346
  characteristics of authentic problems, 18
  design considerations, 213-214
  designing a test as a test, 212-213
  designing short tasks, 127-140
  elementary, conceptual content in, 66
  essay questions, 212
  for general goals, 226-230
  in curriculum reform, 8-9, 362-363, 380
  inference networks in, 311
  manufacturers of, 429
  multiple choice, 9, 180, 182, 205, 214, 286, 322-323, 397
  open-ended questions, 320-321
  single answer, vs. model-eliciting tasks, 37
  standardized, 6, 9, 183-185, 270-272, 310, 379-398, 399-401, 430-435
  systemically valid, 5, 362
  to achieve higher-order goals, 205
  two-stage, 211-213, 242
Textbooks
  development time and curriculum change, 122
  elementary, conceptual content in, 66
  improving, 321-322
  mathematical systems in, 53-56, 302
Tools, problem-solving, 18
  in model-eliciting activity, 41, 44-45
  in transition to realistic mathematics, 236-237
  realistic, 48-49
  role of, 402-404
  teaching process and use of, 20
Training, teacher
  depth of understanding in, 20
  in model-eliciting activities, 40-47, 324-325
  research project on, 409-410, 419-421
  role in improving assessment, 406-408

U

United Kingdom
  assessment in, 120, 178, 180, 182, 185, 192
  cost of assessment in, 187
  curriculum development in, 122
  grading in, 182-183
  testing in, 178-179
  use of extended tasks in, 145, 146

V

Validity
  decision-making, 5
  in portfolio assessment, 5
  of constructs, 5
  predictive, 5
  reliability vs., in standardized tests, 183-185
  systemic, 5, 362
Vocational education, in Netherlands, 207